Th...

WITHDRAWN

ETHICS IN FORESTRY

ETHICS IN FORESTRY

Edited by

Lloyd C. Irland

TIMBER PRESS
Portland, Oregon

PHOTO CREDITS
Jacket courtesy U.S. Forest Service. *Frontispiece* courtesy U.S. Forest Service. *Section I* courtesy U.S. Forest Service. *Section II* copyright Chris Ayres. *Section III* courtesy U.S. Forest Service. *Section IV* copyright Chris Ayres. *Section V* copyright Chris Ayres. *Section VI* copyright Chris Ayres. *Section VII* courtesy U.S. Forest Service.

ISBN 0-88192-281-1
Printed in Singapore

TIMBER PRESS, INC.
The Haseltine Building
133 S.W. Second Ave., Suite 450
Portland, Oregon 97204-3527, U.S.A.

Library of Congress Cataloging-in-Publication Data

Ethics in forestry / edited by Lloyd C. Irland.
   p. cm.
 Includes bibliographical references and index.
 ISBN 0-88192-281-1
  1. Foresters--Professional ethics. 2. Forests and forestry--Moral and ethical aspects. I. Irland, Lloyd C.
SD387.E78E87 1994   93-38153
174'96349--dc20   CIP

Contents

For Connie

Foreword

When Lloyd Irland asked if I would be willing to write a fore-
word to his collection of readings on ethics in forestry, I was both
honored and enthusiastic. Lloyd is a respected colleague who has
served with distinction in both the public and private sectors, and in
this book he addresses a topic that has been neglected for too long.

Today we live in a world in which the population is increasing at
the astounding rate of one billion people every eleven years. The
physical and emotional stresses that result from this explosion could
lead to chaos if we do not constantly attempt to understand and
practice the rules of civilization. These rules, otherwise known as
ethics, are the very cornerstone of any civilized society because they
require the honest and equitable treatment of our fellow human
beings. They are especially important for those of us who, as natural
resource managers, have responsibility for the long-term health of
the land on which all civilization depends. This realization has lead to
the development of a relatively new body of thought: the need for an
environmental ethic.

The readings in this volume encourage us to think through some
very complex and difficult issues. They are not designed to provide us
with the right answers; they are designed to teach us to ask the right
questions. I think this is a very important book that deserves to
become a foundation reference for the practice of forestry.

William H. Banzhaf
Executive Vice President
Society of American Foresters

11

Preface

In our technological and commercial society, the modes of doing business are constantly changing, sometimes moving faster than our own awareness of potential new ethical issues. Professional practice cannot exist in a modern society without public confidence in the propriety of all dealings. Therefore, ethical maxims are as much concerned with preventing the *appearance* of inappropriate conduct as with proscribing certain types of behavior. For these reasons, careful study of ethics is always timely in forestry.

For this book, I divided the field into several categories—professional ethics, business ethics, environmental ethics, and the ethics of government service. These general areas, though sometimes overlapping, define primary ethical areas of concern to foresters practicing in government, research, industry, or in a private capacity. I would note that in many instances the author affiliations have not been updated.

The professional ethics section includes several readings that introduce the basic dimensions of professional ethics as they relate to private and public forestry practice. Professional ethics includes areas of ethics that are particularly important in professional practice, including relationships with clients, other professionals, and society at large. That this material originates primarily in forestry publications is testimony to the profession's enduring concern with ethics.

Since foresters customarily represent clients or employers in sizable financial transactions, an awareness of the principal issues in business ethics is important. Therefore, the second section addresses business ethics proper, as distinct from professional ethics. This includes the important topics of conflict of interest and loyalty to one's employers. In this area, the problems confronted by foresters are similar to those faced by realtors, so I have included two articles by real estate appraisers. A moment's reflection will reveal that business ethics is important to foresters in public service as well.

Foresters attend almost daily to situations concerning the effects of forest practices on the environment and human health. They sense

13

that there is an environmental ethic that must be considered in managing forestland and all its resources. At times, foresters are bombarded with free advice from groups with their own ideas of environmental ethics. How should foresters respond to such claims? To what extent should an environmental ethic constitute an integral part of the forestry profession's ethical standards? Just what should such an ethic contain? The third section of this book offers a few classic readings to help sharpen your thinking in this area. Reflection on environmental ethics should be part of any forester's training and continuing education.

Forest management is continuously affected by changing public policies. Foresters in all lines of employment are involved in advocacy, formulation of policy, policy implementation, and compliance with public policies. For this reason, some attention must be paid to the particular ethical issues that arise in public service and in formulating public policy.

The volume concludes with a number of cases and questions for reflection to assist the reader in applying the message from the readings. This book attempts no catechism, no list of do's and don'ts. Its purpose is not to codify rules of conduct but rather to stimulate and support self-conscious reflection on the many ethical questions that arise in the daily practice of forestry. My hope is to assist foresters in maintaining a confident watchfulness about potential ethical problems. Most ethical problems arise not from a decision to do harm but from a lack of foresight and caution in anticipating a situation. In many cases the lines separating acceptable from unacceptable behavior are indistinct. Foresters who develop a sensitivity to these divisions are more likely to resolve ethical conflicts in a responsible manner.

I would like to thank Albert C. Worrell, Edward C. Stuart, Jr., Archie Patterson, Jim Bethune (now deceased), Keville Larson, Greeley McGowin, and Bill Siegel for their assistance and encouragement in this project, as well as the publications staff of the Society of American Foresters, who assisted in earlier stages. Rondi Doiron kept a constantly changing manuscript and bulky permissions file in excellent shape. I owe a great debt to Timber Press and its staff, especially Karen Kirtley, Suzanne Copenhagen, and Christina Conklin, for undertaking to publish this book and for working with a busy editor who was rarely on schedule and often cantankerous.

Lloyd C. Irland

Introduction:
Developing Ethical Reflection

LLOYD C. IRLAND

If we wish to improve the ethical standards of the forestry profession, we need to stimulate and reward the exercise of ethical reflection—that is, reflective discussion and thought about upcoming issues, in the context of ethics, that help a person make right choices. A great deal is gained by studying ethics codes. Reviewing and discussing cases is also important. But these are only aids in developing a core professional skill, the skill of ethical reflection.

This critical skill enables professionals to make sense of the general and formal rules, and of the case examples offered in courses, readings, or discussions. Ethical reflection focuses on developing a sensitivity to the kinds of situations that can lead to unethical decisions; it helps detect potential problem situations early; and it develops the ability to recognize available options that can assist in resolving a problem in a satisfactory manner.

The ethical choices that foresters face generally fall into several basic categories: business or policy situations that develop over time in unforeseen ways; ethical claims or canons that conflict; ethical rules that are very general and subject to legitimate debate as to how they apply in specific situations; and positions that offer strong temptations.

The only way to confront these facts fruitfully is to treat ethics as a key professional skill for ongoing use, not as a periodic obeisance to canons in a rulebook.

Dr. Irland is President of The Irland Group, forestry consultants, in Winthrop, ME. Reprinted from *Journal of Forestry*, Volume 11, by permission of the Society of American Foresters.

Cultivate Awareness

One way to develop an ethical basis for decisions is to read, attend meetings, and cultivate an awareness of the kinds of situations that tempt people to behave unethically. An excellent starting point is to periodically review applicable codes of ethics (such as those of the Society of American Foresters or Association of Consulting Foresters) and apply them to recent situations in your experience. Discuss these situations with associates and friends in informal settings.

As you develop awareness, you will also develop foresight. You will learn to recognize situations early for their dangers. When you sense such a circumstance arising, stall for time or find a gracious way to defer a decision until you can think the matter over and consult with others. Always think ahead. Where could this situation go? Will it place me in a position where my ethical standards will be challenged?

Learn to see marginal or ambiguous situations more clearly. Frequently ask yourself: "Is this ethical? Could it lead me to act unethically in the future?" Ask certain questions during the early stages. "How would it look to someone else? Who else is affected by this decision and how? Is there an angle I'm not seeing?"

Choose a Mentor

Cultivate one or more friendly but tough mentors—not yes-people. Your mentor need not be your boss, though you are fortunate if your boss would be tough-minded but helpful.

You may need to try a few people out for their advice before finding those with the necessary depth of experience and understanding, and an interest in your development. A mentor who always gives you the easy answer—"no, there's no problem here"—is only teaching excessive regard for expediency. Find another one. They who walk with the wise shall be wise.

Don't just bring up the tough problems. Talk over the routine ones too, asking if you're overlooking something. I always prefer to disclose any minor conflict of interest myself, and I'm surprised at how often I overlook one. Asking for advice will prevent minor lapses. Not only that, it gets you into a good habit.

Construct Options

When a situation raises an ethical concern, think of the options. Are there ways to deal with this situation openly with all parties so that any appearance of improper behavior can be avoided? What is the most straightforward approach? Many of the borderline concerns can be addressed by merely disclosing the situation to all affected parties and asking their advice.

For example, a potential client calls asking if you would accept an assignment. Your first response is, "We are delighted to be on your list. But we have worked for Company X, one of your competitors. Would you see that as creating a conflict in this situation?" Many times they will say no. Before accepting the assignment, it may be wise to consult Company X too. They will appreciate being consulted and will remember your thoughtfulness.

Act Courageously

Recognize and support sound ethical decisions by others, even when they cost you convenience, money, or opportunities. Be tough-minded with yourself; What is your real motivation for this decision? Is it ethical standards? The expense? Convenience or expediency? To spare yourself embarrassment?

Be ready to admit a mistake—to yourself, to others, to the injured party. This can take uncommon courage. The most severe punishment that can be administered to a child is to be compelled to apologize. Some children never outgrow this. Admitting an ethical mistake will not damage your reputation—it will improve it.

Client confidence can be influenced by appearances—that is often all that people have to go by. Many of us react with disgust when we read about members of Congress junketing around in the private jets of high-rollers who are later found to be savings and loan looters. But we ourselves are not always careful to avoid or to carefully manage situations that might give rise to an appearance of a conflict of interest.

Develop Skills

A valuable aid in facing ethical situations can be found in Arch Patterson's (1984) four questions:

- What does my conscience say?

- What would it be like if everyone did this?

- How would it feel if everyone knew about this?

- How would I feel about this tomorrow?

Gaining wisdom is a journey, not a destination. A forester who can develop habits of ethical reflection will be best equipped to handle the ever more complex and murky choices that life brings. Gaining skill in ethical thinking is no different than gaining skill in estimating timber, playing poker, or fly-fishing. It requires expenditures of time, energy, interest, and learning. If you are really serious about fishing, you learn something even when you come home with no fish. This is also true of developing your skill in dealing with ethical questions.

When we treat ethics as a professional skill to be developed, and when we develop commitment to the highest personal and organizational standards of ethical behavior, then forestry will be well on the way to realizing the promises in its ethics codes.

Literature Cited

Patterson, A. 1984. Ethics in forestry. Speech presented at the Association of Consulting Foresters Practicing Foresters Institute IX, Athens, Georgia, October 7–11.

I. CODES OF ETHICS

The codes of ethics reproduced here include those of the two major organizations of professional foresters, the Society of American Foresters (SAF) and the Association of Consulting Foresters (ACF). The ACF is a smaller organization, comprised of foresters working as consultants. Many of its members are also members of the SAF. The third code is a short one adopted as a Code for Government Service by the United States Government. It is intended to cover a broad, diverse group of professions involved in government. To Chapter Three I have added a fourth code, that of scientific ethics for the United States Department of Agriculture (USDA), Agricultural Research Service.

Additional Reading

A more thorough guide to the SAF Code of Ethics has been published by the Society:

> *SAF Ethics Guide*
> Society of American Foresters
> 5400 Grosvenor Lane
> Bethesda, MD 20814
> (Price: $3.00)

Other professional groups have adopted ethical codes, some of imposing complexity, which you may read for the additional guidance they provide.

> *Code of Ethics and Guidelines to Practice*
> American Society of Civil Engineers
> 345 E. 47th St.
> New York, NY 10017

> *Professional Ethics and Standards*
> The Appraisal Institute
> 875 N. Michigan Ave., Suite 2400
> Chicago, IL 60611-1980

Principles of Appraisal Practice and Code of Ethics
American Society of Appraisers
P. O. Box 17265
Washington, DC 20041

Code of Ethics
Institute of Certified Financial Planners
7600 E. Eastman Ave., Suite 201
Denver, CO 80231-4397

Statement of Ethical Principles for Planning
and *AICP Code of Ethics*
American Planning Association
1776 Massachusetts Ave., NW
Washington, DC 20036

CHAPTER ONE

Code of Ethics for Members of the Society of American Foresters

Preamble

Stewardship of the land is the cornerstone of the forestry profession. The purpose of these canons is to govern the professional conduct of members of the Society of American Foresters in their relations with the public, their employers, including clients, and each other as provided in Article VIII of the Society's Constitution. Compliance with these canons demonstrates our respect for the land and our commitment to the wise management of ecosystems, and ensures just and honorable professional and human relationships, mutual confidence and respect, and competent service to society.

These canons have been adopted by the membership of the Society and can only be amended by the membership. Procedures for processing charges of violation of these canons are contained in Bylaws established by the Council. The canons and procedures apply to all membership categories in all forestry-related disciplines, except Honorary Members.

All members upon joining the Society agree to abide by this Code as a condition of membership.

Canons

1. A member will advocate and practice land management consistent with ecologically sound principles.

2. A member's knowledge and skills will be utilized for the benefit of society. A member will strive for accurate, current and increasing knowledge of forestry, will communicate such knowledge when not confidential, and will challenge and correct untrue statements about forestry.

Effective November 2, 1992.

3. A member will advertise only in a dignified and truthful manner, stating the services the member is qualified and prepared to perform. Such advertisements may include references to fees charged.

4. A member will base public comment on forestry matters on accurate knowledge and will not distort or withhold pertinent information to substantiate a point of view. Prior to making public statements on forest policies and practices, a member will indicate on whose behalf the statements are made.

5. A member will perform services consistent with the highest standards of quality and with loyalty to the employer.

6. A member will perform only those services for which the member is qualified by education or experience.

7. A member who is asked to participate in forestry operations which deviate from accepted professional standards must advise the employer in advance of the consequences of such deviation.

8. A member will not voluntarily disclose information concerning the affairs of the member's employer without the employer's express permission.

9. A member must avoid conflicts of interest or even the appearance of such conflicts. If, despite such precaution, a conflict of interest is discovered, it must be promptly and fully disclosed to the member's employer and the member must be prepared to act immediately to resolve the conflict.

10. A member will not accept compensation or expenses from more than one employer for the same service, unless the parties involved are informed and consent.

11. A member will engage, or advise the member's employer to engage, other experts and specialists in forestry or related fields whenever the employer's interest would be best served by such action, and members will work cooperatively with other professionals.

12. A member will not by false statement or dishonest action injure the reputation or professional associations of another member.

13. A member will give credit for the methods, ideas, or assistance obtained from others.

14. A member in competition for supplying forestry services will encourage the prospective employer to base selection on comparison of qualifications and negotiation of fee or salary.

15. Information submitted by a member about a candidate for a prospective position, award, or elected office will be accurate, factual, and objective.

16. A member having evidence of violation of these canons by another member will present the information and charges to the Council in accordance with the Bylaws.

CHAPTER TWO

Code of Ethics of the
Association of Consulting Foresters

These canons formulate the guiding principles of professional conduct for ACF Consulting Foresters in their relations with each other, their employers, the public and with other foresters. Observance of these canons secures decent and honorable professional and human relationships, establishes enduring mutual confidence and respect, and enables the profession to give its maximum service.

In order to apply these canons to the diverse circumstances arising in practice, guidelines are issued from time to time by the Executive Committee. They are incorporated by reference and made a part thereof.

Professional Life

1. ACF Consulting Foresters will utilize their knowledge and skill for the benefit of society.

2. An ACF Consulting Forester will cooperate in extending the effectiveness of the forestry profession by interchanging information and experience with other foresters and by contributing to the work of forestry societies, associations, schools and publications.

3. An ACF Consulting Forester will advertise only in a dignified manner, setting forth in truthful and factual statements services offered prospective clients and the public.

4. Professional work should come to ACF Consulting Foresters on the basis of their experience, competency and reputation. Solicitation by criticism of competitors, self laudation or lobbying is degrading to the profession and is unethical.

Effective May 31, 1991.

Dealing with the Public

5. An ACF Consulting Forester will strive for correct and increasing knowledge of forestry and the dissemination of this knowledge and will discourage and condemn the spreading of untrue, unfair, and exaggerated statements concerning forestry.

6. ACF Consulting Foresters will not issue statements, criticism, or arguments on matters connected with public forestry policies, without indicating, at the same time, on whose behalf they are acting.

7. When serving as an expert witness on forestry matters in a public or private fact finding proceeding, an ACF Consulting Forester will base testimony on adequate knowledge of subject matter, and render opinions based on honest convictions.

8. ACF Consulting Foresters will refrain from publicly expressing opinions on a technical subject unless informed of the facts relating thereto, and will not distort or withhold data for the purpose of substantiating a point of view.

9. Forestry plans and reports should be definite and specific and should have no double meaning.

Dealing with Clients, Principals, and Employers

10. ACF Consulting Foresters will be loyal to their clients and to the organization in which they are employed, and will faithfully perform their work and assignments.

11. ACF Consulting Foresters will clearly present the consequences expected if their professional forestry judgment is overruled by non-technical authority when they are responsible for the technical adequacy of forestry or related work.

12. ACF Consulting Foresters will not voluntarily disclose information, received in confidence, concerning business affairs or their employers, principals or clients, unless express permission is first obtained.

13. An ACF Consulting Forester must avoid conflicts of interest or even an appearance of such conflicts.

14. ACF Consulting Foresters will not, for the same service, accept compensation of any kind other than from their

clients, principals, or employers, without full disclosure, knowledge and consent of all parties concerned.

15. ACF Consulting Foresters will engage, or advise their clients or employers to engage, other experts and specialists in forestry and related fields whenever the clients' or employers' interests would be best served by such actions, and will cooperate freely with them in their work.

16. An ACF Consulting Forester should not undertake work at a fee that will not permit a satisfactory professional performance.

17. ACF Consulting Foresters will not use association with a nonforester, a corporation or partnership as a cloak for unethical acts, but must accept responsibility for their acts.

18. Under no circumstances should ACF Consulting Foresters undertake to make an appraisal when their employment or fee is contingent upon the amount of their estimate of value.

19. It is unethical for an ACF Consulting Forester to pay or claim a fee or commission for the purpose of obtaining or referring employment.

Dealing with Professional Foresters

20. An ACF Consulting Forester will at all times strive to protect the forestry profession collectively and individually from misrepresentation and misunderstanding.

21. An ACF Consulting Forester will aid in safeguarding the profession against the admission to its ranks of persons lacking good moral character or adequate training.

22. In writing, or in speech, ACF Consulting Foresters will be scrupulous in giving full credit to others, in-so-far as their knowledge goes, for procedures and methods devised or discovered, ideas advanced, or aid given.

23. An ACF Consulting Forester will not intentionally and without just cause, directly or indirectly injure the reputation or business of another forester.

24. If ACF Consulting Foresters have substantial and convincing evidence of unprofessional conduct of another forester, they will present the information to the proper authority for action.

25. An ACF Consulting Forester will not attempt to supplant another forester in a particular employment, after becoming aware that the latter has been definitely engaged.

26. An ACF Consulting Forester will base all letters of reference, or oral recommendation, on a fair and unbiased evaluation of the party concerned.

27. An ACF Consulting Forester will not solicit or collect financial contributions from subordinates or employees for political purposes.

28. An ACF Consulting Forester will uphold the principle of appropriate and adequate compensation for those engaged in forestry work, including those in subordinate positions, as being in the public interest and maintaining the standards of the profession.

CHAPTER THREE

A Code of Ethics for Government Service

PL 96-303 Signed by the President on July 3, 1980

1. Put loyalty to the highest moral principles and the country above loyalty to persons, party or government department.

2. Uphold the Constitution, laws, and regulations of the United States and of all governments therein and never be a party to their evasion.

3. Give a full day's labor for a full day's pay; giving earnest effort and best thought to the performance of duties.

4. Seek to find and employ more efficient and economical ways of getting tasks accomplished.

5. Never discriminate unfairly by the dispensing of special favors or privileges to anyone, whether for remuneration or not; and never accept, for himself or herself or for family members, favors or benefits under circumstances which might be construed by reasonable persons as influencing the performance of governmental duties.

6. Make no private promises of any kind binding upon the duties of the office, since a Government employee has no private work which can be binding on public duty.

7. Never engage in any business with the government either directly or indirectly, which is inconsistent with the conscientious performance of governmental duties.

8. Never use any information gained confidentially in the performance of governmental duties as a means of making private profit.

9. Expose corruption wherever discovered.

10. Uphold these principles, ever conscious that public office is a public trust.

Code of Scientific Ethics for the United States Department of Agriculture, Agricultural Research Service

- I dedicate myself to the pursuit and promotion of beneficial scientific investigation, consistent with the mission of the Agricultural Research Service.

- I will never hinder the beneficial research of others.

- I will conduct, discuss, manage, judge, and report science and that of my colleagues in a manner that fosters harmony and quality amid scientific debate.

- I recognize past and present contributors to my science and will not accept unwarranted credit for the accomplishment of others.

- I will maintain and improve my professional skills and be a mentor to others.

- I will ensure safety and humane treatment of human and animal subjects and will prevent abuse of research resources entrusted to me.

Source: United States Department of Agriculture, Agricultural Research Service Directive 129.0, "Procedures for Reporting and Dealing with Possible Misconduct in Science."

II.
PROFESSIONAL
ETHICS

The distinguishing mark of a profession is that its members are supposed to exercise their expert judgment not only in the service of clients but also in the broader interests of society. Clients expect from professionals that their proprietary information will remain confidential, that loyalty to their interests will be undiluted by conflicts of interest, and that technically sound methods and procedures will be employed. These concerns are explored in the readings and cases in this book. Social considerations most often arise in the forestry profession in regard to the land ethic, considered in a later section.

A professional claims specialized expertise that can be the basis for admission as an expert in a court of law. In forestry, professional expertise is frequently relied upon by clients for guidance in expensive decisions, such as the sale of property, construction or closing of mills, or the conduct of harvesting operations. In addition, professionals are usually presumed to be credible by the press and the general public.

Personal ethical standards, of course, go far beyond the specific mandates of professional codes. Archie Patterson, an early writer on forestry ethics, suggests four questions that will help foresters make the ethical decisions: What does my conscience say? What if everyone knew? What if everyone did it? What will it look like tomorrow? These questions do not depend on any particular ethics code. They do not even depend on the questioner having a particularly well-developed ethical sense or large store of ethical knowledge. Yet, asking them and reflecting on them more often could lead to a noticeable improvement in general ethical standards.

Though it was founded in 1900, the SAF did not adopt a formal code of ethics until 1948. Joe Lammi's 1963 paper summarizes the historical development of this code, providing the necessary background for the essays that follow. Professional codes of ethics must, however, be applied in the day-to-day decisions and plans made by all resource managers or they mean nothing. Foresters are often required by organizational policy to report ethical violations they have observed and to comply with rules with which they may not individually agree.

The essays by Flanagan and myself illustrate some of the ethical questions that arise in the areas of land use and pesticide use. For example, imagine a forester for a public lands agency who is working with a private landowner to obtain a right of way to gain access to public land. How can this forester expect to report the landowner's apparent violations of forest practice rules while negotiating with that same landowner on other issues?

One field in which foresters claim specialized knowledge is in assessing the productivity of forests and determining their potential for sustained yield production. Foresters regularly advise small landowners, industrial corporations, and government agencies as to the sustained yield potential of particular forest tracts. It could be said that the concept of sustained yield is the most basic concept underlying the professional practice of forestry—but is sustained yield an ethical mandate? The SAF Code of Ethics certainly suggests in more than one place that it is. Klemperer suggests, however, that sustained yield of timber is not a socially significant concern and is therefore not an ethical mandate for professional foresters.

A thoughtful paper by a military officer, Kermit Johnson, on ethical issues in military service offers useful insights on the personal conflicts encountered when serving in any large organization. That members of the military are increasing their ethical self-evaluation and reflection illustrates a positive trend in society at large.

It is useful in analyzing matters of consequence to seek out conflicting views. John Ladd's essay provides a valuable contrast to the opinions expressed in other essays about the importance and role of ethics codes. He suggests that professional ethics codes contribute little of value and even create a potential for mischief.

Many major government policy decisions are now being made on the basis of facts and projections that are strongly contested within the scientific community. Interest groups regularly conduct their own research or syntheses of knowledge in order to support their own policy preferences. There are many temptations for analysts to selectively edit and interpret facts to support their personal views or to support outside groups in ways that undermine official policy. Additionally, there are instances in which government agencies seek to prevent individual scientists from making their conclusions and views more widely known, as in the case of Howard Wilshire. The excerpt from the report *On Being a Scientist* by the National Academy of Sciences reviews a number of ethical issues, including human error, outright fraud, apportionment of credit, credit and responsibility, and plagiarism. All these issues arise in the professional practice of forest land management.

The papers in this section show how ethical conflicts can affect foresters in their relationships with each other and with their superiors in working with clients and in dealing with citizens and other groups outside their organizations. Sometimes these conflicts concern technical concepts like sustained yield, which are at the center of a forester's technical expertise. More often, they deal with policy matters or questions of prudence for which technical training confers no special wisdom. Themes that emerge from these readings include

- the importance of being sensitive to potential problems before they occur;

- the likelihood that coping with a problem will be easier and less costly if detected at an early stage;

- the importance of honest self-scrutiny;

- a range of alternative solutions exists for most ethical conflicts;

- in many ethically problematic situations, candid consultation with the affected parties is often helpful;

- many ethics issues are connected, and conflicts between legitimate loyalties are frequent. For example, Howard Wilshire, discussed in Eliot Marshall's first article, faced a conflict between his loyalty to his employer and his professional concern for what he considered wise resource management.

As Roger Merchant's essay suggests, relying on a mentor early in your career can assist greatly in developing skills in ethical reflection.

Additional Reading

Anderson, C. Survey tracks misconduct, to an extent. *Science* 262(1993):1203–1204. Notes reactions and concerns about methods used in Swazey et al. article.

Branscomb, Lewis M. Integrity in science. *American Scientist* 73(1985):421–423.

CBE Style Manual Committee. *CBE Style Manual. Fifth Edition.* Bethesda, MD: Council of Biology Editors, Inc. 1983. Ch. 1, Ethical conduct in authorship and publication, should be read by anyone before submitting work for publication.

Chalk, Rosemary; Frankel, Mark S.; and Chafer, Sally B. *AAAS Professional Ethics Project: Professional Ethics Activities in the Sciences and Engineering Societies.* Washington: American Association for the Advancement of Science, 1980.

Frankel, Mark S. Professional codes: why, how, and with what impact? *Journal of Business Ethics* 8(1989):109–115. A thoughtful review of the problems in developing and enforcing professional ethics codes.

_____ , ed. *Science, Engineering, and Ethics: State of the Art and Future Directions.* American Association for the Advancement of Science

(AAAS), Office of Scientific Freedom and Responsibility, 1988. Includes many citations concerning ethics in engineering, and useful comments by Robert Swartz, p. 23 ff. on what he calls "values reasoning."

Hamilton, David P. A shaky consensus on misconduct. *Science* 256(1992):604–605.

_____. In the trenches, doubts about scientific integrity. *Science* 255(1992):1636.

Hazard, Geoffrey C., Jr. *Ethics in the Practice of Law.* New Haven: Yale University Press, 1978. (especially ch. 5, conflict of interest)

Schachtman, H. K. What is misconduct in science? *Science* 261(1993):148–149, 183. Discusses issues encountered in enforcing rules on fraud in science funded by government agencies.

Sigma Xi. *Ethics, Values and the Promise of Science. Forum Proceedings.* Research Triangle Park: Sigma Xi, 1993. Valuable readings on ethics in science.

Swazey, J. P., M. S. Anderson, and K. S. Lewis. Ethical problems in academic research. *American Scientist* 81(1993):542–553. Major survey that documents perceptions among graduate students and faculty about unethical behavior.

US Congress. House. *Forest Resource Management and Personnel Practices: Values in Conflict.* Comm. on Post Office and Civil Service, 101st Cong. 2nd Sess. Ser. No. 101-80, 1991. Excellent source on the conflicts of values and generations within federal land managing agencies.

The Vietnam experience illustrated many of the challenges of coming to grips with professional ethics during particularly momentous events. Readers seeking ethical reflections in those events should consult the following:

Kinnard, (Gen.) Douglas. *The War Managers.* Hanover, NH: Univ. Press of New England, 1977. (especially section in ch. 5 on "Professionalism")

Lewy, Guenter. *America in Vietnam.* New York: Oxford University Press, 1978. (especially ch. 7, "American military tactics and the laws of war;" ch. 9, "Atrocities: fact and fiction;" and the "Epilogue")

The volume from which the Johnson article was selected contains other excellent pieces on military ethics:

Matthews, Lloyd J.; and Brown, Dale E. *The Parameters of Military Ethics.* Washington, D.C.: Pergamon-Brassey's International Defense Publishers, Inc., 1989.

These three volumes deal with the problems of policy and senior command rather than with day-to-day issues faced at lower levels.

Archie E. Patterson was an early contributor to the subject of professional ethics in forestry. In 1949, he produced a syllabus on professional ethics that remains relevant today. In this shorter contribution he offers a more informal overview of professional ethics, including four questions that will help readers in thinking about ethics questions. You will find it helpful to keep these four questions in mind as you read later articles. Patterson is Professor Emeritus of Forest Resources at the School of Forestry, University of Georgia, Athens, GA.

CHAPTER FOUR

Ethics in Forestry:
Four Self-Help Questions

ARCHIE E. PATTERSON

Few professions in the United States can look back upon their historic origins with such justifiable pride as can the profession of forestry. Since the days of Fernow, Pinchot and Graves our profession has been marked as one committed to altruism and the moral law. As a profession we have striven unceasingly to further ourselves within the bounds of these moral obligations. We have made progress; slowly, no doubt, at times, but by the relentless effort of putting one foot before the other we have begun to attain a fruition of our efforts. In so doing we have accepted the responsibility offered us, a responsibility which we must maintain out of our maturity and moral fibre. This acceptance of responsibility is what the public expects of us; the protection of their inherent forest rights, and their forest property, is what they demand.

Certain witness trees and boundary markers have already been established for ·us. Each of us who is a member of the Society of American Foresters or the Association of Consulting Foresters has subscribed to a forester's code of ethics. In my opinon, no greater step forward has ever been taken by the profession of forestry than the adoption of these codes. In addition to these codes, in several states we have taken a further step and introduced licensing of foresters. This adds weight to our professional standing, and although it does not circumscribe our actions beyond the moral obligations of a code of ethics, it serves us well in the eye of the public.

None doubt the need, wisdom, and value of a code of ethics, and those who adhere to the principles of a code have no fear or valid objection to licensing. I fear, however, that among some, a feeling

This article was presented at The Association of Consulting Foresters' Practicing Foresters Institute IX held in Athens, GA, on 7–11 October 1984. Used by permission.

exists that a code of ethics and licensing set forth the only professional obligations we have, and that all else is right and acceptable. Nothing could be further from the truth. Both a code and licensing are of the utmost value, but no code and no form of licensing can cover all existing situations, nor anticipate all those which will occur in the future.

What, then, shall be our guide? What shall be the restraint that we all need? May I propose that, in order for our profession to properly serve the public, we ask ourselves the following four questions concerning our daily manner of professional conduct?

First: *What does my conscience say?* This is the old reliable standard; the one that has been impressed upon us since childhood, and for some of us it may be enough. Conscience, however, is a faculty, or power, or principle conceived to decide as to the quality of our thoughts or acts. It enjoins what is good, but unfortunately we are not born with it; it must be learned. It was taught to us at our mother's knee, and has been bolstered throughout our life by various institutions. But the very fact that it must be learned makes it susceptible to being forgotten, or altered, especially if it has become corroded by disuse. We must also recognize that we can argue with our conscience, and we would not be human if we did not admit that from time to time we believed we had won the argument. I challenge you, therefore, to ask yourselves: Can I wholly rely upon the infallibility of my conscience? Is it the same all of the time? Or does it change from time to time, and from place to place?

If you are not satisfied with your answers, ask yourself this question concerning your day-to-day professional dealings with the public: *What if everyone knew?*

Are you willing to lay your professional dealings of the past week open to the glare of publicity? Are you willing to have them examined openly by the other foresters in this room? If not, how long do you think you can keep them hidden? Another week? Another month? Another year?

I recently read of an ancient Asian temple which was being repaired. When it was built, nearly one thousand years ago, its teak roofbeams had been sunken into the hillside wall of the temple. Supposedly, none would ever see or examine the ends of those beams. Yet, as repairs were made, it was discovered that the ends of the sunken beams had been as carefully and beautifully carved as those that were exposed to the view of every passerby. The long-dead artisans and craftsmen who built that temple acquired the respect of all who examined their work. What if everyone knew? Doesn't everyone, who really matters, know, eventually?

Let us move to a more practical question: *What if everyone did it?* Would you be able to conduct your various businesses if all your competitors operated in the same manner as you do? Examine your professional dealings of the past few days, and try to list the almost innumerable times your actions were based upon honesty, trust, faith, and the spoken assurance. "Put it in writing," is a byword of business conversation, but only a minute fraction of business is conducted in writing. Consider your very livelihood: I will wager that less than ten percent of those in this room will go to work tomorrow morning under a written contract. Your entire business structure is based upon the honesty and faith in those with whom you deal. Would you have it otherwise? I think not. You desire integrity in all of your professional aspects, but there are times when the line of demarcation is very fine; when the proposed action is not covered by code or conscience. When this situation occurs, ask yourself: *What if everyone did it?*

One final question: *What will it look like tomorrow?* We are all human, and one of our common failings is to be emotionally aroused to the extent that we sometimes fail to reason clearly. The excitement of the moment, or the expected results, may lead us down the paths of regret; cause us to make rash promises which we cannot fulfill; stimulate us to actions which we ordinarily would shun. How will these actions appear in the glare of tomorrow's retrospection? Can we afford to build our professional reputation upon our present day actions? Whether we like to face it or not, every professional action we take is another block in our pyramid of professional stature, and the taller we build the easier it is for others to see.

Answers to these four questions are, I believe, the answer to the question of how we can best serve the public. Not by any materialistic listing of doing this or doing that—the materialistic is expected of us. It is a day's work for a day's pay; it is the application of the technical know-how; it is the expected service rendered for fee paid.

As Joe Lammi writes, a formal code of ethics for the Society of American Foresters was many years in the ripening, originating in suggestions by some of the profession's most distinguished founders. Lammi discusses the goals of society as well as the goals of the forestry profession, concluding that ethics is conduct that facilitates "freedom and responsibility."

Lammi served with the US Forest Service and then taught for many years in the forestry department at North Carolina State University, where he gave much time to fostering the growing specialty of forest economics in the South.

CHAPTER FIVE

Professional Ethics in Forestry

J. O. LAMMI

A profession has been defined as "the practice of an art, based on knowledge and skill acquired through rigorous education at the college or university level, and conducted in accordance with high ethical standards of performance and service" (13). Forestry is generally accepted as a profession and its code of ethics, adopted by the Society of American Foresters in 1948 (1), is an indication of professional maturity and acceptance of responsibility. Practicing foresters can benefit from the code in that it serves to guide them in their relationships with colleagues and clients. Students will find the code a useful orientation and an introduction to some of the philosophical aspects of their chosen profession.

History of the Code

The foresters' code did not develop spontaneously—it was the culmination of a great amount of thought and effort put forth by many members of the Society of American Foresters. The needs for it were discussed and debated by those who recognized that rules of conduct, clearly stated and formally accepted by foresters, would add to their professional stature. Over 30 years of the maturing process were needed, however, before a written foresters' code of ethics became a fact.

The first mention of an honor code is credited (5) to B. E. Fernow in 1914. He suggested the establishment of an "Honor Committee" and eventually a "suggestive code" for the guidance of the profession. In 1931, F. E. Olmstead (23) declared that destructive forest

The ideas in this article resulted from discussions with the Forestry Club at Duke University, Durham, NC. Reprinted from *Journal of Forestry*, Vol. 66 (2), February 1968, pp. 111–114, by permission of the Society of American Foresters.

practices were unethical, and discussed the ethical problems of private forestry employment in considerable detail. Olmsted contended that ethical forestry was that which was in the public interest, and "in the long run . . . private and public welfare are one and the same thing."

The decade following World War I was a period of expansion in private forestry with consequent new recognition of ethical problems and needs. The first written code of ethics was presented to the 1923 annual meeting of the Society of American Foresters. A member of the drafting committee for the code, Theodore S. Woolsey, Jr., in commenting on the code (26) defined forestry practices as ethical "if in the long run they make for the well-being of the human species and for normal human relations." Woolsey went on to say that "if there is friction and social loss, it is a sign of unethical practices."

A few years later, during the great expansion of forestry work under the Roosevelt "New Deal," the need for attention to ethical standards was again vividly apparent. The Civilian Conservation Corps and the growth of state forestry brought on problems of "friction and social loss" such as Woolsey had mentioned. Political influences in forestry had never before been as great nor as obnoxious. The professional society was duly concerned (4, 7, 9, 12) and a code of ethics was demanded. It is not clear from the record why no action was taken on the code of 1923; a new effort to draw up some rules was started in 1931 with studies by Guthrie (22) followed by the committee work of Kahn, Chapman, Clepper, Evans, and Moir (11).

Many voices were raised in support of the movement toward higher levels of ethics. In an eloquent plea for a "conservation ethic," Aldo Leopold (19) argued the need for "dealing with man's relationship to land and to the nonhuman animals and plants which grow upon it." [see "The Conservation Ethic" reprinted in this volume] Leopold defined an ethic as a differentiation of social from antisocial conduct. Another socially conscious forester, Bernard Frank, followed Leopold in the pages of the Journal [of Forestry] (14) with the statement that ". . . if we are to rise above trade levels we must be interested in more than technical achievement. Membership in the Society should carry . . . ethical obligations . . . and every member should be expected to fulfill them."

Results were not immediately apparent from the study of ethics which Guthrie began in 1931, but Chapman recalled later (22) that, although a general code of ethics could not be finalized during the New Deal days, a special code dealing with civil service and political appointments was successfully formulated and agreed upon in 1935. This code was beneficial to foresters, although outside the Society framework, Chapman claims credit for drawing up the initial draft of the foresters' code which was ultimately adopted, its basis being the code of ethics of the Society of Mechanical Engineers (*11, 22*).

The present foresters' code was enacted by a vote of the Society membership in 1948 (*1*). It is available in a form suitable for framing, and a syllabus intended especially for student instruction was issued in 1949. The Council of the Society authorized a standing Committee on Ethics in 1950 (*2*) to replace the former Committee on Maintenance of Professional Standards. The full text of the code appeared in the *Journal* [*of Forestry*] of January 1953 (*3*).

In looking back at the history mentioned in the preceding paragraphs, it is evident that foresters have been strongly motivated, in their struggle for ethical improvement, by the ideal of public service. At the turn of the century nearly all American foresters were public servants. As the size of the private employment sector increased, the attitude of foresters did not greatly change. They still considered themselves the custodians of a public trust and continued to subscribe, as Pinchot suggested (*24*), to "Theodore Roosevelt's motto: the public good comes first."

The Meaning of Ethics

The enactment of a code of ethics is a milestone in the growth of the forestry profession. Foresters will want to study the code for the guidance it gives to the maintenance of effective communication and collaboration between colleagues, and the improvement of relations with employers and the ever-increasing number of clients and forest resource users. Ethics are important also in the delicate relations that must be maintained with the vast majority of the public who influence forest policy through their contact with legislators and who have very little knowledge or appreciation of forest resource management.

An understanding of ethics and an improvement of the professional performance mentioned above can be helped by answers to two questions: (1) What are ethics, and (2) how are decisions made on ethical questions?

To answer the first question requires a sorting out of the three related concepts: ethics, morality, and religion. Chapman pondered these questions when he was working on the foresters' code and concluded (*8*) that religion could be distinguished from ethics by the fact that the "tenets of religion relate broadly to human life rather than specifically to professional conduct."

Morals and morality can be categorized as being concerned with the rules and practices of the conduct of an individual within a society. Ethics, on the other hand, relate to individual conduct and group activity with respect to contributions to the goals of a particular profession or human society as a whole.

Ethical practices of a professional man can then be described as those which contribute to progress toward the goals of the profession or the human society. Unethical practices are those which hinder this progress.

Goals are mentioned in the above definition; how can these goals be identified? Of first importance are the goals of society as a whole and to these can be related the objectives sought by the forestry profession.

Goals of Society

A review of some of the thinking of the philosophers who have influenced our civilization may be helpful in identifying goals.

One of the most powerful of ideas has been that of hedonism, the belief that the ultimate purpose of living is happiness or pleasure. Among the famous hedonists can be listed Aristotle and Plato (4th century B.C.), Saint Thomas Aquinas (13th century A.D.) and, more recently, Jeremy Bentham and John Stuart Mill. The philosopher who perhaps influenced the modern world most profoundly, Immanuel Kant (1724–1804), was also greatly interested in the role of happiness in human affairs (17).

Happiness is a happily vague idea and one with which most people can agree. In an attempt to make the idea of happiness less vague, Bentham (6) reasoned that the ultimate goal of civilization was the greatest happiness of all conscious beings, with the total happiness constituting the aggregate of the happiness of the mass of people rather than the intensity of happiness of a few.

Gifford Pinchot may well have had Bentham in mind when he wrote the letter for the signature of the Secretary of Agriculture which included the subsequently famous forestry platitude of "the greatest good for the greatest number." Pinchot's "greatest good" is Bentham's "greatest happiness."

The idea of happiness as a general goal of living has caused many conceptual difficulties. Happiness to one person means something different than it does to another: individual happiness may be incompatible with the happiness to a group or to society: frequently, this notion applied to a country is different than it is in the World context. Ultimately, it is the happiness of the World that takes priority over a country or a group or an individual. This is the thought that motivates much of the international political and economic activity, the search for peace and a better life.

In the context of the world as a whole, the idea of happiness as a goal of human society appears to be a vague and dim goal indeed. What relationship and what useful meaning is there between happiness to an Amazonian aborigine and happiness to an individual in the top stratum of the affluent society? Dag Hammarskjold, the late Secretary-General of the United Nations, a statesman and philosopher, must have often thought about this matter. In his memoirs (16), he gave a meaningful expression of an ethical goal when he said "To become free and responsible. For this alone was man created, and he who fails to take the way which could have been his shall be lost eternally."

The goal which Hammarskjold sets out—*freedom and responsibility*—is based on the justice and equity of Kant (18) and the liberty which Mill (21) emphasized. Hammarskjold's goal has its roots in the philosophy of the past but its appeal is to the present and the future. It expresses a realistic goal, to be free but to be free only with the protection of responsible conduct.

Goal of Professional Forestry

As mentioned in a previous paragraph, it is desirable to relate professional objectives to the goals of the society of which the profession is a part.

If *freedom and responsibility* are acceptable as goals of society can they also be the goals of the forestry profession? Are the goals of the profession identical with those of the society of which it is a part? Foresters are familiar with the many goals or objectives that have been expressed at one time or another for the practice of forestry. A recent issue of the *Journal* [*of Forestry*] (15), for example, lists as forestry objectives the production of human satisfactions (Gould), welfare and needs of the public (Connaughton), wise management of natural resources (Dana), greatest good for society as a whole (Pomeroy), wise management of wildland resources (Orell), and public welfare (Zivnuska). Pinchot's greatest good for the greatest number has already been mentioned. Other familiar goals are sustained yield of forest goods and services and preservation of the nation's heritage. Additional goals are undoubtedly mentioned in the literature.

Public benefits appear foremost in the above expressions of professional goals. Can these be related to the overall goal of *freedom and responsibility*? The relation of forestry activities to *freedom* will be looked at first, then the clear ethical issues of *responsibility* will be pointed out.

Freedom occurs in many forms—political, social, economic, religious, intellectual—but Franklin Roosevelt expressed the idea effectively in his 1941 speech to the U.S. Congress. The four freedoms of Roosevelt were Freedom of Speech and Expression, Freedom of

Worship, Freedom from Want, and Freedom from Fear.

The goals of forestry, whether implicit or explicit, contribute to the Four Freedoms: resource management has a large share in the Freedom from Want, the resources of the forest provide national strength that builds Freedom from Fear, the profession actively supports ideals of free expression and the protection of minority groups. As citizens and as professional men and women, foresters stand for the Four Freedoms.

The second part of the overall goal—responsibility—is the heart of the ethical problem. Responsibility separates the professional man from the craftsman. Responsibility to the public transcends selfish and narrow motivations. The professional man accepts responsiblity when he enters the profession, and he carries out his duties with a high sense of responsibility ever in mind. Personal gain, political advantage, short-run interest, all are eschewed for the permanent public benefit!

The principle of responsibility to society as a whole is a basic difference between a professional society and a labor union. It is also the basis for the foresters' concept of conservation in that it rejects the irresponsibilities of the extremists, those who advocate the very low level of resource use (preservationists) and those who look only at immediate money returns (despoilers). Precisely this same idea was advanced by Chapman over 43 years ago (10).

The foregoing discussion has answered the question of "what is ethics" by indicating that it is professional conduct and practice that facilitates progress toward a goal. The goal was shown to be *freedom and responsibility.*

Remaining to be answered is the second query: "how are decisions made on ethical questions?" As background to the answer, some problems of forestry ethics will be briefly described.

Some Problems of Forestry Ethics

The objectives of forestry have been expressed several ways, as noted in an earlier paragraph. All of them are included in the following statement (5) of the purpose of the profession: "to apply our specialized knowledge toward making the country's forests yield their fullest contribution to the economic and social welfare of the nation."

The statement fits the goal of *freedom and responsibility.* Ethical forestry contributes toward the goal, unethical practice hinders progress toward this goal.

Types of ethics problems which have occurred were listed by Chapman (8) as: unprofessional conduct, false and derogatory statements, questionable dismissal of employees, and professional advice to an employer of a forester without informing the latter.

Unethical practices also include participation, without full professional efforts to resolve conflicts, in the disputes over land use such as the wilderness proposals. Similarly, a forester who promotes destructive timber harvesting or advocates an anti-social land use practice is violating professional ethics, even though he may be within the rules of the written code.

Are the goals of the profession compatible with certain foreign aid projects, regional economic development proposals, reservoir construction programs and taxation schemes? What is the professional judgment on expansion of wood manufacturing plants in areas of increasingly scarce timber supply? Where does the forester stand on the recreation reservation questions that have a great impact on local and regional economies?

The answers to all the above questions require ethical judgments.

Not all violations of ethical standards are deliberate ones, they may stem from lack of knowledge or may even arise from insufficient means to accomplish objectives.

Unethical conduct can thus be classified into: category 1, the deliberate choice cases; category 2, the lack of knowledge; and category 3, the lack of means.

The most abhorrent violations of ethics are the deliberate ones. The penalties which the professional Society can impose include expulsion from professional ranks.

Category 2, the lack of knowledge situations, receives remedial action through ·the normal course of professional development. Forestry schools have a prominent place in overcoming these obstacles to progress toward professional goals.

Category 3, ethics problems which arise from lack of means, are particularly prominent in small-tract forestry and in international forestry development. Public policies, political changes, education— all can be influential in attracting resources that will speed up progress toward goals.

The above categorization of ethical problems is different and more comprehensive than most foresters have used in the past. All three categories are limitations to progress, and therefore unethical. Only category 1, however, is morally despicable.

With the above in mind, the answer to the question "how are decisions made on ethical questions" can be approached.

Action by Foresters

An employer may be unethical but a professional man's loyalty to his employer should not influence condoning the unethical practice. If the employer cannot be persuaded to change his ways, the forester may be well advised to seek other employment.

The question of ethics in "moonlighting" (20, 25) can be resolved when better knowledge is available about the circumstances of this activity. If the moonlighter can provide better services to his employer *and* his client, his work is ethical whereas, in such instance, the complainer who is concerned with his selfish monetary interest would be clearly unethical.

The code of ethics helps the individual guide his personal conduct in this profession. The more general ethical problems require the attention of the professional society through committee or special study action. The Society's National Policy Committee (15) may prove to be the best vehicle to deal with the broad problems which Leopold would have included in the "conservation ethic."

The individual forester will look to the Society for continued expansion of the activity in the "conservation ethic." For decisions regarding his own conduct he may find some help in answers to the following questions: Does the proposed decision or action contribute to the goal of forestry? Is it likely to violate one of the 25 items of the foresters' code? Does one's conscience remain clear after the decision? What are the opinions of trusted colleagues who are familiar with the circumstances of the decision?

The Society of American Foresters and each of its members will need to apply their best efforts to move the profession ahead. American forestry has a past record of distinguished progress sparked by some lively controversy. The future is likely to be even faster-moving, more controversial and certainly more complex. Much of the increased complexity of forestry stems from the changing emphasis in the direction of human relations and away from the traditional major preoccupation with timber growing. Professional ethics will have a greater role in the new age of forestry than they had in 1914 when Fernow began this discussion.

Literature Cited

1. Anon. 1949. Code of ethics adopted by foresters. Jour. Forestry 47:70–71.
2. _____ . 1950. Committee on ethics. Jour. Forestry 48:370.
3. _____ . 1953. Foresters' code of ethics. Jour. Forestry 51:71.

4. _____ . 1934. Society protests removal of State Forester Jackson. Jour. Forestry 32:1031–1032.
5. _____ . (Editorial). 1937. The foundation of professional ethics. Jour. Forestry 35:237–239.
6. Bentham, Jeremy,1948. An introduction to the principles of morals and legislation. Hafner Publ. Co., N.Y. 278 pp.
7. Chapman, H. H. 1936. Appointment of Georgia state forester questioned, Jour. Forestry 34:431–432.
8. _____ . 1947. Does the "profession" of forestry need a code of ethics? Jour. Forestry 45:61–64.
9. _____ . (Editorial). 1934. Politics versus efficiency in government. Jour. Forestry 32:521–523.
10. _____ . 1924. The relation of the Society of American Foresters to the profession of forestry. Jour. Forestry 22:9–15.
11. Chapman, H. H. *et al.* 1948. Report of committee on canons of ethics. Jour. Forestry 46:625–627.
12. _____ . 1934. "Voluntary" political contributions in Indiana. Jour. Forestry 32:874–876.
13. Dana, S. T. and E. W. Johnson. 1963. Forestry education in America, today, and tomorrow. Soc. of Amer. Foresters. Washington, D.C. 402 pp.
14. Frank, Bernard. (Letter to the Editor). 1933. The profession and the Society. Jour. Forestry 31:991–992.
15. Glascock, H. R., Jr. 1966. A symposium—what scope for S.A.F. policy activities. Jour. Forestry 64:517–526.
16. Hammarskjold, Dag. 1964. Markings. Alfred A. Knopf, N.Y. 222 pp.
17. Kant, Immanuel. 1963. Analytic of the beautiful. The Bobbs-Merrill Co., Inc., N.Y. 141 pp. (Trans. by Walter Cerf).
18. _____ . 1963. Lectures on ethics. Harper and Row, N.Y. 253 pp. (Trans. by Louis Infield).
19. Leopold, Aldo. 1933. The conservation ethic. Jour. Forestry 31:634–643.
20. Mason, Howard F. R., Jr. 1964. Comments on "moonlighting"—a professional problem. Jour. Forestry 62:344–345.
21. Mill, John Stuart. 1936. Utilitarianism, liberty and representative government. E. P. Dutton and Co., Inc. N.Y. 393 pp.
22. Montgomery, David. 1962. Evolution of the Society of American Foresters. 1934–1937, as seen in the memoirs of H. H. Chapman. For. History. Vol. 6, No. 3. pp. 2–9.
23. Olmsted, Frederick E. 1922. Professional ethics. Jour. Forestry 20:106–112.
24. Pinchot, Gifford. 1941. The public good comes first. Jour. Forestry 39:208–212.
25. Vardaman, James M. 1964. Moonlighting—a professional problem. Jour. Forestry 62:48–49.
26. Woolsey, Theodore S., Jr. 1924. A forester's code of ethics. Jour. Forestry 22:59–61.

Professional growth in ethical awareness is a major theme of Roger Merchant's essay. He emphasizes the importance of having a strong mentor who can "listen, share, and teach." Too many of us learn from people who model *poor behavior*—business or bureaucratic leaders whose values more resemble J. R. Ewing's than those of Merchant's mentor. Merchant suggests that each of us, as we rise in our professions, has a responsibility to act in ways that model ethical behavior.

Merchant is an extension agent in Maine.

CHAPTER SIX

Personal and Professional Ethics

ROGER MERCHANT

To learn and grow as a person and a professional, it helps to cultivate an attitude of openness. The thoughts and recollections shared in this presentation reflect my experiences with ethics and codes of personal and professional conduct.

When I was a "greenhorn" forester the meaning of ethics was remote to my practices in the field of forestry. Occasionally, I would hear of a forester who had been censured for misconduct and would think: "Well now, there must be a bad apple." I had no idea of the particulars surrounding this type, so I judged the person in ignorance. Back then, communication seemed impersonal and nothing like what is needed today. In my view, open, supportive, thoughtful, and constructive communication is a necessity.

As a company forester, I took pride in my successes, yet stumbled and tripped over my ignorance-caused mistakes. Sometimes my errors were in technical forestry matters, more often they involved relationships with people.

Today's well-indoctrinated work ethic—"success is what is most important no matter how you get there"—was usually responsible for my "stumbling." What I needed most was support and help in dealing with problem areas. Hiding from the truth was never easy because my conscience was intact. Due to my personal ethics and honesty, I was eventually able to recognize problems and get help from those around me.

My personal and professional growth was enhanced by having a mentor. A mentor has knowledge, experience, and a willingess to listen, share, and teach. My mentor was Oric O'Brien. He listened, was supportive, and advised in a way that was neither overbearing nor arrogant. He had been a greenhorn once and understood both the

This article was presented at the First Annual Conference for Licensed Professional Foresters held at the University of Maine, Orono, on 14 May 1987. Reprinted by permission from *Forestry Notes*, Vol. 2 (1), 1988, pp. 1–4.

human and technical aspects of the art and science of forestry. I grew in many ways from our relationship.

Today, I believe that mentors are an important bridge to help come to grips with our current age of high-tech solutions and dehumanization. I propose that we speak with a *human voice,* as well as a *technical voice.* If you don't have a mentor, think about it in terms of what you have mastered, as well as what you haven't. Seek one who is suitable for your needs. Consider a mentor outside of your organization/employment circle. Learning does not end or begin with an AAS, BS, MS or PhD; it is on-going and lifelong.

While I was selling sawlogs from company lands to a sawmill owned by the same company, a conflict of interest became apparent. At that time I documented, with impeccable technical skill, a consistent 15% difference between check and mill scale. I presented a compelling case which illuminated a 15% loss of revenue to woodlands department, loggers, truckers, and cutters.

Like Lancelot, my cause was noble, truthful, and technically accurate. Of course, communications were frequent if not forceful, yet this problem was persistent enough to not go away. Oh, we got a stumpage rebate once, but the perceptions of truth and excellence remained 15% apart. Loggers, truckers, and cutters received little or no rebate because that's the way it was supposed to be; yet each of them were involved with those products, and the one-time stumpage rebate seemed discriminatory and unfair in my view.

Did I blow the whistle, run up the professional flag? No, I accepted this unresolvable conflict with *ethical cynicism.* In retrospect, profession had little to do with the issue. Personal perceptions and personal codes of ethics were operating by choice.

I think that it may be safe and easy for any one of us to hide behind a professional shield, but in the end it is *you as a person* that is on the front line with your various arrangements and involvements. Perhaps it is more comfortable to "hang out in the forest on the stump of our feelings;" the alternative is more risky. In my view it may be difficult, even threatening, to stand and put yourself in the light of day among your peers, professionals, society; yet, if you are committed to your own integrity and growth as a person and professional, there is no better way to go.

I explored social environments through graduate studies in social work. By moving beyond the safe, unconfrontive confines of the forest, I was seeking to fulfill learning needs about people, relationships, and self-in-society. My earliest studies in the field of human behavior directed me to pay attention to my unexamined personal code of ethics.

A type of forum called "case review" made a significant difference to my ethical learnings. In case review, you presented to your peers the issues, problems, and needs that were involved in the clients' human and social environment. Then the variety of plans and actions to be taken by you and the client, as well as the client alone, were presented.

Inevitably, the discussions openly addressed the counselor's self-interest in working with the client, thereby enabling both parties to work on those areas in need of attention. This helped keep the counselor's agenda on a helping, supportive relationship. The discussions in case review were often personal, rather than only technical. Discussions centered on how the client felt, as well as on the effectiveness of the consultant. The client/consultant relationship was extremely important.

I believe that this procedure strengthens personnel effectiveness and helps prevent conflicts of interest and other ethical problems. Open, constructive communications about how we are doing, as well as what we are doing, makes a difference in fostering personal and professional growth.

After three decades of career development, I put my forestry and social work experience together in public service as a county agent. As an extension/forestry professional, I have multiple codes which address knowledge, application, intellectual honesty, and benefit to society. The integrity and honesty in my personal code still prevails as both a *strength and a weakness*.

Each profession wears its own particular crown of special knowledge, skill, and practice for the benefit of the individual, organization, client, profession, society, yet what makes an honorable and credible profession is honorable and credible people. A professional code of ethics can and does encourage this, yet it is our personal code of ethics, values, and beliefs which governs our conduct day by day, month by month, year by year.

In my view, this calls for some self-analysis and discussion on a regular basis. The exercise and learning to be found in this arena can only strengthen us as people and professionals.

Today we wear many hats, serve many interests: self, family, employer, client, profession, society, if not the global environment of human and natural resources. Our personal code of ethics is ever present, choosing and acting every step of the way and in every involvement. There are things *you* can do to address ethical issues:

1. Explore, become self-aware, and learn more about your own personal code of conduct; motives, choices, consequences. Opportunities for this kind of learning are available every day through your job and family experiences.

2. Society of American Foresters Canon 2 states that your knowledge and skill are to be used to benefit society. Examine your

own position on what is harmful and beneficial to the individual, to the family, to the organization, to the community, and to society. Study each Canon and act out each in the role of forester, employer, client, and society. Pay attention to what fits and feels OK, as well as what doesn't fit or feel so good.

3. Recognize and learn to accept as fact that human and technical error from ignorance is a given. We are imperfect beings. Strive to learn and grow as much from your mistakes as you would from your accomplishments. Seek help, support, and constructive feedback for difficult and uncertain situations.

4. Acknowledge your levels of mastery and accept each one as a personal strength. Recognize those areas in which you are weak and strive to improve. Mentors can make a difference in your learning. Seek them from inside and outside of your immediate working group or organization. Putting yourself openly on the line with another means finding the courage to take risks. Nothing ventured, nothing gained.

5. Finally, I believe that the healthiest thing we can do as people and professionals is to openly discuss what we are about, involved in, and doing. Case review studies have the greatest potential to be constructive in working groups, as well as being a good way to prevent ethical uncertainties from becoming conflicts of interest. There is so much to be learned from each other.

In this paper, David Flanagan outlines reasons for ethical codes, and their relation to protecting consumers and to compliance with regulatory programs. He raises the tantalizing possibility that ethical codes could be construed to create binding legal obligations. While an assistant attorney general, Flanagan served as the first attorney for the Maine Board of Licensing for Professional Foresters. When he wrote this paper he was a senior aide to Governor Joseph E. Brennan. He also worked with the Maine Department of Conservation, an experience that gave him a thorough understanding of practical forestry matters. Mr. Flanagan is now President of the Central Maine Power Company and serves on the Board of Trustees of the University of Maine and of the American University of Bulgaria.

CHAPTER SEVEN

Legal Considerations of Professional Ethics

DAVID T. FLANAGAN

It is a privilege for me to be here to participate in your seminar on professional ethics. I am honored—but also a little puzzled—as the chairman noted; I am an attorney and I am also legal counsel for a governor, a position with obvious political implications. I think in repeated national surveys of leading occupations the ones that consistently rank at the botom in public esteem are, last—politicians; and next to last—lawyers.

I am afraid that episodes such as Watergate, with the involvement of so many lawyers, have seriously eroded public confidence in the ethics of my profession. But perhaps that has heightened our awareness of ethical considerations and led to more introspection than has been true of many of the professions. Since Watergate, in fact, over 90% of the law schools in this country have instituted mandatory courses in professional responsibility.[1] I hope that my remarks today, from the perspective of an attorney, with both public and private experience, and as the first lawyer for the Maine State Board of Registration for Professional Foresters, will be of some use to you.

Today, I would like to address the following issues: First, whether you should have a code of professional ethics at all. Second, what legal liability arises because you do have a code of ethics. Third, what the impact of violation of Governmental Land Use laws may be on individual professionals. I understand all of these to be issues of concern to many of you right now.

Let me address first the question of whether you should have a

This article was presented at the New England Section Meeting of the Society of American Foresters held in Portland, ME, on 12 March 1981. Reprinted by permission from *The Consultant*, July 1981, pp. 59–64.

code of professional ethics at all. Many professions have voluntarily adopted codes of ethics to set standards for the on-the-job behavior of those practicing. Lawyers probably have the most elaborate code, but doctors, accountants, architects, engineers, and others also have rules of professional conduct. As you know, the Society of American Foresters has a code of ethics consisting of 16 separate rules, or canons, of conduct.

There are several reasons such codes of behavior have been developed. Those reasons include:

1. *Promoting the pride of practitioners in their occupations.* The general rule of business is that whatever minimum standards for conduct exist are those imposed by government. Thus, the SEC's Insider-Information Disclosure Rules, the Federal Anti-Kickback Statute, and the National Labor Relations Act will prescribe standards for what businessmen can and cannot properly do in certain situations. But when a professional voluntarily adopts standards of behavior that are higher, and more stringent, than the minimums imposed by government, they are saying to themselves that they are special, out of the ordinary, and self-sacrificing, all of which tend to create more pride in their specialties and the quality of the work they do. That is one reason.

2. Another reason is for *the protection of the consuming public.* This was stated concisely in the Maine Bar Rules which provide:

> A proceeding brought against an attorney under these rules shall be an inquiry to determine the fitness of an officer of the court to continue in that capacity. The purpose of such a proceeding is not punishment *but the protection of the public* and the courts from attorneys who by their conduct have demonstrated that they are unable, or unlikely to be able, to discharge properly their professional duties. (Rule 2A)

Thus, if professional canons and disciplinary rules are fairly and consistently applied, they can serve to protect the public from unfit practitioners and increase public confidence in the quality of the service consumers receive.

3. Finally, a code of professional conduct can *help guide the professionals' own decision-making process in deciding difficult issues of professional conduct.* We are all confronted with moral choices in our work: whether we should blow the whistle on improper conduct by superiors; whether we should bend a legal rule for our personal or business advantage; whether we should take advantage of a careless client; how far we can go in advertising; and on and on. In the abstract, those questions may seem to have easy answers, but in practice they can be very difficult indeed. And certainly, as our organizations,

society and technology grow more complex, the questions become more difficult.

A good code of ethics can be useful to the practitioner in sorting his way through this maze of choices—and give him some parameters as to what conduct may be acceptable. But for all these desirable features, codes of ethics obviously have some drawbacks. They restrict what you can do—they may have an impact on your income or profit, or even job opportunities—and they impose on you some duties that can lead to legal liability if you disobey a rule.

Let me explore this topic of legal liability next. It is my understanding that there have been very few malpractice suits brought against professional foresters in the past. My legal research revealed none which had ever reached a state supreme court on that issue.

Usually foresters are sued for breach of contract—failing to perform the promised services for the landowner for which he suffered damages through loss of sales or profits. I am sure that as a general rule that will continue to be the usual course of action against foresters. But even in breach of contract suits, your code of ethics can be relevant.

Let us say, for example, that a forester is employed by a landowner to cruise his land, mark the trees for their highest and best use, and arrange for sale to mills. And let us further suppose this same forester has an on-going contract with a paper mill to identify and secure sources of pulp. This forester could, in his cruising and marking, treat trees suitable for higher value veneer and lumber products as pulp. This will result, of course, in a lower return to the landowner and a higher profit to the forester.

All right, this guy has obviously done something wrong—and someone else has suffered as a result. He has clearly breached his contract with his client—the landowner. But he has also breached the code of ethics, §4 of which requires "unqualified loyalty to the employer," §8, "a member must avoid conflicts of interest," and §9 [now SAF canons 5, 9, and 10—Ed.], which bans compensation from more than one employer for the same service.

What is his legal liability as a result of this double breach? As I noted, ordinarily he will be sued for breach of contract, and he will be liable for the difference between what the landowner actually got and what he should have gotten.

In helping to make his contract case, the landowner can seek to establish that he *relied* on the expectation that a person who called himself a *professional* forester would also adhere to the *professional* code of conduct and that became an implied term of the contract. Then he could introduce evidence of the existence of the code, and the forester's breach, as a part of his contract case, but he doesn't have to since he can make out his contract case just on the failure to fetch the price promised.

On the other hand, I must point out that if the landowner did not in fact know of or expect that there existed a professional code to which foresters adhered, this would be irrelevant. By using the word "professional" in your titles, however, you certainly open yourselves up to more people having the expectation that you *do* have a code of ethics. But, that is not the end of our friend's troubles. The landowner can also make another claim against the forester as well. That is for malpractice.

Most malpractice cases have been against doctors. Let me tell you the test for physicians' malpractice: The Plaintiff must prove:

1. the existence of a duty to the patient;

2. violation of the applicable standard of care;

3. a compensable injury; and

4. a causal connection between the violation of the standard of care and the injury.

Now let's apply these tests to your problems. First, did the forester owe a duty to the landowner? Arguably, he did—when he held himself out to the landowner as a *professional,* and took his money, he *did* owe a duty to act in a professional manner. If he *said* he would so act then there would be no question but that a duty was created.

Second, did he violate the applicable standard of care? Ordinarily this question refers to whether the doctor used the most up-to-date, proven medical technique. But, it can also be whether the doctor breached an ethical obligation. Here again, our forester meets the test. Not only can breach of a provision of the code of ethics itself constitute a breach when it is the cause of damages, but the code also can be evidence of the minimum standards of care set by the forestry community.

Finally, was the breach of the ethical rule the proximate cause of the injury? Disobeying SAF canons 4, 8, and 9 [now SAF canons 5, 9, 10—Ed.] was clearly a part of the underlying problem here which caused the landowner's damages. Again, I must stress there is no law on this yet, but by analogy to the medical profession, there would appear to be grounds for a kind of malpractice action against a forester, based in part on your canons of ethics. This possibility is significant to you, because malpractice has a different measure of damages from contracts. In malpractice cases, it is not just the amount that will make up the difference between actual profits and expected profits, but also allows for punitive damages in excess of the actual loss.

By the way, if I were the landowner's lawyer, I'd also throw in a claim that the forester trespassed on my land under these circumstances.

Taking this same fact situation one step further, there is another consequence that concerns the state. In Maine, and some other states, the forestry profession is regulated by a state agency. In Maine, it is the Board of Registration for Professional Foresters. This board has adopted all of the SAF standards for professional conduct as rules of the board. Thus, in Maine your ethical code now has the force of laws and all professional foresters have a mandatory legal duty to obey those rules. If the landowner, or another person, brings the double-dealing of this guy to the attention of the Maine Board, it can investigate independently. If it finds he has violated a rule, the board can refer charges to the Maine Administrative Court. If the court finds the forester breached the code of ethics in this case, the judge could revoke his license to practice forestry—regardless of whether the landowner suffered any damage at all.

Let me stress that much of this legal theory is new and has not been tested in the courts. But I can assure you that violation of your code of ethics can have immediate financial consequences for you should you breach the rules it establishes—especially if they have been incorporated into state regulations, as has been the case in Maine.

So you can see the tradeoff—having professional rules of conduct has benefits—professional esteem, consumer protection, and guidance for proper conduct. But there's no such thing as a free lunch—it exposes you to professional liability in excess of that of the ordinary businessman—to malpractice damages and loss of license to practice. I have no doubt personally that Maine foresters have chosen the right alternative and should be commended. But these are the considerations you all should be aware of.

As promised, there is one other subject I would like to address. That is the violation of state and other laws relating to forest practices. As you may know, here in Maine, we have amended the SAF code of ethics to introduce a new element. Here, it is also required that a professional forester "at all times in the performance of his services, to the best of his ability and knowledge, abide by federal and state laws and municipal ordinances relating to the protection and environmental improvements of the air, land and waters of the state of Maine." In other words, Maine professional foresters are obliged to follow the law relating to land use as a professional requirement.

For other professions, this would hardly be unusual. It is inconceivable that a professional board for lawyers would sanction fraud, or that a physician's review board would tolerate illegal dispensation of drugs. But land use laws are a relatively new phenomenon—and one that has provoked considerable controversy. These laws are neither so well accepted nor so stable as rules against fraud and drug use, for example.

In the outline of issues the Society [of American Foresters] suggested I discuss, they referred specifically to the issues of "intent, negligence, and competence." I take it that these words were to be taken in this context.

First, must a forester *intend* to violate a state land use law to get into trouble? I don't think so. In Maine, like most states, we have two kinds of legal violations. Technically, they are known as crimes which are called, in Latin, *malum in se*, and others which are called *malum prohibitum.*

Malum in se—literally, "bad in itself," is a crime such as murder, rape or theft—which our moral code calls intrinsically evil behavior. These crimes obviously and automatically bring harm to the victim. Under our legal code, to be guilty of these crimes, they must have *deliberately* intended to commit the offense. For example, you must deliberately intend to kill someone to be guilty of murder. If you kill someone, without intending to do so, as in an auto accident, then society will not characterize that as murder, but rather as reckless homicide, or even just "wrongful death."

But there is another class of crime, which is conduct that we do not regard as intrinsically evil, but simply as contrary to society's best interests. These offenses are called *malum prohibitum*—forbidden conduct. The classic case is a parking violation. There's nothing wrong with parking in any particular place in and of itself. But it would be against society's best interest to park in front of a fire hydrant or near an intersection.

So this conduct is prohibited—and even if the parker did not intend to park in an intersection—just wasn't paying attention—the driver is still liable to a ticket.

Well, land use laws fall into the latter category. It is in society's interest—the legislature has concluded—that we shouldn't have roads wash into streams or clear cut the trees on riverbanks or on mountain tops. There's nothing intrinsically evil about sloppy roads or bare streambanks. But it is now against society's interest—according to the people responsible for representing society.

As a result of all this, intent is *not* a test—it's simply whether you did or did not *in fact* commit the prohibited act. Your exposure to liability for violation of this canon is somewhat complicated by the fact that the professional forestry regulation says that you must have acted "to the best of his ability and knowledge" to abide by the law. This softens the impact of a strictly result-oriented test. But it does not diminish the responsibility of the forester much, as it may be presumed the professional forester *knows* which laws apply and what he must do to comply with them. That is part of being a professional. So I would not expect the board or the Administrative Court to exonerate a professional forester from responsibility just because he did not intend to violate land use law. On a practical level, of course, the boards and the courts are very likely to be more forgiving of an unintentional violation, especially if the forester did not act with wanton disregard of the law or if no great damage resulted.

You have also asked about negligence. The legal definition of negligence is:

> The omission to do something which a reasonable man, guided by those ordinary considerations which ordinarily regulate human affairs, would do, or the doing of something which a reasonable and prudent man would not do.

For a professional, the standard is a little higher—it is not what the ordinary person would do, but what the professional community to which he belongs would do. This concept relates not only to taking reasonable action to comply with the land use laws, but also to malpractice itself. I don't think there's any question but that professional foresters are held to a duty to avoid negligently breaching land use laws, but also to maintaining a minimum level of professional knowledge—and if they ignore that standard, or fall behind in their knowledge of recent scientific advances in forestry, or ignore the natural and probable consequences of their decision, that can be grounds for liability both to clients and to the board when it involves violations of standards.

Finally, there is the question of "competence." This term does not have a precise legal definition. Rather, I take it to mean acting professionally, without violating laws or committing negligent acts.

Let me tie all of this together.

There are real advantages to having a code of professional ethics. It will increase your sense of having a legitimate, specialized profession. It will offer consumers some protection, and it will give you some parameters to deciding what you can and cannot do in good conscience. But having a code also gives you some responsibilities beyond that of ordinary businessmen. These responsibilities mean additional rights for your clients and additional liability when a state adopts these rules as its own. The legal consequences involve liability not only for breach of contract, but also possibly for malpractice and loss of the rights to practice.

Finally, in testing whether you have violated the code of ethics in a state that has incorporated responsibility for land use law violations into the code, you will be judged not only for your specific intent, but also for the standard of care you exercised and for the natural and probable consequences of your acts.

Notes

1. Applied Ethics. *Carnegie Quarterly.* Vol. 28, No. 2 & 3 (Spring and Summer, 1980).

The use of pesticides in forests often leads to controversy and not infrequently to litigation. Because outbreaks of insects or diseases are episodic, the effects of treatment on the forest and on nontarget organisms are often not well measured. Health and environmental concerns can arise. Rarely, however, has pesticide use been as extensive as in the Maine spruce budworm program implemented between 1975 and the mid-1980's. Administering this program for three years gave me an intensive short course in political controversy and conflicting ethical demands.

CHAPTER EIGHT

Pesticides: Ethical Problems for Foresters

LLOYD C. IRLAND

Pesticide use in forestry has generated a large share of our profession's collisions with other groups in our society. I administered Maine's Spruce Budworm Control Program during an especially controversial period, and later faced pesticide use decisions in forest management on the state's Public Lands. These experiences led me to some thoughts on ethical problems faced by pesticide users that I would like to share with you. I'm trying not to preach, but to generate a useful debate.

Foresters expect to be held responsible for how they balance competing values in forest pest management. Choices to spray—or not spray—can affect the forest, the environment, and may affect human health. As foresters, we lack the expertise to be the sole judges of all of these effects. The radical uncertainty about technical facts, the divergent interpretations of those facts, the differing incidence of risks and benefits, and conflicting values bring the use of forest pesticides into the realm of ethical, as opposed to purely technical, decision making.

In my view, the key issues are: (1) obstacles to pesticide decision making; (2) duties of professional foresters; and (3) rights of affected citizens, which I will discuss in turn.

Obstacles to Pesticide Decision Making

A list of formidable obstacles prevents us from solving the problem of forest pesticide use as a straightforward technical

Dr. Irland is President of The Irland Group, forestry consultants, in Winthrop, ME. This article is condensed from a talk presented at the Maine Biological and Medical Sciences Symposium held in Farmington, ME, on 27–28 May 1982. Reprinted by permission from *The Consultant*, January 1983, pp. 17–20.

problem. It is these obstacles that force us to recognize an inescapable ethical dimension to this problem.

These obstacles can be summarized generally as:

1. uncertain knowledge of direct treatment benefits and costs;

2. diverse, incomplete, and often conflicting data on environmental effects;

3. conflicting scientific interpretations of the data;

4. uncertain knowledge of human health effects (even for the best-studied chemicals, gaps remain, or the experts dispute the meaning of the findings); and

5. conflicting values on how the data should be interpreted, how risk-averse decisionmakers should be, and on environmental purity.

This short list of obstacles demonstrates that grave difficulties hinder any attempt to devise a rational, technically based pesticides policy. The uncertainties and conflicting values render it almost impossible. The only course for us is informed, ethically conscious decisions.

Duties of Professional Foresters

Legal Duties

Pesticide users have obligations that go beyond mere compliance with the letter of federal and state regulations. One reason is that such regulations rarely are specific enough to guide a spray project. The user, then, bears a responsibility to develop sound application procedures.

Pesticide users are engaged in what the law regards as an ultrahazardous activity. This is an activity, like blasting, which, even when properly handled, may damage another person's health or property. In the eyes of our courts, even "best behavior" does not excuse a person engaged in such an activity from legal liability for damage to others. Negligence is not necessary to create liability. This is sound policy. The policy is that persons so engaged shall bear a heavy duty of care to protect the interests of others. When they proceed, it is in the knowledge that they bear full responsibility for the consequences.

One of the effects of government operation of insect control projects is to insulate the benefitting landowners from liability exposure under this doctrine. While this is a significant factor, private owners have proposed to spray against budworm in New Hampshire and Newfoundland (they were denied permits by the pesticide authorities). In Maine, in 1982, J. D. Irving Co. treated an area of its land in northern Maine. Herbicide applications on private lands are always carried out by the owners themselves.

Ethical Duties

Foresters who are members of the Society of American Foresters subscribe to a brief code of professional ethics that is probably similar to those adopted by most other professions. It consists of broad guidance on how to balance the often competing claims of the forest itself, the landowners or client, and society as a whole. As Lowrance notes, professionals bear a series of ethical obligations as a result of society's expectations about their behavior.

Four basic problems emerge from professional ethics.

Determining the Necessity of Pesticide Use

This is a basic professional function, just as is prescribing aspirin, choosing beams for a bridge, or doing a client's taxes. Unfortunately, there are instances of unprofessional conduct in the prescription of pesticides. Though 20-20 hindsight is powerful, I have heard experienced foresters denounce DDT applications in the West as based on emotion, panic, and a simple desire for revenge. These colorful caricatures may be unfair, but they convey a conclusion that people didn't do homework.

Professionals owe to their clients and to society a serious duty to assure that a proposed pesticide use is, in fact, necessary. This is so if only for legal and financial reasons, but is also an ethical matter, since interests other than the landowner's are affected. Sloppy, hipshot judgments, based on cliches or unexamined generalities, are simply not professional. Have alternatives been fully considered? Is the treatment in fact cost-effective in the planned conditions?

Judging Acceptability of Health Effects: Independent Determination vs. Limits of Expertise

It would seem that forest managers have an obligation to fully inform themselves on the impacts of pesticides on health and the environment. Courts have offered guidance, saying that, at least in the federal government, they do.

They must make, in an Environmental Impact Statement, an independent and documented determination of safety. It won't do to simply point to the EPA label and say, "they say it's O.K." The sorry state of EPA's credibility today is a source of frustration to many foresters who need to use pesticides in their work.

The obligation to make an independent determination of safety runs counter to another important one—the duty to recognize the limits of professional expertise. As a forester, I'm taught about trees. What do I know about carcinogenesis, inhalation of aerosols, sensitive medical conditions, or allergies?

There is no way out of this dilemma. Clearly, a credible, independent regulatory authority should make the final judgments about safety and rules of application. Foresters should fully inform themselves, but should refrain from pretending to have expertise on health and environmental safety that they cannot possibly possess.

A related professional obligation is one of conduct. It is incumbent on professionals to hold, and to display, a decent respect for the conflicting views and values of others.

Obligations to the Future

As foresters, our professional duty to the future is to pass on a productive forest. This means that we are not to recommend, tolerate, or implement the short-term liquidation of forest capital in disregard of the claims of the future.

Spray opponents argue that the sprayers are only serving short-term interests. *Clearly, no one interested only in the short term would ever spray for budworm or apply herbicide to accelerate tree growth.* For the real benefits of these outlays appear only in the timber supply several decades into the future. Would a short-sighted individual ever consider such investments?

If your time horizon is ten years, there is no reason to spend a nickel on budworm control, far less to release a young stand that will be harvested in 2030. Even in the worst conceivable budworm outbreak, followed by fire and hurricane, we could run our mills on surviving green timber and deadwood for ten years or so. This is enough time to fully write them off for tax purposes.

Company managers or public land managers clearly have a more distant time horizon in view when they authorize outlays for forest pesticide use. They are implementing their duties to the future as they see them. Some could argue that foresters have an ethical duty to argue for the maximum application of intensive management practices.

The Land Ethic

As Aldo Leopold wrote,

A thing is right when it tends to preserve the integrity, stability, and beauty of the biotic community. It is wrong when it tends otherwise.

The land ethic is far from widely accepted as a practical mandate for land managers. Does it provide guidance on pesticides?

Based on the quoted lines, an argument could be made that use of pesticides in forests is wrong. I think this carries the idea too far. At times, pesticides are applied (budworm in parks, gypsy moth) to sustain the beauty of the forest. It can be argued, though it shouldn't be taken too far, that insect control improves short-run stability of at least some forest characteristics. But it would be hard to argue that even biological insecticides or herbicides augment the integrity of the forest.

To me, the Land Ethic must be taken seriously. I think it argues for the elimination of long-lived "hard" pesticides, like DDT, for the utmost judiciousness in prescribing and using chemicals, and for maximum use of biologicals. I think that properly managed plantations are not inconsistent with the Land Ethic, so I would find careful herbicide use acceptable. The Land Ethic does allow other interpretations than mine, and its actual application depends on judgments about environmental impact, about which informed persons disagree.

Rights of Affected Citizens

Chemical Trespass—The Citizen's Right to Clean Air

More and more people today arrange their lives so as to minimize their exposure to toxic chemicals. They do this for many reasons: practical, religious, ethical. Their values should be respected. It is sometimes argued that fears of pesticides are irrational, or are based on anti-social or even destructive motives. I don't agree. We are dealing with legitimate and serious value conflicts here.

These people argue that they have a right to clean air—to be free from the trespass of forest chemicals on their persons and property. This view is taken by many statements of "Environmental Bills of Rights" and by texts on environmental ethics.

What obligation do these values create on the part of forest managers? I have strong sympathy for these values, but I am forced to conclude that they create obligations of degree and of procedure only.

There is clearly an obligation to avoid chemical trespass that imposes actual damage—to paint, to bees, to plants. This is only legal.

To assure purity, however, is not within the power of forest managers. Everyone drives through farm areas and is exposed to pesticide drift there; our food contains residues; the air itself contains low level traces of pesticides. How can the managers of our forests be asked to assume a burden that cannot be sustained by the rest of society?

The Citizen's Right to Informed Consent

In 1980, an advisory panel to the Maine Forest Service advised that unconsented exposure to carbaryl should not occur in the state's budworm control program. This was a significant watershed in ethical terms.

State officials had long conducted public information activities about their spray program. Over the years, active press attention was given to health risks. In 1979, a strong health warning was employed. But the 1980 recommendation was a radical break with this, since it asserted that citizens had a right to informed consent. I understand that in medical law and medical ethics the concept of informed consent is a very difficult one, encountered in research and in the use of novel or risky procedure.

My personal conclusion is that it is not clear what specific obligations are imposed on foresters by the citizen's right to informed consent. We need more help from the medical profession on this point.

Should it be necessary, it is clearly possible in most situations to simply exclude recreationists and woodsworkers from the forest for any desired length of time. In view of our obligation to future forest productivity, we should seriously consider this option before giving up the use of pesticides entirely.

Agonies, Dilemmas, and Recommendations

In a way, it doesn't help to take an ethical view of pesticide problems. It makes you more conscious of duties you can't fully meet, and makes you aware of legitimate claims that you cannot grant.

But professional practice is always a matter of, in Lowrance's words, "being, and being held, responsible."

Experience suggests several practical guidelines for resolving the ethical dilemmas described here.

1. Assure that all uses of forest pesticides are in fact necessary, measured against reasonable alternatives and reasonable

management goals. Of course, what is "reasonable" is what most of the arguing is about.

2. Assure that off-target drift is minimized.

3. Assure that exposure to humans and their water is minimized, even if this requires material sacrifice of timber values. Always err on the side of reducing human exposure. Exploit to the maximum practicable extent the use of small aircraft, biological insecticides, salvage, and harvesting to minimize human exposure to pesticides.

4. Support with project funding, the study of drift, human exposure, and environmental impacts. Be prepared to use the results, if necessary, to modify operating practices.

5. Assure that no forest visitor or worker is unaware of the fact that a forest spraying treatment is under way.

6. Consult closely with local medical authorities in developing advisories to address the need for informed consent, and to make them fully aware of the planned project.

7. Operate in an ultra-conservative manner with respect to all project employees likely to come into contact with insecticides.

8. Provide amply for contingencies involving possible accidental spills of insecticide.

9. Where use of pesticides in settled areas is required, establish a process where legitimate local government forums can exercise regulatory control over the program, if they so desire.

My exploration of this topic grew from my conclusion that pesticide use rules cannot be based solely on technical facts. The knowledge is too sparse and controversial for that. There is a clear mandate to seriously consider the values of people who do not want to be exposed to pesticides. In the end, we won't make everyone happy, including ourselves.

These considerations led me into a search for guidance in the literature of ethics. That search so far has only helped me develop a list of dilemmas—answers to practical problems remain elusive. Nonetheless, we will be held responsible for how we balance the competing claims of forest productivity, of rights to pure air, of a healthy environment, and the value of others. Still, I think that a heightened awareness of these ethical dilemmas is an aid to responsible professional performance in using pesticides in forestry.

Kermit Johnson's essay is based on his military experience, and is relevant to this book because it speaks to the realities of life in large, bureaucratic organizations. His emphasis on the destructive side of the drive for success rings true throughout our profession and throughout society. His discussion uncovers some of the roots of compromise that can leach away ethical sensitivity in people as they seek or gain authority. This article was published in 1974, as the Vietnam War was slowly ending, but it remains relevant to this day. Johnson retired after serving as Chief of Chaplains of the U.S. Army.

CHAPTER NINE

Ethical Issues of Military Leadership

KERMIT D. JOHNSON

Several years ago, I awoke at 0500 hours thinking about an ethics talk I was scheduled to give at the U.S. Army War College Memorial Chapel. As I allowed my mind to wander in free association, I got more than I bargained for. I started out with a flashback of Vice President Nixon's visit to the heavy mortar company I commanded on Okinawa in 1954.

It was pleasant to recall that my company had been selected for the vice president's visit because we consistently had the best mess on the island. However, this triggered a thought about my mess sergeant. For some unknown reason, he would come up with juicy steaks whenever they were needed, whether they were on the menu or not. I recalled that he had some contacts with the Air Force and apparently was involved in trading, but I never bothered to look into it.

My next thought was that trading in steaks wasn't much different from trading in bullet-proof vests. This brought to mind the supply sergeant of another company I commanded during the Korean War. He had no administrative ability whatever, but he always had a good supply of bullet-proof vests. The only thing that helped me out of Korea without supply shortages was those bullet-proof vests— valuable trading materials.

These uncomfortable thoughts, dredged from the semi-subconscious at five in the morning, formed the starting point for my thinking about the ethics of military leadership. But still another question forced itself upon me: "Is this the sort of thing which forms the substance of Watergate and mini-Watergates?"

With this as background, I can't pose as a flaming prophet or crusader in the ethical area. Maybe this is just as well. Perhaps in order to have an ethical consciousness we should be aware of our per-

This article reprinted by permission from *The Parameters of Military Ethics* edited by Lloyd J. Mathews and Dale E. Brown (Washington, DC: Pergamon Press, 1989), pp. 73–78. Copyright © 1989 by Pergamon Press.

sonal fallibility. In recent reading, I've noticed this awareness in
Abraham Lincoln's life. He was constantly at odds with puritanical
moralists and idealists who he could never please. Yet Lincoln knew
very intimately what we are like as human beings. It came out in a
comment he made about our judicial system as he quoted Thomas
Jefferson, with approval: "Our judges are as honest as other men, and
not more so. They have, with others, the same passions for party, for
power, and the privilege of their corps."[1]

At the outset, I must admit that I am probably as silent, as tactful,
as self-protective, and as non-risk-taking and gutless as anyone else.
Yes, I have been forced to take some clear-cut goal-line stands—those
Martin Luther deals where you say, "Here I stand. I can do no other,"
whether it's to the detriment of efficiency report, career, or whatever.
However, this is exceptional.

On a day-to-day basis, the tightrope is a better metaphor. I believe
that we walk a tightrope, constantly oscillating between the extremes
of crusader and chameleon: both roles are difficult and we burn up a
lot of energy attempting to walk the tightrope between these two
positions. The crusader, to use a phrase of J. D. Salinger, seems to
"give off the stink of piousness" or self-righteousness.[2] On the other
hand, the chameleon is so non-principled that if you told him "A" was
right one week and then that "non-A" was right the next week, he'd
dutifully and loyally click his heels together and say, "Yes, sir."

My own self-understanding, then, in discussing this matter of
ethics is that of a tightrope walker caught alternately between the
positions of crusader and chameleon—in one instance, donning the
uniform of a pure knight in shining armor and, at the other times,
crawling into my chameleon skin of comfort and compromise. To the
extent that readers have felt this ethical tension, I hope this article will
encourage fellow crusader-chameleons to surface those ethical issues
with which we all struggle from day to day.

In the December 1973 issue of *Worldview,* Josiah Bunting, a former
Army officer and a crusader type who wrote *The Lionheads,* refers to
"the tyranny of the dull mind," which, he says, "one so often
encounters in the military." But he's objective enough to speak also of
"the tyranny of the gifted mind" and he says these types are more
dangerous because they withhold their true judgments lest they
jeopardize the hopes for success which their ambitions have carved
out for them.

He quotes B. H. Liddell Hart, discussing British officers, at this
point:

> A different habit, with worse effect, was the way that ambitious
> officers, when they came in sight of promotion to the general's
> list, would decide that they would bottle up their thoughts and

ideas, as a safety precaution, until they reached the top and could put these ideas into practice. Unfortunately, the usual result, after years of such self-repression for the sake of their ambition, was that when the bottle was eventually uncorked the contents had evaporated.[3]

What Hart is saying should not be limited to promotion to general. The process starts much earlier. I would have to agree that if we don't *now* expose the relevant ethical issues that affect our daily lives, when we become Chief of Staff or Chief of Chaplains and open up the bottle, we're going to find that there isn't any carbonation left, no zip. It will be gone. It simply can't be saved that long.

I would like to emphasize four pressing ethical issues for leaders in the military establishment to consider. The first is the danger posed by the acceptance of various forms of *ethical relativism,* or the blurring of right from wrong. It appears obvious that the erosion of a sense of right and wrong in favor of a "no-fault" society poses a threat to sound ethical judgments.

A brilliant young major, now out of the Army, once told me that we can never say anything is right or wrong. He said very blatantly, "Everything is relative. There is no right or wrong." I then asked him if the killing of six million Jews in World War II was wrong and whether the actions of an Adolph Eichmann were wrong. He said, "Well, it depends on what was going on in Eichmann's mind." What basis does this man have for making ethical judgments with his belief that all is relative?

Less blatant but equally devastating to ethical judgments is a subtle and disguised form of ethical relativism practiced frequently in the military setting. It comes out of the tendency to have a functional or pragmatic attitude. I've heard Army officers say impatiently, "Hell, don't give me all that theory. I just want to know what works."[4] This, of course, *is* a theory—"what works is right." Such a hazardous ethical position is made worse by emphasis on getting the job done, no matter what. Performance of the mission is everything; therefore, the question of what is right often gets lost in the shuffle of practicality and necessity, if indeed ethical questions are even raised.

A second ethical issue every military leader should face is what I call the *loyalty syndrome.* This is the practice wherein questions of right or wrong are subordinated to the overriding value of loyalty to the boss. Loyalty, an admirable and necessary quality within limits, can become all-consuming. It also becomes dangerous when a genuine, wholesome loyalty to the boss degenerates into covering up for him, hiding things from him, or not differing with him when he is wrong.

General Shoup, a former Marine Corps commandant, once said something like this: "I don't want a 'yes' man on my staff, because all he can give back to me is what I believe already." Now for a leader to honestly say this and to attempt to carry it out, I would think he would have to be very secure. To turn it around, the less secure a leader is, the greater his need for pseudo-loyalty, that is, for fewer ideas that threaten his position. The simplest and quickest way he can get this type of loyalty is through fear. There is little doubt in my mind that fear is often a motivational factor in Army leadership, and also a major trouble spot in terms of ethical practice. This is confirmed in a study titled *The United States Army's Philosophy of Management*, done by eight officers in the Army Comptrollership Program at Syracuse University. With reference to a survey of officers and civilians on managerial practices in the Army, the report said:

> From the statements concerning fear, one can conclude that the use of fear is perceived by a majority of respondents, especially the lower-ranking respondents, to deeply pervade the Army's organization structure. Lower-ranking respondents generally believe that managers are unwilling to admit errors and are encouraged to stretch the truth because of how fear operates within the system. They believe that fear itself and the life-and-death power of efficiency reports are the primary means used by their superiors to motivate subordinates' performance. When lower-ranking officers are afraid to tell superiors about errors, embarrassing situations for the individual, the manager, and the organization can arise when the errors are finally disclosed. The persistence of fear as a stimulator of performance can have repercussions.[5]

This report says that "when lower-ranking officers are afraid to tell superiors about errors" it is an "embarrassing situation." More than this, the use of fear to guarantee a sterile form of loyalty contributes to an environment where suppression of truth is guaranteed.

Concern about what might turn out to be an "embarrassing situation" leads into a third ethical trap on which we've been particularly hung-up for years in the Army, namely, the anxious worry over *image*. We frequently run scared; instead of acting upon what is right, we often hear: "You know, if we do this, it'll be embarrassing to the Army's image."

Whereas with the loyalty syndrome people are reluctant to tell the truth, with the image syndrome they aren't even interested in it. What becomes important is how things are perceived rather than how things really are. Thus, a dream world of image is created which is often different from the world of reality.

Let's look at some quick examples:

- The old recruiting poster: not "Join The U.S. Army" but "The Army wants to join you." How true is it?

- A general at his new duty station who tells his information officer: "You're going to make me my next star."

- A unit commander who says: "This is the best unit in the U.S. Army," and then refuses to seriously consider negative input.

- And what about our craze for "innovation?" How much of it is based on a desire for good publicity or catching our rater's eye with "dash and flash," and how much of it is based on the desire for quality and solid achievement in the unglamorous "bread and butter" items of our daily job?

As you read this, add examples from your own experience and you will probably arrive fairly close to my conclusion: at times, the obsession with image in the U.S. Army borders on institutional paranoia.

A fourth ethical trouble spot in our military experience involves *the drive for success.* This is the masochistic whip by which, sometimes, we punish ourselves and by which we sometimes are beaten sadistically by others.

In Vietnam, I escorted a speaker who was sponsored by the Department of Defense. I took him to see some of the best and the brightest of our leadership. On one occasion, I heard a high-ranking officer tell our visitor about a field grade officer who objected to the body count and to the wisdom of some current operations. The general to whom we were talking repeated gruffly what he told this field grade officer's superior: "Give 'em some candy and send 'em back up." In other words you can buy off his ethical sensitivity—give him some medals and ribbons and send him back to his unit.

Compare this with a comment by one of the respondents in the section on "Integrity" from the *Study on Military Professionalism* done by the U.S. Army War College in 1970: "One of the most violent reactions we got was from the body count, particularly from the young combat arms officers recently back from Vietnam . . . basically being given quotas, or if not given quotas, being told that their count wasn't adequate—go back and do it again."[6] "Give 'em some candy and send 'em back up." But at what price success or even survival?

The internally generated drive for success that we all possess is compounded by the externally demanded results which signal success. In one word this adds up to *pressure.* We have this in common with other professions. While reading a study of 1,700 executive leaders entitled "How Ethical Are Businessmen?" conducted by *Harvard Business Review,* I found the following comments under the title "Pressure":

A controller resents "repeatedly having to act contrary to [his] sense of justice in order to 'please.' In upper middle management, apparently, one's own ethical will must be subordinated to that of interests 'at the top'—not only to advance, but even to be retained."

The sales manager of a very large corporation phrases his views most bluntly: "The constant everyday pressure from top management to obtain profitable business; unwritten, but well understood, is the phrase, 'at any cost.' To do this requires every conceivable dirty trick."

A young engineer testifies that he was "asked to present 'edited' results of a reliability study [he] refused, and nearly got fired. [He] refused to defraud the customer, so they had others do it."[7]

It may be small comfort to realize that business leaders also experience pressures to buy off ethical sensitivity, through jeopardy of career advancement or retention. Yet one would hope for better standards in the military services where profit motive demands are absent, and where its members are dedicated to a lifetime of service to their country.

Interestingly enough, the *Harvard Business Review* study also indicated that there were pressures from bosses which helped employees to act ethically. The study concluded: *if you want to act ethically, find an ethical boss.*[8]

Fortunately, there are a great many leaders in the Army who, by personal example, offer this ethical encouragement to others. However, while the Army neither compels its personnel to compromise their ethical principles nor condones their unethical behavior, the importance of an institutional drive to push ethical leaders to the fore becomes significant since individuals cannot always choose their commanders. It also means building into the institutional structure and leadership training process such emphasis on ethics that leaders who use unethical methods will be exposed.

The task of building an ethical environment where leaders and all personnel are instructed, encouraged, and rewarded for ethical behavior is a matter of first importance. All decisions, practices, goals, and values of the entire institutional structure which make ethical behavior difficult should be examined, beginning with the following:

First, blatant or subtle forms of ethical relativism which blur the issue of what is right or wrong, or which bury it as a subject of little or no importance. Second, the exaggerated loyalty syndrome, where people are afraid to tell the truth and are discouraged from it. Third, the obsession with image, where people are not even interested in the truth. And last, the drive for success, in which ethical sensitivity is bought off or sold because of the personal need to achieve.

Before being sentenced for his Watergate role, Jeb Stuart Magruder testified: "Somewhere between my ambition and my ideals I lost my ethical compass. I found myself on a path that had not been intended for me by my parents or my principles or by my own ethical instincts."[9] In the Army, we must ensure that the ambition of the professional soldier can move him along the path of career advancement only as he makes frequent azimuth checks with his ethical compass.

Notes

1. Elton Trueblood, *Abraham Lincoln, Theologian of American Anguish* (New York: Harper and Row,1973), p. 123.

2. J. D. Salinger, *Franny and Zooey* (Boston: Little Brown, 1955), p. 158.

3. Josiah Bunting, "The Conscience of a Soldier," *Worldview* (December, 1973), p. 7.

4. Scientific research by James W. Tyler in *A Study of the Personal Value Systems of U.S. Army Officers and a Comparision with American Managers*, an unpublished University of Minnesota thesis, August 1969, has shown "first-order" values to be pragmatic ones such as high productivity, organizational efficiency, one's boss, and achievement. "Second-order" values are ethical and moral values such as trust, honor, dignity, equality, etc. See U.S. Army War College, *Study on Military Professionalism*, Carlisle Barracks, Pa., 30 June 1970, pp. B-6, B-7.

5. Management Research Center Report, *The United States Army's Philosophy of Management,* Syracuse University, August 1972, p. 77.

6. *Study on Military Professionalism* pp. B-1 to B-10.

7. George A. Smith, Jr., *Business, Society, and the Individual* (Homewood, Ill.: Richard D. Irwin, 1962), pp. 59–60.

8. Ibid., p. 52.

9. *New York Times,* 22 May 1974, p. 37.

The concept of sustained yield is said to have originated in Germany, a stable society accustomed to taking a long view of policy and concerned about its ability to provide its own raw materials. American conservationists seized upon this idea in the late 19th century as a counter to the massive destruction of forests they saw in this country. Sustained yield is enshrined in law as a goal of the National Forests and could be said to be one of the core concepts of professional forestry. In this essay, however, Klemperer challenges the notion that sustained yield of timber is in society's best interest, suggesting that society is really interested in the "total utility" or total benefit of all resources. He, therefore, concludes that sustained yield of timber cannot be an ethical mandate for foresters.

Klemperer is a professor at Virginia Polytechnic Institute in Blacksburg, Virginia, and is a leader in applying mathematical analysis to forestry problems.

CHAPTER TEN

Is Sustained Yield an Ethical Obligation in Public Forest Management Planning?

W. DAVID KLEMPERER

Introduction

Is sustained yield an ethical obligation in public forest management planning? Dictionaries define ethical as that which relates to morals or principles that are right and proper. We might then ask whether sustained yield is intrinsically "good" and "right" in the same sense that we view the pursuit of truth of the sanctity of human life. Many among us would probably answer "yes" and some would answer "no."

My basic thesis is that *in the broadest sense, planners have an ethical obligation to maximize the present value of society's satisfactions from all its resources combined.* This view recognizes that forests are but one aspect of our resource base. Following this line of thought would suggest that, depending on the values of forest outputs relative to other outputs, optimal forest yields for any given region could fluctuate between zero and some positive level. Yields from individual resource systems would be managed so that total utility from all systems combined could be maximized.

When is total utility maximized? Resource allocation can be considered optimal when no re-allocation could bring about any net gains (the economist's "Pareto optimum"). We would accept changes—for example, reduced forest outputs—as long as resulting gainers could overcompensate losers (ignoring income distribution for the moment).

This article reprinted by permission from the Proceedings of the XVII International Union of Forest Research Organizations World Congress, Japan, 1981, pp. 421–432.

It may appear that the thesis has been framed in such a way that sustained yield in forestry, however defined, could not be optimal. But is that necessarily true? Hidden in this thesis are significant questions: How do we value resource inputs and outputs? What is the social rate of discount that will help us determine the optimal allocation of outputs over time? And how do we handle the major uncertainties in forest management planning? Add to this the fact that physical input-output relationships are often poorly defined, and the problem becomes still worse.

After defining sustained yield and two of the major rationales for the policy of even-flow, this paper evaluates the even-flow doctrine, discusses economic harvest optimization, and briefly mentions the problems of including nontimber values.

Due to space limitations, I will concentrate primarily on timber output, recognizing that other forest outputs are often at least as important as wood, but also recognizing that the fundamental questions of whether or not to sustain, and at what level, are similar and relevant for all outputs.

Sustained Yield Defined—Wood Output Only

In general, sustained yield forestry involves a commitment to continued long-term wood output. Public agencies have typically defined sustained yield as:

1. Annual harvest equal to potential long-run annual growth, i.e. "even-flow."

Private forest industries sometimes define sustained yield as:

2. Ensuring regeneration after harvest but not necessarily even-flow.

Between these extremes lie many variations on the same theme. For consistency, my initial discussions will consider the first definition.

Obviously the harvest pattern over time depends partly on the size of the forest area. If the forest area is small enough, annual yields are impossible. Barely above some minimum area, it is physically possible to attain annual even-flow through uneven-aged management. And various forms of regulation with even-aged management can yield the same result. However, the latter is highly impractical on small areas.

Under an even-flow policy the ultimate goal will be a "regulated forest." Let this be defined as a forest in which an equilibrium has been attained where annual harvest is stable and equal to annual growth. This includes but is not limited to the "normal" forest with a perfectly even distribution of age classes. In forests with large mature timber inventories, even-flow dictates cutting no more than the long-run sustainable yield, even when accelerated harvest would be physically possible. On understocked forests, definition 1 must be modified to allow a harvest gradually increasing until regulation is attained.

The larger the developed forest region, the more likely we are to attain a fairly even sustained yield, whether or not it is consciously planned. For example, in recent decades in the United States as a whole, annual timber harvest has been gradually increasing but remaining below growth (U.S.F.S. 1980a). However, smaller regions show marked harvest instability and uneven relationships between growth and cut.

If transportation costs were zero and we ignored nontimber considerations, the rationale for annual harvests in smaller regions would be weakened. Under these assumptions one could support the notion of a huge regulated or "normal" forest the size of the United States with one enormous wood processing complex in the center of the country. One might harvest the first 20 age classes in the Northeast, the next 20 in the South, 20 more in the Lake States, and the last 20 in the West, after which we could again cycle through the regions. (This pattern may not be too far from what actually happened!) Far-fetched, yes—because it totally ignores economic and social dislocations and non-market goods—but the example drives home the relevance of transportation costs. As the latter increase, the optimal size of regulated forest will decrease, other things equal.

This question of optimal size of even-flow planning unit must underlie any discussion of sustained yield. If the unit is large enough—international if you wish—we will almost always have a fairly even flow of timber. And if the area is small enough, we will *never* have it. Where between these two extremes is the optimal sized planning area? I submit that this issue has not been adequately addressed in forestry.

The Rationale for Even-flow

Community Stability

As a rationale for even-flow, community stability is at best an elusive goal. Forest policy documents do not usually define clearly how large an area a "community" encompasses. In addition, what is to be stabilized? Employment? Income? Prices? Lifestyle? Rates of growth?

In spite of these ambiguities, stability of timber-dependent communities continues to be one of the major rationales for the U.S. Forest Service's even-flow policy (Waggener 1977).

Even with a stable regional timber harvest, some timber-dependent communities are likely to experience declining employment as timber processing becomes more automated (Schallau 1974). On the other hand, forest-dependent communities having strong ties with bouyant regional centers often experience economic growth despite stable or declining timber harvests (Schallau et al. 1969). Thus there is no assurance that even-flow policies can foster community stability, although they apparently have in some cases (e.g., see Beuter and Olson 1980).

In addition, even-flow in the face of dynamic demand for wood products can exacerbate timber price fluctuations—a form of instability. Downs (1974) has suggested a price-responsive public harvest pattern to alleviate such problems.

Throughout this paper, the size of sustained yield region will be a major factor. For example, two adjacent federal forests with uneven-flow can often provide even-flow (and community stability?) when combined. In addition, uneven-flow from public lands can offset irregular harvests from private lands. So we are back to the previous point: some form of even-flow (and economic stability) is likely to be attained on a large national scale. How far should we move in compressing the geographic scale of our planning view?

One might also question the desirability of community stability as a goal in itself, particularly within one economic sector. With economic diversity one can plan smooth transitions from decreasing timber processing activity to increasing nontimber enterprises (or vice versa). Such instability can be desirable if it meets with citizen approval and elevates living standards.

It is true that communities can adjust to a wide range of harvest options, and there is no way that even-flow can guarantee community stability. However, this should not suggest that community stability is such an elusive goal that we can ignore employment impacts of harvest options. We need only think of my far-fetched example of the entire United States as one regulated forest to be convinced that impacts of harvest patterns on communities *do* matter. Perhaps we could achieve greater consensus if we stopped using the term "community stability" and attempted in some way to *optimize* the economic dislocations from changes in harvest patterns. We could accept changes in harvest levels as long as the resulting gains exceeded the resulting costs of economic dislocations.

It is difficult to predict employment and income impacts of alternative harvest levels, although attempts have been made (see Waggener 1972, Bell 1977, and Darr and Fight 1974). Efforts along these lines must be continued if we are to search for the optimal level of instability in forest based communities.

Uncertainty

If one knew with certainty all future forest and nonforest benefits and costs at various output levels, one could compute optimal departures from even-flow. It is sometimes argued that since we are so uncertain about these future benefits and costs (as well as all the variables which determine them), we should therefore opt for even-flow.

For example, in order to decide whether to change from timber to agricultural production on a given area, for cases where the decision is not immediately obvious, one must estimate highly uncertain values of wood and farm crops several decades in the future. Analogous information is required in deciding whether to accelerate or decelerate timber harvests on a given area.

A related problem is the "tyranny of small decisions" (Kahn 1966) in which a series of willingly made small changes may accumulate to an unpalatable result. Again the problem is imperfect information. The dilemma is magnified if decisions are made by the living for the not-yet-born—the ultimate form of dictatorship.

There was at one time reason to be somewhat unconcerned about future generations. Historically, later generations usually were better off than earlier ones. Under that view, saving resources for future generations was distributing wealth from the poor to the rich. The argument is less compelling today, given our projected energy crisis. Rawls (1971), Solow (1974 and 1974a), and Ramsey (1928), among others, have wrestled with the question of intergenerational equity. No clear consensus exists except for the nagging discomfort that use of the private market interest rate doesn't answer all the questions when we deal with irreversibilities, personal utilities, imperfect information, and non-renewable resources. The extent to which these problems are relevant in forestry demands more attention.

The uncertainty issue has led Duerr (1975 and 1981) to suggest that a driving force behind sustained yield forestry has been "faith" or "conviction." In this vein, Duerr (1974) has maintained that timber supply is not particularly price-responsive. My own bias is that expected wood price-increases do influence private forest investment levels and should also be considered in public planning. Rising stumpage prices and expectations of further increases have been accompanied by major expansion in United States industrial forestry investment and by industry efforts to stimulate management of small woodlands. However, I will not attempt to prove whether that indicates causality.

If we plead near-total ignorance about the future, then even-flow might seem a defensible posture. Some argue that it allows us to maintain diversity and options for future wood consumption, and it prevents irreversible forest destruction. However, such a view implies that the status quo is optimal, which may be far from true. In addition, intelligent (if not perfect) estimates of future conditions, with frequent re-evaluations, could suggest a variable yield policy that provided greater benefits than even-flow.

Analysis of Even-flow

Even-flow sustained yield, especially when combined with maximization of growth, has been widely criticized on economic grounds (e.g., see Waggener 1969, Nelson and Bennet 1965, Smith 1969, Thompson 1966, Keane 1972, Hirshleifer 1974, Roberts 1974, Ledyard and Moses 1974, Samuelson 1976, Walker 1977, Dowdle 1974). In general, the growth-maximizing rotation does not maximize present value. Furthermore, even if rotations were economically optimal, an even-flow policy would not permit the harvest fluctuations often needed to meet social demands.

Let us first examine the combined conditions under which even-flow would maximize present net worth of society's resources: (1) if the forest were already regulated at the present value maximizing rotation; (2) if the current allocation of land to different uses were optimal, i.e. no change in land use could bring about an increase in present value of benefits to society; (3) if population were projected to be stable and to have unchanging preferences; (4) if technology were unchanging over time; (5) if projected real interest rates and relative prices of all commodities were stable.

If the above conditions held for all time, no fluctuations in harvest level could improve social welfare. On the other hand, if any of the conditions do not hold, welfare could be improved by changing harvesting and investment levels and/or changing the amount of land devoted to timber production. Consider the unlikelihood of each condition: In most regions, forests are not fully regulated, and the costs of attaining regulation may exceed the resulting benefit.

As for condition 2, cases about where changes in land use from forestry to, say, agriculture or vice versa will increase social benefits. For example, consider . . . [the] declining forest area and increasing cropland area between 1800 and 1920 in the United States—a typical and desirable trend during settlement of forested regions. To have arbitrarily frozen this land allocation at any point during the 1800's could have reduced social welfare. In recent decades the land use balance to a shift from agricultural to forest land use in the north-eastern United States during the 1950's and 60's. The reverse is presently occurring in portions of the southern U.S. These are desirable changes as long as the resulting benefits exceed the costs, and an even-flow policy would impede such gains. . . .

Often forest destruction has progressed far beyond optimal levels for reasons such as poorly defined property rights, limited technology, natural disasters, and low timber prices. In these cases public forest policy may aim to substantially increase the forested areas. Again even-flow is not optimal. Current examples include Haiti and Nepal.

Certainly condition 3 does not generally hold. A changing population with varying preferences over time will demand changes in the mix of forest outputs as well as changes in land use. And if technology changes, as it continuously does, the useable sizes and species of trees will change as will the availability of substitute products.

Most of the above dynamics have caused continuing changes in condition 5. Variations in expected real interest rates will change the optimal path of resource use over time, and changes in relative prices of all commodities will alter the optimal mix of forest and non-forest outputs.

An added problem with the even-flow constraint is an "allowable cut effect" which can occur in regions with large old growth reserves. In such cases, current investments which increase long-term timber growth can immediately boost the allowable cut, thereby apparently generating tremendous rates of return. Artificially attributing such returns to current investment can foster decisions to spend large sums whose actual distant-future incomes may represent very poor rates of return (Teeguarden 1973 and Klemperer 1975). In addition, the allowable cut effect in reverse could lead to under-investment in protecting excess old growth inventories, the loss of which would not affect annual harvest under a non-declining even-flow policy (Bell et al. 1975).

Given the unlikelihood of the restrictive conditions under which regional even-flow would be optimal, probably some form of uneven-flow will best serve society's interests.

Economic Harvest Scheduling

Harvest scheduling models dearest to many economists' hearts are those which determine the present value-maximizing harvest pattern at some interest rate, allowing stumpage price to be a function of output in any given period (e.g., Walker 1976 and Johnson and Scheurman 1977). However, forest planners are often distressed by the widely fluctuating harvest patterns sometimes yielded by such economic models. For example, Johnson (1976) found that projected harvest levels often varied by more than fourfold from one decade to the next when the criterion was to maximize present value on several national forests in the western United States. Some schedules showed harvest peaks over ten times higher than troughs, with oscillating cycles spanning 30 to 80 years (Johnson 1976). Results depended on the initial distribution of age classes and site-types, and oscillation did not always occur. However, in general, the present value maximization models yielded results that were far from even-flow.

If the rational economic model yields harvest fluctuations that displease us, perhaps the fault lies not in the model but in our failure to properly incorporate the costs of harvest fluctuations. Where wood output peaks and troughs are accompanied by negative side-effects (e.g., economic and social dislocation and environmental damage), then this should be reflected in the present value calculations. . . .

The above might tempt even-flow proponents to tinker with output prices until economic models yielded the desired timber flow. That would of course defeat the model's purpose. However, it could be instructive to adjust prices until even-flow becomes economically optimal. One could then examine if these adjustments were in some sense "reasonable." If they were not, then even-flow would be suboptimal.

Sometimes it is suggested that an economic model could dictate no harvesting at all in cases where stumpage prices are expected to rise at a rate exceeding the interest rate. . . . That may indeed be the case in some instances, but present value calculations must include the negative effects (if any) of not harvesting; i.e., reductions in processing income and employment, where relevant. These could be very important on large planning areas.

There are extreme dangers in following the harvest dictates of an apparently logical economic model which ignores crucial input data. It is equally as foolish, however, to reject potentially useful models because we dislike their outputs under a partial analysis.

Because of the major uncertainties about the future, outputs from economic harvest scheduling models must not be the basis for rigid long range harvest plans. Such models should be frequently rerun using revised predictions. It is unlikely that subsequent re-evaluations would match earlier plans. In essence, this is a plea for flexibility. Rigid harvest scheduling plans have no place in a dynamic uncertain system. . . .

Total Forest Output

In general, the foregoing questions of wood output levels apply also to forest benefits such as dispersed and developed recreation, fish and wildlife, water quality and quantity, scenic beauty, and soil protection. The optimal mix of these outputs will vary as the factors change which determine demand for them.

Since market prices are absent for most nonwood forest outputs, their optimization is left to public forest planners and regulators of private forests. To determine the present value-maximizing pattern of nonmarket goods over time, some effort must be made to derive their values. While such valuation is difficult, recent techniques show promise and are continuing to improve (Dwyer et al. 1977, Charbonneau and Hay 1978).

Even where valuation of nonmarket goods appears impossible, one can use economic harvest scheduling models to calculate the loss in timber income when harvesting is reduced to gain nonmarket benefits (Walker 1978). This at least defines trade-offs and allows planners to ask important questions: Are the added nonmarket benefits from any harvesting restriction worth the resulting costs of foregone income?

The above question attempts to measure citizens' willingness to pay for nonmarket forest benefits such as increased wilderness. A problem is that this value tends to be less than the compensation they would require when losing the same benefit. Under the latter legal framework, the above question would be rephrased: Is the potential added harvest income sufficient to compensate those who would lose nonmarket benefits? The choice of legal framework in this decision model can have a significant impact on what is deemed to be an optimal level of harvest-induced damage to nonmarket goods (Klemperer 1979).

One of the thorniest problems in total forest output planning is that of benefit distribution. We often find that several alternative management plans may have similar present values and capital requirements but vastly different flows of outputs over time (e.g., see USFS 1980). Among alternatives with equal present values, one plan may produce more recreation, the other more timber, and the distribution of outputs may differ between income groups, regions, and over time.

Who should receive what forest benefits and when? Economic analysis cannot readily answer such distribution questions which are generally left to the political sphere. But there the problems are no easier. As Thurow (1980) points out, the economy is a zero-sum game in which policy changes involve gainers and losers, and no one wants to be a loser.

Simple voting solutions to resource problems have the draw-back that intensity of preference is not registered in a vote. The indi-vidual with a flaming passion for or against a program has the same power in the voting booth as one who is nearly indifferent about the same issue. Thus there is no assurance that simple majority voting for resource management issues will necessarily maximize social satis-factions (Buchanan and Tullock 1962). The interfaces between economics, social welfare, and politics need greater attention in multiple use forestry. How can welfare-maximizing public choices be made most effectively in the absence of well-functioning markets?[1]

Conclusions

I'd like to close with a touch of what may seem like heresy. Humans have an uncanny ability to adapt to new situations, forget old ones, substitute products and experiences, and develop new tech-nologies. From one generation to another, this ability is truly awe-some. Historic and prospective cases of learning to appreciate changing resources have been widely discussed (Munley and Smith 1976, Michael 1973, Schechter 1977).

In response to the doomsday models (Meadows et al. 1974), Goeller and Weinberg (1976) make a case for the "age of substi-tutability" in which technology provides tremendous opportunities for developing substitute products. In the long run (especially across generations), most consumers can rather happily change their consumption patterns ... from lumber to plywood to flake-board, from pristine wilderness to less "natural" wilderness, from rivers to lakes, from softwoods to hardwoods, from wood to concrete, or from hula hoops to frisbees. Even those who espouse the "steady-state economy" recognize that in such a system, product mixes and tech-nologies will change (Daly 1980, p. 325). So the steady-state economy does not necessarily imply even-flow sustained yield forestry.

All this is not to suggest that nothing matters or anything goes. But it does imply that within a wide range of options, it would be impossible to prove that society would be better off in the long run with one resource management scenario than with another!

Prediction and distribution problems aside, the foremost ethical obligation lies in my opening thesis of maximizing the present value of satisfactions from all resources combined. This may or may not mean sustained yield forestry. In the final analysis, what we should sustain is the land's capability to provide benefits—forest or nonforest—to society. But even here, flexibility is advised. On any given area, one must not spend more to maintain productivity than that productivity is worth.

Notes

1. See Buchanan and Tullock 1962, Downs 1957, and Dahl and Lindblom 1953 for examples of work in this area.

Literature Cited

Bell, E. F., 1977: Estimating effect of timber harvesting levels on employment in western United States. U.S. Forest Service reseach note INT-237, Ogden, UT, 11 pp.

Bell, E., R. Fight, and R. Randall, 1975: ACE the two edged sword. *Journal of Forestry*, Vol. 73, No. 10, pp. 642–643.

Beuter, J. H. and D. C. Olson, 1980: Lakeview federal sustained yield unit, Freemont National Forest: a review, 1974–79. Report to the U.S. Forest Service, Oregon State University, Corvallis.

Buchanan, J. M. and G. Tullock, 1962: The calculus of consent. Univ. of Michigan Press, Ann Arbor, 361 pp.

Charbonneau, J. J. and M. J. Hay, 1978: Determinants and economic values of hunting and fishing. Transactions of 43rd N. Am. Wildl. and Nat. Res. Conf., Wildlife Mgt. Institute, Wash., DC, pp. 391–403.

Clawson, M., 1979: Forests in the long sweep of American history. *Science*, Vol. 204, pp. 1168–1174.

Dahl, R. A. and C. E. Lindblom, 1953: Politics, economics, and welfare. Harper and Brothers, New York.

Daly, H. E. (ed.) 1980: Economics, ecology, ethics—essays toward a steady-state economy. W. H. Freeman & Co., San Francisco, 372 pp.

Darr, D. R. and R. D. Fight, 1974: Douglas County, Oregon: potential economic impacts of a changing timber resource base. U.S. Forest Service Res. Pap. PNW-179, Portland, OR, 41 pp.

Dowdle, B., 1974: Comments and viewpoints. In: Timber policy issues in British Columbia. Univ. of British Columbia Press, Vancouver, pp. 226–320.

Downs, A., 1974: Sustained yield and American social goals. Proceedings of symposium on the economics of sustained yield forestry. University of Washington, Seattle. In press.

Downs, A., 1957: An economic theory of democracy. Harper and Row, New York, 310 pp.

Duerr, W. A., 1974: Timber supply: goals, prospects, problems. *American Journal of Agricultural Economics*, Vol. 56, No. 4, pp. 927–935.

Duerr, W. A. and J. Duerr, 1975: The role of faith on forest resource management. In: Social Sciences in Forestry: A Book of Readings, W. B. Saunders & Co., Philadelphia, pp. 30–41.

Duerr, W. A., 1981: Productivity as a theme. Paper presented at the Society of American Foresters Appalachian Section Meeting, Roanoke, VA. January 29, 7 pp.

Dwyer, J. F., J. R. Kelley, and M. D. Bowes, 1977: Improved procedures for valuation of the contribution of recreation to national economic development. Univ. of Illinois at Urbana-Champaign Wat. Res. Cent. Rept. No. 128, Urbana, 218 pp.

Goeller, H. E. and A. M. Weinberg, 1976: The age of substitutability. *Science,* Vol. 191, pp. 683–689.

Hirshleifer, J., 1974: "Sustained yield" versus capital theory. Proceedings of symposium on the economics of sustained yield forestry. University of Washington, Seattle. In press.

Johnson, K. N., 1976: Consequences of economic harvest scheduling procedures on five national forests. Background report for Timber Harvest Scheduling Issues Study, U.S. Forest Service, Wash., DC 55 pp.

Johnson, K. N. and H. L. Scheurman, 1977: Techniques for prescribing optimal timber harvest and investment under different objectives— discussion and synthesis. *Forest Science Monograph 18,* 31 pp.

Kahn, A. E., 1966: The tyranny of small decisions: market failures, imperfections, and the limits of economics. *Kyklos.* Vol. 19, No. 1, pp. 23–47.

Keane, J. T., 1972: Even flow—yes or no. *American Forests,* Vol. X, No. 6, pp. 32–35 & 61–63.

Klemperer, W. D., 1975: The parable of the allowable pump effect. *Journal of Forestry,* Vol. 73, No. 10, pp. 640–641.

Klemperer, W. D., 1979: On the theory of optimal forest harvesting regulations. *Journal of Environmental Management,* Vol. 9, pp. 1–13.

Ledyard, J. and L. N. Moses, 1974: Dynamics and land use: the case of forestry. Proceedings of symposium on the economics of sustained yield forestry. University of Washington, Seattle. In press.

Meadows, D. H., D. L. Meadows, J. Randers, and W. W. Behrens III, 1974: The limits to growth. Second ed., New American Library, New York, 207 pp.

Michael, R. T., 1973: Education in nonmarket production. *Journal of Political Economy,* Vol. 81, No. 2, pp. 306–327.

Munley, V. G. and V. K. Smith, 1976: Learning-by-doing and experience: the case of whitewater recreation. *Land Economics,* Vol. 52, No. 4. pp. 546–552.

Nelson, T. C. and F. A. Bennet, 1965: A critical look at the normality concept. *Journal of Forestry,* Vol. 63, No. 2, p. 107.

Ramsey, F. P., 1928: A mathematical theory of saving. *The Economic Journal,* Vol. 38, pp. 543–559.

Rawls, J., 1971: A theory of justice. Harvard University Press, Cambridge, MA.

Roberts, M., 1974: Sustained yield and economic growth. Proceedings of symposium on the economics of sustained yield forestry. University of Washington, Seattle. In press.

Samuelson, P. A., 1976: Economics of forestry in an evolving society. *Economic Inquiry,* Vol. 14, pp. 466–492.

Schallan, C. H., 1974: Can regulation contribute to economic stability? *Journal of Forestry,* Vol. 72, No. 4, pp. 214–216.

Schallau, C., W. Maki and J. Beuter, 1969: Economic impact projections for alternative levels of timber production in the Douglas-fir region. *Annals of Regional Science,* Vol. 3, No. 1, pp. 96–106.

Shechter, M., 1977: Open-access recreational resources: is doomsday around the corner? In: Outdoor recreation—advances in applications of economics. U.S. Forest Service Gen. Tech. Rept. WO-2, Wash., DC, pp. 35–41.

Smith, J. H. G., 1969: An economic view suggests that the concept of sustained yield should have gone out with the crosscut saw. *Forestry Chronical*, Vol. 45, No. 3, pp. 167–171.

Solow, R. M., 1974: Intergenerational equity and exhaustible resources. *Review of Economic Studies*, Symposium on the Economics of Exhaustible Resources, pp. 29–45.

Solow, R. M., 1974a: The economics of resources or the resources of economics. *American Economic Review*, Vol. 64, No. 2, pp. 1–14.

Teeguarden, D. E., 1973: The allowable cut effect: a comment. *Journal of Forestry*, Vol. 71, No. 4, pp. 224–226.

Thompson, E. F., 1966: Traditional forest regulation model: an economic critique. *Journal of Forestry*, Vol. 64, No. 11, pp. 750–752.

Thurow, L. C., 1980: The zero-sum society. Penguin Books, New York, 230 pp.

United States Forest Service, 1980: A recommended renewable resources program—1980 update. Publ. FS-346. U.S. Gov't Printing Office, Wash., DC, 540 pp. plus append.

United States Forest Service, 1980a: An analysis of the timber situation in the United States—1952–2030. U.S. Dept. of Agriculture, Wash., DC, 541 pp. plus append.

Waggener, T. R., 1969: Some economic implications of sustained yield as a forest regulation model. Univ. of Washington Contemporary Forestry Paper No. 6, 22 pp.

Waggener, T. R., 1972: Estimating the economic impact of changes in the supply of timber. *Oregon Business Review*, Vol. 31, Nos. 3–4, pp. 1–5.

Waggener, T. R., 1977: Community stability as a forest management objective. *Journal of Forestry*, Vol. 75, No. 11, p. 714.

Walker, J. L., 1976: ECHO: solution techniques for a nonlinear economic harvest optimization model. In: Systems analysis and forest resource management (Meadows, Bare, Ware and Row, eds.), Society of Am. Foresters, Wash., DC, pp. 172–188.

Walker, J. L., 1977: Economic efficiency and the national forest management act of 1976. *Journal of Forestry*, Vol. 75, No. 11, pp. 715–718.

Walker, J. L., 1978: Land use planning applications of the economic harvest optimization (ECHO) model. In: Proceedings of IUFRO symposium on simulation techniques in forest operational planning and control, Wageningen, the Netherlands.

Ethical conflicts often occur when a consultant or government official is privy to trade secrets or other proprietary data. The many temptations that can arise in this situation demand the highest level of professionalism and integrity. Getting and keeping secrets is a two-way street. As a forester, you may be asked for information about your organization that is secret by policy. Is it legitimate to share such information? In this paper, I discuss a number of the ethical issues involved in both obtaining and keeping information that others would like to keep secret.

CHAPTER ELEVEN

Getting and Keeping Secrets: Some Ethical Reflections for Consultants

LLOYD C. IRLAND

Why We Want Other People's Secrets

Why do we need people's secrets? We need proprietary and confidential information to serve our clients' needs for accurate and timely information. We can benefit from a moment's reflection on the ethical issues raised by the need to obtain and keep secrets in the course of our work.

We frequently need the *client's own proprietary information* in order to have all the facts about production volumes, marketing patterns, estimated accounting costs, and prices. In land management, we need inventories, cutting records, and a variety of other data that land owners customarily hold closely. We may also need to know the details of business contracts. Finally, we may need to know extensive details about the client's tax position, business plans, strategy, or other facts that might affect decisions that we have under study.

In addition, we frequently need to know *other people's secrets* as well. Now, none of us are in the business of industrial espionage. We are not being retained to steal other people's secrets. But in many kinds of work we can't function without at least a general knowledge of matters which others would prefer to keep confidential. An example is in fiber supply work. Consultants are asked for detailed information and forecasts regarding the cost of delivered wood. This requires information on stumpage and delivered wood prices in the marketplace, which is frequently not well documented in published sources. It also requires that we gain at least a rough understanding of

This article reprinted by permission from *The Consultant*, Summer 1990, pp. 11–14.

the components and likely trends in logging costs. So we need to get the secrets of the loggers.

In other instances, a company may wish to know whether others in its timbershed are considering expansions. Its officials need to assess future competing demand for wood to make a wise decision. And certainly, there is a social interest in seeing that companies do not overexpand through ignorance of each other's plans.

The final reason we need to know people's secrets is both self-serving and client-serving. Only when we know the nitty-gritty details of our industry are we well placed to assist our clients with sound advice that is well grounded in facts. Now, the only way to really know the facts of life in an industry is to study costs, prices, marketing patterns, technology, and resource procurement patterns as they are actually changing in the marketplace. The only way to learn these things is to work with confidential, private data.

So here we are presented with a dilemma. We are aware of and may possess in our files confidential and proprietary data from a variety of clients. Our value to clients depends on our awareness of and access to that information. If it were felt that we might make inappropriate use of that information, we would never be granted access to it. So our business demands a high degree of judgment and discretion in the ability to get and to keep secrets. This is true for consultants working on large corporate projects, for foresters managing and appraising woodlots, and not infrequently for public servants as well.

Why Do People Have Business Secrets?

It is useful to wonder for a moment why people have secrets in the first place. There are a variety of reasons. Many companies, from individual proprietorships to the most immense corporations, place a high value on privacy in general. This is understandable. The ability of companies and landowners to enjoy a high degree of privacy is a widely noted feature of the American economy and is one reason for its attractiveness to foreign investors.

So there is a general cultural orientation placing high value on business privacy for its own sake, quite apart from any business gain or risk that might be involved. But there are many situations where people are competing in thin markets for labor or for timber, or where product markets are highly fragmented. In these cases, people are sensitive about revealing their hiring and compensation policies, or their wood procurement prices and practices, or their customer lists.

Some firms possess or believe that they possess superior pro-prietary business practices or engineering processes which contribute to their profitability. We have all been in mills where cameras are not allowed for this reason.

Many businesses are inclined to believe that if specifics of their costs and prices were known to their competitors it would result in a disadvantage. Whether they are correct in this is not the issue. But it is one more reason for businesses to closely guard many kinds of information.

Finally, firms are constrained by antitrust laws (and consent decrees) from discussing with one another many aspects of costs and prices. While this is a different matter than our principal interest here, it can affect willingness to share information.

Getting Secrets

In laying out your plans for a client project, you will quickly see whether it will be necessary to obtain information that people may not want to give you. There are several simple approaches. Here is a short list of possibilities.

1. To check an existing source. We maintain an extensive library of industry directories for the region in which we do most of our work. This gives us a starting point for information as to wood consumption, product volume, capacity, or employment on many different companies. When we call an individual and ask to check an item of directory information, at times they will tell us information that we suspect we would not get at all had we not come into the conversation with something.

2. Alternative contacts. In many situations there may be more than one individual or organization which can give you at least a rough estimate of the information that you need. An important skill in building a consulting practice is to develop as wide a network of people like this as you possibly can.

3. Other side of transactions. When we need information about production volumes or prices, we may have the opportunity to check with the other side of the transaction to elicit an estimate of the amounts that we require.

4. The wild guess. If there are no directories or other sources of estimates, it may be possible to obtain a response by making your own estimates of what the figures should be and seeking a response from the interviewee.

5. Index the data. When you are dealing with information that is

extremely sensitive, but for which you need a precise estimate, it may be possible to ask respondents to give you not the
raw data but an index of the change. It may be that you can use
this knowledge of change to update some previous published
benchmark. Or perhaps simply knowing the magnitude of
change tells you what you need. If a way can be found to
"purify" a proprietary number in this manner, it may improve
people's willingness to help.

6. The managed exchange. There are occasions when it is
   desirable to obtain information that the sources are extremely
   reluctant to divulge. If you can obtain a client agreement to
   share the average or sum of aggregate information on an
   anonymous basis this frequently improves respondents'
   willingness to assist. In many instances a client will be happy to
   authorize such an approach since it may be the only way they
   have of obtaining the information. Some clients recognize that
   they do not give away anything of value in doing this because
   of the shared interest in having this information available.

So there are a number of ways to elicit proprietary information
when it is needed. But none of these methods will work unless the
respondent trusts your assurances that the individual data and the
names of respondents will never be seen by the client or by a competitor.

Keeping Secrets

Keeping secrets would seem to be a good deal easier than getting
secrets. But in fact, as you move ahead from year to year in the
consulting business, it will demand care, discretion, and conscious
effort to make certain that you don't inadvertently reveal some of the
things you know. You have to bring to bear your general knowledge
on problems. But you must make certain that even by innocently
mentioning an apparently trivial detail you don't tip off to a competitor, for example, that Company A is doing a major expansion
study.

When stumpage and log market prices and volumes are involved,
the clients' own attorneys should specify that information reported to
the client must be summarized and presented in a manner that the
individual competitors involved cannot be discerned by the client.
This is for obvious antitrust reasons and may lead to a certain amount
of difficulty in presenting results when the market is highly concentrated. Nonetheless, it would be all too easy to inadvertently create an
antitrust infraction for yourself and your client by failure to observe

due caution. Once you get deeply into information of this kind and begin casually discussing its various dimensions and implications, you may accidentally reveal information that you shouldn't. In doing so you not only create a potential antitrust violation, but you violate the trust placed in you by the person who provided the information.

In social settings among our peers, it can be tempting to use our inside knowledge to impress others with how important we are. This happens all too frequently; it would be encountered a good deal less often if people thought for a moment that such actions abuse another person's trust in them.

In many cases, the client will insist on anonymity. This leads to the difficulty often encountered when respondents casually inquire why you want to know, what you're asking for, and who wants to know it. We usually do not tell our interviewers who the client is or the detailed nature of the project. If asked by an interviewee, they can then honestly respond that they don't know. Most respondents will still help us with what we're looking for.

Another technique for keeping secrets is to exercise care in your custody of client materials, and, in particular, to be careful about what kinds of material you cite when using that information in work for other clients. One way to ensure against accidental disclosure would be to generate your work sheets and data sheets with numbers or letters to designate the sources of particular bits of information instead of using company names. A final precaution is to make sure that sensitive and proprietary information is simply only seen or used by those who absolutely must see it. In litigation work, you will often be required to sign and comply with stringent and detailed confidentiality agreements covering such matters.

Some Ethical Questions

Getting and keeping secrets inevitably raises questions in areas where we may not see clear rules to guide our behavior. Here are a few of them. I would like to get your responses as to how you would handle these questions.

1. How far can we conceal or misrepresent the purpose for which we are seeking information? Is it permissible to obtain information on false pretenses from the respondents in the interest of getting the data we need to serve our client?

2. How much can we tell about the need for the information? This is the reverse of the previous question. We have to abide by our duty to the client. But when the client's need for anonymity conflicts with the client's needs for information,

what should we do? Probably the best practice is to consult the client for guidance when people are withholding needed information. What other options might we have?

3. Can we go around a reluctant source? If we call the normal cus-todian of a certain item of information and our request is declined on grounds of policy, is it appropriate to call other people we may know in the same organization to seek an esti-mate by going around an official who has just declined us?

4. What limits must be respected in using client-owned informa-tion? Data produced for an appraisal or other project may be legally the client's property. What rules should guide us in protecting or using such information in other work?

Conclusion

Getting and keeping secrets as a consultant is a complex matter. It grows more complex the more different problems you deal with and the more clients you serve. Many ethical gray areas will test your discretion and judgment. When those are encountered two simple rules to fall back on are: (1) How would I feel about it if somebody did this to me? and (2) What would the effect on my business be if my clients knew I was doing this? These rules may seem self-centered and self-serving. But they may concentrate your mind on the matter in a way that more general and often vague moral principles or ethics codes might not.

It hardly needs to be mentioned that we do not utilize client information for purposes of personal gain. We do not convey or sell to others this information without proper authorization, and we do not offer bribes or other incentives to respondents to induce them to provide information. Yet how often do we read in the newspapers about people who've been unable to resist temptations to violate these ordinary rules of business ethics and courtesy?

As consultants, we deal with matters that depend on client trust and confidence every day. Our ability to serve them, to survive as a business, and to feel good about ourselves, depends upon main-taining a reputation for judgment and discretion. One way to destroy a reputation is to be sloppy or inattentive in the all-important area of getting and keeping secrets.

Among professional foresters, the arguments in support of professional ethics codes are generally accepted. Yet, as General Patton was fond of saying, "if everyone's thinking the same, then nobody's thinking." In this challenging essay, John Ladd, a philosopher at Brown University, offers a valuable alternative view. In fact, virtually all the arguments made earlier in this section receive rough handling here. This essay is included because of its cogent reasoning and thorough handling of the issues; confronting its argument fully and honestly will assist professional foresters in clarifying the role of ethical codes in their working lives.

CHAPTER TWELVE

The Quest for a Code
of Professional Ethics:
An Intellectual and Moral Confusion

JOHN LADD

My role as a philosopher is to act as a gadfly.... My theme is stated in the title: it is that the whole notion of an organized professional ethics is an absurdity—intellectual and moral. Furthermore, I shall argue that there are few positive benefits to be derived from having a code, and that the possibility of mischievous side effects from adopting a code is substantial....

1. To begin with, ethics itself is basically an open-ended, reflective, and critical intellectual activity. It is essentially problematic and controversial, both as far as its principles are concerned and in its application. Ethics consists of issues to be examined, explored, discussed, deliberated, and argued. Ethical principles can be established only as a result of deliberation and argumentation. These principles are not the kind of thing that can be settled by fiat, by agreement, or by authority. To assume that they can be is to confuse ethics with law-making, rule-making, policy-making, and other kinds of decision-making. It follows that ethical principles, as such, cannot be established by associations, organizations, or by a consensus of their members. To speak of codifying ethics, therefore, makes no more sense than to speak of codifying medicine, anthropology, or architecture.

2. Even if substantial agreement could be reached on ethical principles and they could be set out in a code, the attempt to impose such principles on others in the guise of ethics contradicts the notion of ethics itself, which presumes that persons are autonomous, moral

This article reprinted by permission from the proceedings of the American Association for the Advancement of Science Workshop on Professional Ethics held in Washington, DC, on 15–16 November 1979; pp. 154–159. Copyright © American Association for the Advancement of Science.

agents. In Kant's terms, such an attempt makes ethics heteronomous; it confuses ethics with some kind of externally imposed set of rules such as a code of law which, indeed, is heteronomous. To put the point in more popular language: ethics must, by its very nature, be self-directed rather than other-directed.

3. Thus, in attaching disciplinary procedures and methods of adjudication and sanctions, formal and informal, to the principles that one calls "ethical," one automatically converts them into legal rules or some other kind of authoritative rules of conduct, such as the bylaws of an organization, regulations promulgated by an official, club rules, rules of etiquette, or other sorts of social standards of conduct. To label such convention, rules, and standards "ethical" simply reflects an intellectual confusion about the status and function of these conventions, rules, and standards. Historically, it should be noted that the term "ethical" was introduced merely to indicate that the code of the Royal College of Physicians was not to be construed as a criminal code (i.e. a legal code). Here "ethical" means simply non-legal.

4. That is not to say that ethics has no relevance for projects involving the creation, certification, and enforcement of rules of conduct for members of certain groups. But logically it has the same kind of relevance that it has for the law. As with law, its role in connection with these projects is to appraise, criticize, and perhaps even defend (or condemn) the projects themselves, the rules, regulations, and procedures they prescribe, and the social and political goals and institutions they represent. But although ethics can be used to judge or evaluate a disciplinary code, penal code, code of honor, or what goes by the name of a "code of ethics," it cannot be identified with any of these, for the reasons that have already been mentioned.

Some General Comments on Professionalism and Ethics

5. Being a professional does not automatically make a person an expert in ethics, even in the ethics of that person's own particular profession—unless of course we decide to call the "club rules" of a profession its ethics. The reason for this is that there are no experts in ethics in the sense of expert in which professionals have a special expertise that others do not share. As Plato pointed out long ago in the *Protagoras,* knowledge of virtue is not like the technical knowledge that is possessed by an architect or shipbuilder. In a sense, everyone is, or ought to be, a teacher of virtue; there are no professional qualifications that are necessary for doing ethics.

6. Moreover, there is no special ethics belonging to professionals. Professionals are not, simply because they are professionals, exempt from the common obligations, duties, and responsibilities

that are binding on ordinary people. They do not have a special moral status that allows them to do things that no one else can. Doctors have no special right to be rude, to deceive, or to order people around like children, etc. Likewise, lawyers do not have a special right to bend the law to help their clients, to bully witnesses, or to be cruel and brutal— simply because they think that it is in the interests of their client. Professional codes cannot, therefore, confer such rights and immunities, for there is no such thing as professional ethical immunity.

7. We might ask: do professionals, by virtue of their special professional status, have special duties and obligations over and above those they would have as ordinary people? Before we can answer this question, we must first decide what is meant by the terms "profession" and "professional," which are very loose terms that are used as labels for a variety of different occupational categories. The distinctive element in professionalism is generally held to be that professionals have undergone advanced, specialized training and that they exercise control over the nature of their job and the services they provide. In addition, the older professions—lawyers, physicians, professors and ministers—typically have clients to whom they provide services as individuals. (I use the term "client" generically so as to include patients, students, and parishioners.) When professionals have individual clients, new moral relationships are created that demand special types of trust and loyalty. Thus, in order to answer the question, we need to examine the context under which special duties and obligations of professionals might arise.

8. In discussing specific ethical issues relating to the professions, it is convenient to divide them into issues of *macro-ethics* and *micro-ethics.* The former comprise what might be called collective or social problems, that is, problems confronting members of a profession as a group in their relation to society; the latter, issues of micro-ethics, are concerned with moral aspects of personal relationships between individual professionals and other individuals who are their clients, their colleagues and their employers. Clearly the particulars in both kinds of ethics vary considerably from one profession to another. I shall make only two general comments.

9. Micro-ethical issues concern the personal relationships between individuals. Many of these issues simply involve the application of ordinary notions of honesty, decency, civility, humanity, considerateness, respect, and responsibility. Therefore, it should not be necessary to devise a special code to tell professionals that they ought to refrain from cheating and lying, or to make them treat their clients (and patients) with respect, or to tell them that they ought to ask for informed consent for invasive actions. It is a common mistake to assume that *all* the extra-legal norms and conventions governing professional relationships have a moral status, for every profession

has norms and conventions that have as little to do with morality as the ceremonial dress and titles that are customarily associated with the older professions.

10. The macro-ethical problems in professionalism are more problematic and controversial. What are the social responsibilities of professionals as a group? What can and should they do to influence social policy? Here, I submit, the issue is not one of professional roles, but of *professional power*. For professionals as a group have a great deal of power, and power begets responsibility. Physicians as a group can, for instance, exercise a great deal of influence on the quality and cost of health care, and lawyers can have a great deal of influence on how the law is made and administered, etc.

11. So-called "codes of professional ethics" have nothing to contribute either to micro-ethics or to macro-ethics as just outlined. It should also be obvious that they do not fit under either of these two categories. Any association, including a professional association, can, of course, adopt a code of conduct for its members and lay down disciplinary procedures and sanctions to enforce conformity with its rules. But to call such a disciplinary code a code of *ethics* is at once pretentious and sanctimonious. Even worse, it is to make a false and misleading claim, namely, that the profession in question has the authority or special competence to create an ethics, that it is able authoritatively to set forth what the principles of ethics are, and that it has its own brand of ethics that it can impose on its members and on society.

I have briefly stated the case against taking a code of professional ethics to be a serious ethical enterprise. It might be objected, however, that I have neglected to recognize some of the benefits that come from having professional codes of ethics. In order to discuss these possible benefits, I shall first examine what some of the objectives of codes of ethics might be, then I shall consider some possible benefits of having a code, and, finally, I shall point out some of the mischievous aspect of codes.

Objectives of Codes of Professional "Ethics"

In order to be crystal clear about the purposes and objectives of a code, we must begin by asking: to whom is the code addressed? Although ostensibly codes of ethics are addressed to the members of the profession, their true purposes and objectives are sometimes easier to ascertain if we recognize that codes are in fact often directed at other addressees than members. Accordingly, the real addressees might be any of the following: (a) members of the profession, (b) clients or buyers of the professional services, (c) other agents dealing with professionals, such as government or private institutions like universities or hospitals, or (d) the public at large. With this in mind, let us examine some possible objectives.

First, the objective of a professional code might be "inspirational," that is, it might be used to inspire members to be more "ethical" in their conduct. The assumption on which this objective is premised is that professionals are somehow likely to be amoral or submoral, perhaps, as the result of becoming professionals, and so it is necessary to exhort them to be moral, e.g. to be honest. I suppose there is nothing objectionable to having a code for this reason; it would be something like the Boy Scout's Code of Honor, something to frame and hang in one's office. I have severe reservations, however, about whether a code is really needed for this purpose and whether it will do any good; for those to whom it is addressed and who need it the most will not adhere to it anyway, and the rest of the good people in the profession will not need it because they already know what they ought to do. For this reason, many respectable members of a profession regard its code as a joke and as something not to be taken seriously. (Incidentally, for much the same kind of reasons as those just given, there are no professional codes in the academic or clerical professions.)

A second objective might be to alert professionals to the moral aspects of their work that they might have overlooked. In jargon, it might serve to sensitize them or to raise their consciousness. This, of course, is a worthy goal—it is the goal of moral education. Morality, after all, is not just a matter of doing or not doing, but also a matter of feeling and thinking. But, here again, it is doubtful that it is possible to make people have the right feelings or think rightly through enacting a code. A code is hardly the best means for teaching morality.

Third, a code might, as it was traditionally, be a disciplinary code or a "penal" code used to enforce certain rules of the profession on its members in order to defend the integrity of the profession and to protect its professional standards. This kind of function is often referred to as "self-policing." It is unlikely, however, that the kind of disciplining that is in question here could be handled in a code of ethics, a code that would set forth in detail criteria for determining malpractice. On the contrary, the "ethical" code of a profession is usually used to discipline its members for other sorts of "unethical conduct," such as stealing a client away from a colleague, for making disparaging remarks about a colleague in public, or for departing from some other sort of norm of the profession. (In the original code of the Royal College of Physicians, members who failed to attend the funeral of a colleague were subject to a fine!) It is clear that when we talk of a disciplinary code, as distinguished from an exhortatory code, a lot of new questions arise that cannot be treated here; for a disciplinary code is quasi-legal in nature, it involves adjudicative organs and processes, and it is usually connected with complicated issues relating to such things as licensing.

A fourth objective of a code might be to offer advice in cases of moral perplexity about what to do: e.g. should one report a colleague for malfeasance? Should one let a severely defective newborn die? If such cases present genuine perplexities, then they cannot and should not be solved by reference to a code. To try to solve them through a code is like trying to do surgery with a carving knife! If it is not a genuine perplexity, then the code would be unnecessary.

A fifth objective of a professional code of ethics is to alert prospective clients and employers to what they may and may not expect by way of service from a member of the profession concerned. The official code of an association, say, of engineers, provides an authoritative statement of what is proper and what is improper conduct of the professional. Thus, a code serves to protect a professional from improper demands on the part of employer or client, e.g. that he lie about or coverup defective work that constitutes a public hazard. Codes may thus serve to protect 'whistle-blowers.' (The real addressee in this case is the employer or client.)

Secondary Objectives of Codes—Not Always Salutory

I now come to what I shall call "secondary objectives," that is, objectives that one might hesitate always to call "ethical," especially since they often provide an opportunity for abuse.

The first secondary objective is to enhance the image of the profession in the public eye. The code is supposed to communicate to the general public (the addressee) the idea that the members of the profession concerned are service oriented and that the interests of the client are always given first place over the interests of the professional himself. Because they have a code they may be expected to be trustworthy.

Another secondary objective of a code is to protect the monopoly of the profession in question. Historically, this appears to have been the principal objective of a so-called code of ethics, e.g. Percival's code of medical ethics. Its aim is to exclude from practice those who are outside the professional in-group and to regulate the conduct of the members of the profession so as to protect it from encroachment from outside. Sometimes this kind of professional monopoly is in the public interest and often it is not.

Another secondary objective of professional codes of ethics, mentioned in some of the literature, is that having a code serves as a status symbol; one of the credentials for an occupation to be considered a profession is that it have a code of ethics. If you want to make your occupation a profession, then you must frame a code of ethics for it: so there are codes for real estate agents, insurance agents, used car dealers, electricians, barbers, etc., and these codes serve, at least in the eyes of some, to raise their members to the social status of lawyers and doctors.

Mischievous Side-effects of Codes of Ethics

I now want to call attention to some of the mischievous side-effects of adopting a code of ethics:

The first and most obvious bit of mischief, is that having a code will give a sense of complaisance to professionals about their conduct. "We have a code of ethics," they will say, "so everything we do is ethical." Inasmuch as a code, of necessity, prescribes what is minimal, a professional may be encouraged by the code to deliver what is minimal rather than the best that he can do. "I did everything that the code requires. . . ."

Even more mischievous than complacency and the consequent self-congratulation, is the fact that a code of ethics can be used as a cover-up for what might be called basically "unethical" or "irresponsible" conduct.

Perhaps the most mischievous side-effect of codes of ethics is that they tend to divert attention from the macro-ethical problems of a profession to its micro-ethical problems. There is a lot of talk about whistle-blowing. But it concerns individuals almost exclusively. What is really needed is a thorough scrutiny of professions as collective bodies, of their role in society and their effect on the public interest. What role should the professions play in determining the use of technology, its development and expansion, and the distribution of the costs (e.g. disposition of toxic wastes) as well as the benefits of technology? What is the significance of professionalism from the moral point of view for democracy, social equality, liberty and justice? There are lots of ethical problems to be dealt with. To concentrate on codes of ethics as if they represented the real ethical problems connected with professionalism is to capitulate to *struthianism* (from the Greek word *struthos*-ostrich).

One final objection to codes that needs to be mentioned is that they inevitably represent what John Stuart Mill called the "tyranny of the majority" or, if not that, the "tyranny of the establishment." They serve to and are designed to discourage if not suppress the dissenter, the innovator, the critic.

By way of conclusion, let me say a few words about what an association of professionals can do about ethics. On theoretical grounds, I have argued that it cannot codify an ethics and it cannot authoritatively establish ethical principles or prescribed guidelines for the conduct of its members—as if it were *creating* an ethics! But there is still much that associations can do to promote further understanding of and sensitivity to ethical issues connected with professional activities. For example, they can fill a very useful educational function by encouraging their members to participate in extended discussions of issues of both micro-ethics and macro-ethics, e.g. questions about responsibility; for these issues obviously need to be examined and discussed much more extensively than they are at present—especially by those who are in a position to do something about them.

Scientists in public service are likely to encounter situations in which their views run counter to official policies. This is occurring with increasing frequency in federal land management. Policies are made by properly elected and appointed officials who do not want their decisions challenged publicly by more specialized underlings. The specialists want to be able to campaign for the policy decisions they want, sometimes choosing loyalties to their own beliefs and those of their professional peers over loyalty to their agency. Interest groups seek allies, and even unauthorized information leaks from sympathetic agency staff, but the public has a right to know the personal views of these experts.

In this paper, the case of Howard Wilshire, a U.S. Geological Survey scientist, illustrates the ethical conflicts experienced by all technically trained individuals serving in large organizations. The author, Eliot Marshall, is a science journalist.

CHAPTER THIRTEEN

Ethics Debate Sends Tremors Through USGS

ELIOT MARSHALL

When a scientist goes to work for the government, must he avoid using his expertise to help a private cause—even if he limits himself to giving advice on his own time?

This question confronted Howard Wilshire, a geologist at the U.S. Geological Survey, when, as a private citizen, he criticized another federal agency's plan to lay out a track for motorcyclists in a national forest.

Wilshire has a passionate interest in defending arid lands from the onslaught of off-road vehicles. His official and personal efforts have intertwined to make him one of the world's leading authorities on this subject. But they also brought him attention of a different sort: earlier this year, he was locked in a widely publicized battle with his boss, USGS director Dallas Peck, who tried to suspend him for misconduct.

Peck's action, which was overturned recently, grew out of what he saw as a conflict of interest. Peck said Wilshire had violated federal criminal law by advising an environmental group called FAWN (Friends Aware of Wildlife Needs) on how to prevent the Forest Service from building a playground for motorcyclists in California's El Dorado National Forest. Wilshire at one point suggested that FAWN subpoena him as a USGS expert, even though federal employees are not allowed to testify as experts against the government.

This case highlights a problem that is likely to arise again and again for scientists in public service. Because of their expertise they may have—or believe they have—the best insight into a scientific issue that has become embroiled in public debate. But if their view happens to clash with the position taken by responsible officials, they face some difficult choices: they may keep silent, speak out at risk of punishment, or resign. The middle option is perhaps the most satisfying, but also the trickiest to negotiate, for the rules are always changing. To what degree public employees may combat the decisions of their bosses in public is a question that the Wilshire case brings into sharp focus.

Peck's attempt to punish Wilshire, combined with a directive to all staffers that they should limit involvement with environmental and other nonprofit groups, caused a storm in USGS over the summer. The directive was later withdrawn and replaced with general advice to use "sound judgment." But many scientists still look uneasily on the Wilshire case, fearing that headquarters has become too ready to sacrifice personal freedom in the interest of conformity.

"I think it's an important issue," says Brent Dalrymple, a colleague of Wilshire's in the USGS office at Menlo Park, California. He claims that taxpayers have a right to hear all manner of testimony and opinion from U.S. experts, regardless of whether it clashes with the official line. "Should the public have access only to the information that an Administration wants to release?" Dalrymple asks.

Environmental groups such as the Sierra Club also see the Wilshire case as having broad implications. They link it with other examples in which they think the government has bottled up information. These include:

- U.S. Fish and Wildlife researchers who allegedly were not permitted to speak out on the threats to wildlife posed by irrigation runoff in California.

- USGS staffers who say that risk data on the nuclear waste burial site at Yucca Mountain, Nevada, are not being presented accurately.

- A U.S. Park Service and USGS researcher whose unfavorable testimony on offshore drilling was canceled by superiors at the last moment.

- A decision by White House staffers to edit controversial testimony on global warming by a National Aeronautics and Space Administration scientist.

- An incident in the 1960s, in which Dalrymple and several other USGS staffers at Menlo Park got in trouble with an Interior Department chief for opposing as too risky a building project that was slated to go on the soft mud of San Francisco Bay. After first ordering the scientists to be quiet, the Department retreated and let them speak.

Peck, on the other hand, sees the Wilshire case narrowly, as one in which a scientist used not just his expertise but his public office to advance private ends. Peck put it as follows in a letter of reprimand to Wilshire: "Your support of FAWN could most certainly result in the loss of your independence or impartiality, resulting in damage to the reputation of the USGS as a scientific agency. . . . We risk some erosion of our organizational credibility if we place ourselves in a position where we can be easily identified not as a private citizen but as a USGS employee in an advocacy role or supporting either side of a controversial issue."

At the heart of this case is a battle between off-road vehicle (ORV) users and those who would like to steer ORVs away from public lands. This debate on where ORVs may travel has raged in California for more than a decade, and Wilshire, a mantle petrologist with a Ph.D. from the University of California at Berkeley and a 30-year veteran at USGS, has been involved from the start. He was drawn in, he says, when scientists in the Apollo program asked him to find out how long disturbances of the moon's surface would remain visible. Using the Mojave Desert as a model, he made studies in the late 1960s that led him to conclude that the scrapes and marks left by the lunar rover, for example, would last a very long time—17 million years. This raised a concern in his own mind about the "least geologically active" areas on Earth.

"This happened at a time when the use of ORVs was expanding tremendously," says Wilshire. "I was doing it myself. Whenever I wanted to go somewhere in the desert, I just drove there. But I could see the tracks left by the recreationists, and the abundance of them bothered me." He began to study the long-term effects of human activity on arid lands, eventually concluding that the use of ORVs does irreparable harm.

Wilshire also began to speak out as a citizen against the disruption of arid ecosystems. Two years ago, Wilshire's reputation as an opponent of ORVs reached Karen Schambach, president of FAWN, a tiny activist group in the Sierra Nevadas. She was preparing to fight the U.S. Forest Service over its plan to build a network of motorcycle trails through the El Dorado Forest near her home in Georgetown, California.

In 1987, at Schambach's request, Wilshire visited the El Dorado Forest on his free time and offered to serve as an expert witness on flaws in the plan for ORV trail maintenance. He pointed out that his testimony would have more impact if he were subpoenaed as a USGS official.

Later, when it looked as though the case might not go to court, he wrote a letter as a private citizen poking holes in the Forest Service's environmental assessment, filing it in the public docket and sending a copy to FAWN.

It was about this time, according to Wilshire and Schambach, that the U.S. attorney handling the case for the Forest Service called Wilshire's superiors. The outraged attorney complained that Wilshire was using his USGS status to foment trouble, putting himself and the Survey in a conflict of interest. Without asking the USGS to investigate on its own, Peck took up the cudgels and many months later, in March 1989, sent a rebuke to Wilshire, threatening to suspend him for 4 weeks.

Peck's rationale had two parts, one based narrowly on the 1879 "organic act" that created the USGS and the other on general codes of conduct that apply to all federal employees. The organic act states plainly that members of USGS "shall execute no surveys or examinations for private parties."

The prohibition's meaning has never been spelled out in court, and USGS staffers have made their own interpretations. E-an Zen, a senior scientist at USGS headquarters who became involved at Peck's request, says, "I did a little digging" through records of the National Academy of Sciences, which proposed the creation of USGS in 1878, and concluded that "there is no way one can tell why these words were put in." He thinks they may have been included in the act as boilerplate.

Wilshire contends the rule has been interpreted lately to mean that no one may write a professional study for a private business. However, the USGS has not interpreted this tradition very strictly. "We'd be violating the law every time we turn around, if that were true," says Wilshire. Properly sanctioned reports benefiting industry are issued often. Furthermore, according to Wilshire, USGS staffers are sometimes asked to visit a private site and interpret the geology.

But Peck didn't let Wilshire's visit to the El Dorado forest slide by in the same way, ruling instead that it violated the organic act. Many senior USGS scientists rose up in protest, sending a flood of outraged letters to the director, warning that he—not Wilshire—was hurting the Survey's integrity. But the director did not back off.

Wilshire appealed the decision to the Department of the Interior, USGS's bureaucratic home, and got a Solomon-like judgment. Deputy assistant secretary Charles Kay ruled in August that, while Wilshire had misbehaved, there was not enough evidence to convict. According to USGS personnel chief Maxine Millard, the Interior Department solicitor—the same one who earlier had recommended punishment of Wilshire—decided that the case would be hard to win. Peck decided not to push any further, agreeing to a rebuke rather than suspension.

Wilshire has filed a grievance anyway, asking that Kay's and Peck's letters be removed from his file, also that the Survey make its policy crystal clear. Since his case blew open, he claims, headquarters has been sending out vague memos on what makes for a conflict of interest. Millard says that those now demanding clearer guidelines may regret it later. In her view, it's better to leave such matters to personal discretion, but, she adds, "They've raised the issue; now we have to deal with it." Consequently, USGS staffer Clifford Nelson has been asked to look into the history of rules limiting off-duty consulting and to help the agency write some definitive new guidelines.

While all this was being debated, Peck added fuel to the fire by issuing last November a directive (Administrative Digest 933 or AD 933) that seemed to prohibit a whole new raft of activities. It named the Sierra Club and the Nature Conservancy as two typical nonprofit organizations that should not be allowed to benefit privately from USGS labors. It also seemed to cut back on working-hours involvement in professional societies.

Coming at a time of low morale and tight budgets, the memo seemed to many in USGS like the final insult.

Even those with no particular interest in the Wilshire case, including many senior USGS scientists, asked Peck to reconsider his words. Peck responded by asking Zen to chair a review committee. The result was the issuance of a new ethics directive (AD 1009) dated 31 July. It claimed to "amend" the offensive sections of AD 933, making the prohibitions more general and appealing to scientists to use "common sense" and "sound judgment" in balancing the desire to speak out as citizens against the need to protect the credibility of USGS.

As for the question of whether Wilshire himself got fair treatment, opinion remains divided. Many senior staffers think the final result was acceptable. For example, Jack Evernden, a USGS geophysicist who had come to Wilshire's defense earlier, says: "In my view, it came out fine. The system worked."

Others remain uneasy, however. For example, Brent Dalrymple says he is "disappointed." Because there had been no definitive interpretation of the 1879 USGS act before, Peck had an opportunity to make one and to make it generous. Instead, the decision on the Wilshire case and AD 933, according to Dalrymple, reveals bad judgment. While most of the damage has been repaired, he regrets that the USGS leadership wavered in the first place. Dalrymple says leaders of strong research institutions must tolerate some confusion and dissonance to retain good people: "That is the price you pay for hiring creative, aggressive scientists."

Professional land managers are regularly called upon to summarize and set forth the scientific basis for decisions, as in Environmental Impact Statements made by many federal agencies. They always face the challenges of minimizing deviations from professional standards while working under tight deadlines and with inadequate funding. They may also need to properly allocate credit for work done and generally uphold the integrity of the process by which they work. This is often done in the face of intense political pressures from groups demanding that particular conclusions be reached.

This selection is excerpted from a short booklet prepared by the National Academy of Sciences that briefs young people on many aspects of being a scientist.

CHAPTER FOURTEEN

On Being a Scientist

COMMITTEE ON THE CONDUCT OF SCIENCE, NATIONAL ACADEMY OF SCIENCES

. . . . Given the progressive nature of science, a logical question is whether scientists can ever establish that a particular theory describes the empirical world with complete accuracy. The notion is a tempting one, and a number of scientists have proclaimed the near completion of research in a particular discipline (occasionally with comical results when the foundations of that discipline shortly thereafter underwent a profound transformation). But the nature of scientific knowledge argues against our ever knowing that a given theory is the final word. The reason lies in the inherent limitations on verification. Scientists can verify a hypothesis, say by testing the validity of a consequence derived from that hypothesis. But verification can only increase confidence in a theory, never prove the theory completely, because a conflicting case can always turn up sometime in the future.

Because of the limits on verification, philosophers have suggested that a much stronger logical constraint on scientific theories is that they be falsifiable. In other words, theories must have the possibility of being proved wrong, because then they can be meaningfully tested against observation. This criterion of falsifiability is one way to distinguish scientific from nonscientific claims. In this light, the claims of astrologers or creationists cannot be scientific because these groups will not admit that their ideas can be falsified.

Falsifiability is a stronger logical constraint than verifiability, but the basic problem remains. General statements about the world can never be absolutely confirmed on the basis of finite evidence, and all evidence is finite. Thus, science is progressive, but it is an open-ended progression. Scientific theories are always capable of being reexamined and if necessary replaced. In this sense, any of today's most cherished theories may prove to be only limited descriptions of the empirical world and at least partially "erroneous."

Human Error in Science

Error caused by the inherent limits on scientific theories can be discovered only through the gradual advancement of science, but error of a more human kind also occurs in science. Scientists are not infallible; nor do they have limitless working time or access to unlimited resources. Even the most responsible scientist can make an honest mistake. When such errors are discovered, they should be acknowledged, preferably in the same journal in which the mistaken information was published. Scientists who make such acknowledgments promptly and graciously are not usually condemned by colleagues. Others can imagine making similar mistakes.

Patent Procedures

In some areas of research, a scientist may make a discovery that has commercial potential. Patenting is a means of protecting that potential while continuing to disseminate the results of the research.

Patent applications involve such issues as ownership, inventorship, and licensing policies. In many situations, ownership of a patent is assigned to an institution, whether a university, a company, or a governmental organization. Some institutions share royalty income with the inventors. Universities and government laboratories usually have a policy of licensing inventions in a manner consistent with the public interest, at least in cases in which federal funds have supported the research.

Scientists who may be doing patentable work have an obligation to themselves and to their employers to safeguard intellectual property rights. Particularly in industry or in a national laboratory, this may involve prompt disclosure of a valuable discovery to the patent official of the organization in which the scientist works. It also entails keeping accurately dated notebook records written in ink in a bound notebook, ideally witnessed and signed by a colleague who is not a coinventor. Data scribbled in pencil on scraps of paper interleaved in loose-leaf notebooks, besides being professionally undesirable, are of no use in a patent dispute.

Under U.S. patent law, a person who invents something first can be granted a patent even if someone else files a claim first so long as witnessed laboratory records demonstrate the earlier invention. Any public disclosure of the discovery prior to filing for a U.S. patent can jeopardize worldwide patent rights.

Mistakes made while trying to do one's best are tolerated in science; mistakes made through negligent work are not. Haste, carelessness, inattention—any of a number of faults can lead to work that does not meet the standards demanded in science. In violating the methodological standards required by a discipline, a scientist damages not only his or her own work but the work of others as well. Furthermore, because the source of the error may be hard to identify, sloppiness can cost years of effort, both for the scientist who makes the error and for others who try to build on that work.

Some scientists may feel that the pressures on them are an inducement to speed rather than care. They may believe, for instance, that they have to cut corners to compile a long list of publications. But sacrificing quality to such pressures is likely to have a detrimental effect on a person's career. The number of publications to one's name, though a factor in hiring or promotion decisions, is not nearly as important as the quality of one's overall work. To minimize pressure to publish substandard work, an increasing number of institutions are adopting policies that limit the number of papers considered when evaluating an individual.

Fraud in Science

There is a significant difference between preventable error in research, whether caused by honest mistakes or by sloppy work, and outright fraud. In the case of error, scientists do not intend to publish inaccurate results. But when scientists commit fraud, they know what they are doing.

Of all the violations of the ethos of science, fraud is the gravest. As with error, fraud breaks the vital link between human understanding and the empirical world, a link that is science's greatest strength. But fraud goes beyond error to erode the foundation of trust on which science is built. The effects of fraud on other scientists, in terms of time lost, recognition forfeited to others, and feelings of personal betrayal, can be devastating. Moreover, fraud can directly harm those who rely on the findings of science, as when fraudulent results become the basis of a medical treatment. More generally, fraud undermines the confidence and trust of society in science, with indirect but potentially serious effects on scientific inquiry.

Fraud has been defined to encompass a wide spectrum of behaviors. It can range from selecting only those data that support a hypothesis and concealing the rest ("cooking" data) to changing the readings to meet expectations ("trimming" data) to outright fabrication of results. Though it may seem that making up results is somehow more deplorable than cooking or trimming data, all three are intentionally misleading and deceptive.

Fraud and the Role of Intentions

The acid test of scientific fraud is the intention to deceive, but judging the intentions of others is rarely easy. The case of William Summerlin illustrates both situations: an instance of blatant fraud and a previous history in which the origins of serious discrepancies are harder to determine.

In 1973 Summerlin came to the Sloan-Kettering Institute for Cancer Research in New York, where he subsequently became chief of a laboratory working on transplantation immunology. He believed that by placing donor organs in tissue culture for a period of some days or weeks before transplantation, the immune reaction that usually causes the transplant to be rejected could be avoided. The work had become well-known to scientists and to the public.

However, other scientists were having trouble replicating Summerlin's work. Another immunologist at Sloan-Kettering was assigned to repeat some of Summerlin's experiments, but he, too, could not make the experiments work. As doubts were growing, Summerlin began a series of experiments in which he grafted patches of skin from black mice onto white mice. One morning as Summerlin was carrying some of the white mice to the director of the institute to demonstrate his progress, he took a felt-tipped pen from his pocket and darkened some of the black skin grafts on two white mice. After the meeting, a laboratory assistant noticed that the dark color could be washed away with alcohol, and within a few hours the director knew of the incident. Summerlin subsequently admitted his deception to the director and to others.

Summerlin was suspended from his duties and a six-member committee conducted a review of the veracity of his scientific work and his alleged misrepresentations concerning that work. In particular, in addition to reviewing the "mouse inci-

Instances of scientific fraud have received a great deal of public attention in recent years, which may have exaggerated perceptions of its apparent frequency. Over the past few decades, several dozen cases of fraud have come to light in science. These cases represent a tiny fraction of the total output of the large and expanding research community. Of course, instances of scientific fraud may go undetected, or detected cases of fraud may be handled privately within research institutions. But there is a good reason for believing the incidence of fraud in science to be quite low. Because science is a cumulative enterprise, in which investigators test and build on the work of their predecessors, fraudulent observations and hypotheses tend eventually to be uncovered. Science could not be the successful

dent," the committee examined a series of experiments in which Summerlin and several collaborators had transplanted parts of corneas into the eyes of rabbits. The committee found that Summerlin had incorrectly and repeatedly exhibited or reported on certain rabbits as each having had two human corneal transplants, one unsuccessful from a fresh cornea and the other successful from a cultured cornea. In fact, only one cornea had been transplanted to each rabbit, and all were unsuccessful.

When asked to explain this serious discrepancy, Summerlin stated that he believed that the protocol called for each rabbit to receive a fresh cornea in one eye and a cultured cornea in the other eye. Summerlin subsequently admitted that he did not know and was not in a position to know which rabbits had undergone this protocol, and that he only assumed what procedures had been carried out on the rabbits he exhibited. After reviewing the circumstances of what the investigating committee characterized as "this grossly misleading assumption," the report of the investigating committee stated: "The only possible conclusion is that Dr. Summerlin was responsible for initiating and perpetuating a profound and serious misrepresentation about the results of transplanting cultured human corneas to rabbits."

The investigating committee concluded that "some actions of Dr. Summerlin over a considerable period of time were not those of a responsible scientist." There were indications that Summerlin may have been suffering from emotional illness, and the committee's report recommended "that Dr. Summerlin be offered a medical leave of absence, to alleviate his situation, which may have been exacerbated by pressure of the many obligations which he voluntarily undertook." The report also stated that, "for whatever reason," Dr. Summerlin's behavior represented "irresponsible conduct that was incompatible with discharge of his responsibilities in the scientific community."

institution it is if fraud were common. The social mechanisms of science, and in particular the skeptical review and verification of published work, act to minimize the occurrence of fraud.

The Allocation of Credit

Fraud may be the gravest sin in science, but transgressions that involve the allocation of credit and responsibility also distort the internal workings of the profession. In the standard scientific paper, credit is explicitly acknowledged in two places: at the beginning in the list of authors, and at the end in the list of references or citations (sometimes accompanied by acknowledgments). Conflicts over proper attribution can arise in both places.

Citations serve a number of purposes in a scientific paper. They acknowledge the work of other scientists, direct the reader toward additional sources of information, acknowledge conflicts with other results, and provide support for the views expressed in the paper. More broadly, citations place a paper within its scientific context, relating it to the present state of scientific knowledge.

Citations are also important because they leave a paper trail for later workers to follow in case things start going wrong. If errors crop up in a line of scientific research, citations help in tracking down the source of the discrepancies. Thus, in addition to credit, citations assign responsibility. The importance of this function is why authors should do their best to avoid citation errors, a common problem in scientific papers.

Science is both competitive and cooperative. These opposing forces tend to be played out within "invisible colleges," networks of scientists in the same specialty who read and use each other's work. Patterns of citations within these networks are convoluted and subtle. If scientists cite work by other scientists that they have used in building their own contributions, they gain support from their peers but may diminish their claims of originality. On the other hand, scientists who fail to acknowledge the ideas of others tend to find themselves excluded from the fellowship of their peers. Such exclusion can damage a person's science by limiting the informal exchange of ideas with other scientists.

It is impossible to provide a set of rules that would guarantee the proper allocation of credit in citations. But scientists have a number of reasons to be generous in their attribution. Most important, scientists have an ethical and professional obligation to give others the credit they deserve. The golden rule of enlightened self-interest is also a consideration: Scientists who expect to be treated fairly by others must treat others fairly. Finally, giving proper credit is good for science. Science will function most effectively if those who participate in it feel that they are getting the credit they deserve. One reason why science works as well as it does is that it is organized so that natural human motivations, such as the desire to be acknowledged for one's achievements, contribute to the overall goals of the profession.

Credit and Responsibility in Collaborative Research

Successful collaboration with others is one of the most rewarding experiences in the lives of most scientists. It can immensely broaden a person's scientific perspective and advance work far beyond what can be accomplished alone. But collaboration also can generate tensions between individuals and groups. Collaborative situations are far more complex now than they were a generation ago. Many papers appear with large numbers of coauthors, and a number of different laboratories may be involved, sometimes in different countries. Experts in one field may not understand in complete detail the basis of the work going on in another. Collaboration therefore requires a great deal of mutual trust and consideration between the individuals and groups involved.

One potential problem area in collaborative research involves the listing of a paper's authors. In many fields the earlier a name appears in the list of authors the greater the implied contribution, but conventions differ greatly among disciplines and among research groups. Sometimes the scientist with the greatest name recognition is listed first, whereas in other fields the research leader's name is always last. In some disciplines, supervisor's names rarely appear on papers, while in others the professor's name appears on almost every paper that comes out of the lab. Well-established scientists may decide to list their names after those of more junior colleagues, reasoning that the younger scientists thereby receive a greater boost in reputation than they would if the order were reversed. Some research groups and journals avoid these decisions by simply listing authors alphabetically.

Frank and open discussion of the division of credit within research groups, as early in the process leading to a published paper as possible, can avoid later difficulties. Collaborators must also have a thorough understanding of the conventions in a particular field to know if they are being treated fairly.

Occasionally a name is included in a list of authors even though that person had little or nothing to do with the genesis of completion of the paper. Such "honorary authors" dilute the credit due the people who actually did the work and make the proper attribution of credit more difficult. Some scientific journals now state that a person should be listed as the author of a paper only if that person made a direct and substantial contribution to the paper. Of course, such terms as "direct'" and "substantial" are themselves open to interpretation. But such statements of principle help change customary practices, which is the only lasting way to discourage the practice of honorary authorships.

As with citations, author listings establish responsibility as well as credit. When a paper is shown to contain error, whether caused by mistakes or fraud, authors might wish to disavow responsibility, saying that they were not involved in the part of the paper containing the errors or that they had very little to do with the paper in general.

However, an author who is willing to take credit for a paper must also bear responsibility for its contents. Thus, unless responsibility is apportioned explicitly in a footnote or in the body of the paper, the authors whose names appear on a paper must be willing to share responsibility for all of it.

Apportioning Credit Between Junior and Senior Researchers

The division of credit can be particularly sensitive when it involves postdoctoral, graduate, or undergraduate students on the one hand and their faculty sponsors on the other. In this situation, different roles and status compound the difficulties of according recognition.

A number of considerations have to be weighed in determining the proper division of credit between a student or research assistant and a senior scientist, and a range of practices are acceptable. If a senior researcher has defined and put a project into motion and a junior researcher is invited to join in, major credit may go to the senior researcher, even if at the moment of discovery the senior researcher is not present. Just as production in industry entails more than workers standing at machines, science entails more than the single researcher manipulating equipment or solving equations. New ideas must be generated, lines of experimentation established, research funding obtained, administrators dealt with, courses taught, the laboratory kept stocked, informed consent obtained from research subjects, apparatus designed and built, and papers written and defended. Decisions about how credit is to be allotted for these and many other contributions are far from easy and require serious thought and collegial discussion. If in doubt about the distribution of credit, a researcher must talk frankly with others, including the senior scientist.

Similarly, when a student or research assistant is making an intellectual contribution to a research project, that contribution deserves to be recognized. Senior scientists are well aware of the importance of credit in the reward system of science, and junior researchers cannot be expected to provide unacknowledged labor if they are acting as scientific partners. In such cases, junior researchers may be listed as coauthors or even senior authors, depending on the work, traditions within the field, and arrangements within the team.

Plagiarism

Plagiarism is the most blatant form of misappropriation of credit. A broad spectrum of misconduct falls into this category, ranging from obvious theft to uncredited paraphrasing that some might not consider dishonest at all. In a lifetime of reading, theorizing, and experimenting, a person's work will inevitably incorporate and overlap with that of others. However, occasional overlap is one thing; systematic, unacknowledged use of the techniques, data, words or ideas of others is another. Erring on the side of excess generosity in attribution is best.

The intentional use of another's intellectual property without giving credit may seem more blameworthy than the actions of a person who claims to have plagiarized because of inattention or sloppiness. But, as in the case of fraud, the harm to the victim is the same regardless of intention. Furthermore, given the difficulty of judging intentions, the censure imposed by the scientific community is likely to be equally great.

Special care must be taken when dealing with unpublished materials belonging to others, especially with grant applications and papers seen or heard prior to publication or public disclosure. Such privileged material must not be exploited or disclosed to others who might exploit it. Scientists also must be extremely careful not to delay publication or deny support to work that they find to be competitive with their own in privileged communication. Scrupulous honesty is essential in such matters.

Even though plagiarism does not introduce spurious findings into science, outright pilfering of another's text draws harsh responses. Given the communal nature of science, the plagiarist is often discovered. If plagiarism is established, the effect can be extremely serious: All of one's work will appear contaminated. Moreover, plagiarism is illegal, and the injured party can sue.

Upholding the Integrity of Science

Perhaps the most disturbing situation that a researcher can encounter is to witness some act of scientific misconduct by a colleague. In such a case, researchers have a professional and ethical obligation to do something about it. On pragmatic grounds, the transgression may seem too distant from one's own work to take action. But assaults on the integrity of science damage all scientists, both through the effects of those assaults on the public's impression of science and through the internal erosion of scientific norms.

To be sure, "whistle-blowing" is rarely an easy route. Fulfilling the responsibilities to oneself discussed earlier . . . will not harm a person's career. That has not necessarily been the case with whistle-blowing. Responses by the accused person and by skeptical colleagues that cast the accuser's integrity into doubt have been all too common, though institutions have been adopting policies to minimize such reprisals.

Accusing another scientist of wrongdoing is a very serious charge that can be costly, emotionally traumatizing, and professionally damaging even if no transgression occurred. A person making such a charge should therefore be extremely careful that the claim is justified. One of the best ways to judge one's own motives and the accuracy of a charge is to discuss the situation confidentially with a trusted, experienced colleague. Many universities and other institutions have designated particular individuals to be the points of initial contact in such disputes. Institutions have also prepared written materials that offer guidance in situations involving professional ethics. In addition, Sigma Xi, the American Association for the Advancement of Science, and other scientific and engineering organizations are prepared to advise scientists who encounter cases of possible misconduct.

Once sure of the facts, the person suspected of misconduct should be contacted privately and given a chance to explain or rectify the situation. Many problems can be solved in this fashion without involving a larger forum. If these steps do not lead to a satisfactory resolution or if the case involves serious forms of misconduct, more formal proceedings will have to be initiated. For this purpose, most research institutions have developed procedures that take into account fairness for the accused, protection for the accuser, coordination with funding agencies, and requirements for confidentiality and disclosure.

Assaults on the integrity of science come from outside science as well as from within. Vocal minorities that call for a halt to whole areas of scientific research or individuals who use a few events to question the entire ethos of science can undermine the public's confidence in science, with potentially serious consequences. Just as scientists need to protect the workings of science from internal erosion, they have an obligation to meet unjustified or exaggerated attacks from without with sound and persistent arguments.

Daniel Koshland, Jr., is Editor of *Science,* one of the nation's most prestigious scientific publications. *Science* publishes numerous articles on conflicts of interest, claims of fraud or faulty allocation of credit, and other ethical issues. Koshland's editorial emphasizes the difficulty of clearly defining conflicts of interest in a scientific context. As he notes, we all have ways to rationalize our own familiar in-group procedures, while distrusting those of others.

CHAPTER FIFTEEN

Conflict of Interest

DANIEL E. KOSHLAND, JR.

Two principles in modern life with wide support are that judges or regulators of a system should be free of conflict of interest and that those judged or regulated deserve to be evaluated by their peers. Yet these two principles are frequently in serious conflict. When asked if the nuclear industry can regulate itself, a biologist would probably say, "Of course not!" A congressman if asked whether biologists can regulate their own ethical behavior might well answer, "How can you expect NIH to evaluate its own grantees?" And a nuclear physicist if asked whether Congress can regulate itself would undoubtedly burst into laughter, and so forth.

Yet in each of these diverse groups, the recipient of a grant or the accused in an ethical inquiry would stoutly maintain that he or she can only be judged by a jury of peers within that same profession. Modern specialization makes it inevitable that those who evaluate complex subjects must have the relevant expertise within that profession to make fair judgments. However, each group is quite willing to say that in somebody else's profession all of the participants are thick as thieves and only outside observers with no axe to grind can protect the public's interest.

Almost all commissions, judges, peer-review panels, and the like, are chosen from within the discipline that is to be regulated. It is not only that expert judgment is required, it is also that individuals will spend time and energy with some unselfishness for their own profession, whereas it is too much to ask them to do this for some other group. Scientists serve on peer-review panels for the National Science Foundation and the National Institutes of Health and on editorial boards of journals, at conditions of pay and hours of labor that would make a sharecropper weep. Lawyers serve on *pro bono*

---

committees of the bar, newspaper journalists on fellowship committees, and so on. It is quite apparent that the time spent on such *pro bono* activities, though offering some reward in the form of recognition and mutual trade-offs, is on the whole not justifiable on a strict cost-benefit analysis. Prominent and busy people are willing to spend the time within their own disciplines because they know the survival of that system depends on that sacrifice.

The systems work, in part because the outside world is always watching. The inside group is needed to provide detailed rules and sophisticated analysis. But inevitably, the big picture can be explained to outsiders, and outside groups do intervene if the insiders' decisions seem unwarranted. Almost invariably, sloppiness or negligence in designing procedures develops into a major scandal in which the outsiders demand reform.

One of the more amusing aspects of conflict of interest morality is how easy it is to be sanctimonious about the ethical systems of other professions. Scientists are utterly confident that the "tiny" honorarium they got from University X does not disqualify them from considering University X's grants, but believe an equivalent small honorarium disqualifies a businessman in a parallel situation. When serving on scientific panels one scrupulously leaves the room while one's own university is being considered. Would the congressman from Arizona leave the room when Arizona appropriations are being considered? Newspaper reporters decry any effort of concealment as *prima facie* evidence of guilt, and yet, asked to give the source of their leaks, discover that confidentiality is essential to their system.

There has to be a reasonable compromise between expertise and conflict. Some cases of conflict are obvious. A businessman cannot serve on a committee to provide a waste disposal license to his own business. A scientist should not be asked to evaluate a colleague's grant. However, firms involved in toxic waste disposal have to be consulted to devise general laws to control toxic waste. Scientists have to be used to evaluate scientific proposals. The line, therefore, must be a compromise. Fame, fortune, and self-interest will tempt anyone, but the idea that one's own profession has a monopoly on virtue is unlikely.

The procedures devised by insiders should always be subject to the scrutiny of outsiders. There will inevitably be some provincialism. Scientists are proud of science and want it to prosper, and they will benefit indirectly if the system prospers. The same is true of businessmen, newspaper reporters, politicians, and public interest groups. That pride, and its concomitant sense of responsibility, is the basis of the *pro bono* sacrifices that allow any system to work. History shows such a system can be destroyed by excessive suspicion or excessive neglect. A spirit of compassionate skepticism is needed to make it work.

Potential conflicts of interest arise when academic research is funded by private corporations. As one university official quoted in this article notes, "It is extremely difficult for the most honest and upright of scholars to acknowledge their own conflicts for what they are." The difficulty of self-appraisal identified by this official is the best reason for foresters to treat ethical self-examination as a critical professional skill. Our acuity in detecting the mote in someone else's eye is often considerably diminished when we look in the mirror. The author is a science journalist.

CHAPTER SIXTEEN

When Commerce and Academe Collide

ELIOT MARSHALL

Consider a scientist about to embark on a clinical trial of what promises to be a blockbuster drug. He happens to have shares in the company that manufactures the drug. Should he decline to participate? Conduct the tests but declare his potential conflict of interest when he reports the results? Sell the shares? Or assume that his objectivity will overcome any possible bias and carry on regardless?

Or take a university researcher who has developed a new biological technique. He sets up a company with funding from a major corporation to exploit his discovery. But the corporation also funds research in the scientist's university lab. How can he respond to a charge that university resources are being used for private gain? And what about the students? How can they be sure that a professor's advice—on such things as choosing a thesis topic—is inspired by academic and not business interests?

Questions like these confront and often haunt faculty members and deans these days—a product of the boom in university-industry partnerships. And one reason they haunt academe is that they have also caught the attention of Congress. Last year, two congressional committees put the spotlight on academic conflicts of interest, focusing on cases in which commercial agreements went sour or violated ethical standards.

Goaded into action by these hearings, the National Institutes of Health (NIH) drafted a set of rules designed to steer its grantees toward a common approach to the problem. The guidelines would have required faculty to disclose their investments, along with those of their children and spouses, to college administrators. In addition, they would have prohibited faculty from having an interest in any company whose products they were testing. The rules sparked a

This article reprinted by permission from *Science,* Vol. 248 (April 1990), pp. 152–156. Copyright © American Association for the Advancement of Science.

storm of protest, and they were withdrawn last December (*Science,*
Vol. 248, 1990, p. 154).

But the federal government hasn't entirely quit the field. There
are rumblings that Representative Ted Weiss (D—NY) may attempt to
add confict-of-interest standards to the NIH authorization bill this
year. And the Department of Health and Human Services may yet
come back with a revised set of rules. Meanwhile, a few schools have
taken the moral high ground by voluntarily adopting tighter standards. Their responses are diverse, however, making for a confusing
array of requirements that vary from one university to another, even
from one department to another. Some codes are quite explicit, like
those approved by the Harvard Medical School in March [1990] but
others treat the subject almost in hypothetical terms.

The problem is anything but hypothetical, however. The two
situations sketched at the beginning of this article, for example, have
real-life counterparts. In 1988 and 1989, it was revealed that
investigators at several clinical centers reviewing the heart-attack
drug TPA were also stockholders in the company that makes TPA.
Several of them later signed a pledge agreeing that in the future they
would not hold stock in a company whose product they were
studying. And in the early 1980s a controversy broke out among the
faculty of the University of California at Davis when it became known
that Allied Chemical Company was funding research on nitrogen
fixation in plant geneticist Ray Valentine's lab and at Calgene, a company Valentine had founded. Valentine resolved the conflict by
dropping out of the university-based research program sponsored by
Allied. The university also adopted more stringent rules asking
faculty to disclose outside commercial activities and established a
standing committee to review potential research conflicts.

A similar conflict now confronts George Levy, a chemist at
Syracuse University. Levy described his own predicament during a
public meeting earlier this year in an attempt to show how the
proposed NIH rules would punish innovators. A decade ago, Levy
had an idea for improving the software for the nuclear magnetic
resonance machines he uses in his research. He advanced the state of
the art, and in 1983 founded a small company called New Methods
Research, Inc. In 1986, NMRI moved off campus, and by 1988 it had
$2.1 million in sales. The following year, NMRI was sold to new
owners. Levy returned to his university lab to resume academic
research full time. Meanwhile, the university lab got caught in the
funding pinch at the National Institutes of Health and discovered this
year that it may lose its grant.

In desperation, Levy says he may ask NMRI (in which he still has
an interest) to spend discretionary funds on research at Syracuse.
This would generate new ideas for the company and keep his lab at

Syracuse going. The profits would be shared between the school and NMRI.

This rescue may succeed, but it makes Levy uncomfortable because, "Here am I, sitting in the middle," trying to negotiate between NMRI and the university. Levy's position is troublesome because he has a direct personal financial stake in the outcome. "I don't like it," he says.

Syracuse has been kept informed at every stage of NMRI's creation and development, says Levy, and he is very much in favor of disclosing the necessary details of academic-industry deals like this.

The university has no formal conflict-of-interest guidelines to cover such situations, according to spokesperson Sandi Mulchonry. "The departments handle it on a case-by-case basis," she says. But Levy says that neither the rescue of his lab nor the creation of NMRI in the first place would have been permissible under NIH's aborted conflict rules. Nor is it likely that the rescue would be allowed under stringent rules being adopted voluntarily by Harvard and several other universities.

Indeed, even a brief survey of major schools reveals, as Carol Scheman of the Association of American Universities says, that "there are a huge number of ways in which institutions approach these issues." Some institutions are taking a laissez faire approach. Caltech, for example, relies on its strong honor code to keep the faculty out of conflict situations, according to vice provost David Goodstein. "There are no requirements for disclosure as far as I know," he says. The only "really explicit rule" is that faculty may not take operational responsibilities outside the school. "We have not had any problems," Goodstein says. In contrast, the Massachusetts Institute of Technology requires everyone, staff and faculty alike, to file full outside interest reports every year.

The latest to adopt strong measures is Harvard University, and many observers believe its rules could be a model for others. Harvard was stung last year by news coverage of a researcher named Scheffer C. G. Tseng at a Harvard-affiliated eye clinic who had a financial stake in an experimental eye medicine he was testing on patients. Before releasing data showing that the medication was ineffective, Tseng cashed in most of his 530,000 shares in a company established to promote the drug. Two other scientists who advised Tseng, one at Harvard and another at Johns Hopkins, also had a financial stake in the company.

Medical school dean Daniel C. Tosteson appointed a committee in 1989 to review conflict-of-interst policies and suggest changes. Tosteson did this, he says, not because of any scandal but because Harvard has encouraged its faculty to spread new ideas to the world through commercial agreements, and many gray areas that were vaguely discussed in the old (1983) rules have now become important.

Barbara J. McNeil, chair of the department of health care policy, headed the rule-drafting group. Their recommended changes were unveiled before a full faculty meeting in February, where they met a noisy and hostile reception. Opponents, who had bused in scores of doctors from the Massachusetts General Hospital, dominated the podium. Many felt, as one professor said, that Harvard was "using a cannon to kill a fly." But a month later, McNeil and the reformers won a quiet victory in the select faculty council, which backed the dean with a lop-sided vote of more than 30 to 1. Says council member Leon Eisenberg, "We sensed the world was watching."

The new rules require all faculty members to make a full disclosure of their potential conflicts of interest to university administrators at least once a year, and they require researchers to get explicit approval before embarking on studies funded by companies in which they or their families have a financial interest. They also put strictures on faculty involvement in the operations of profit-making companies. The rules [went] into effect in May [1990], and faculty members [were] allowed six months to adjust, either by divesting financial holdings or bringing their research into compliance.

Tosteson notes that Harvard's approach is "more explicit" than most. Other institutions that are revising their own conflict-of-interest standards have been looking carefully at Harvard's new rules, but many schools will find some of the specific provisions too Draconian. That, at least, is the view of the Association of American Universities' Carol Scheman, who says that Harvard, with its network of 14 affiliated clinics and hospitals, has a "unique and difficult problem" in trying to keep tabs on its diverse faculty.

Johns Hopkins University went part way toward a stringent code of ethics in November. According to associate dean David Blake, "We didn't get that many disclosures under the old system," so it was revised. "Our assumption is that the problem is mostly one of perceptions, so disclosure is the key." Faculty must report all written agreements involving privately sponsored research and disclose consulting deals that demand more than 26 days a year. Blake himself does not think that clinical research needs special attention because it is already heavily regulated by the federal government. But the medical school does have one "absolute prohibition": you may not own even one share of a corporation that is sponsoring your research at the university.

The Stanford University School of Medicine, according to its dean, David Korn, has begun doing some "spot auditing" of the disclosure forms it requires faculty members to submit each year. In addition, Korn says, the review protocol for human subjects has been

Harvard's Tough New Rules

Over the objections of some faculty members enraged by invasion of their privacy, Harvard Medical School adopted new conflict-of-interest guidelines last month. They are among the toughest yet adopted by a U.S. university. Among the provisions:

• The rules define two problem areas: conflict of commitment and conflict of interest. Commitment issues are simpler, requiring faculty to give "their primary professional loyalty" to the university and to devote no more than 20% (one working day per week) to outside activities.

• Faculty members must now disclose all potential conflicts at least once a year on a new form, which must be updated whenever a new conflict arises. The forms will be collected by the appropriate hospital or dean's office and be kept "strictly confidential." Questionable cases are to be reviewed and settled by a standing committee of the faculty.

• Unless they receive explicit approval, faculty members and their families may not hold stock in or receive consulting fees from a company whose technology they are investigating in clinical trials.

• Without prior approval, faculty members may not do sponsored research at a university facility for a business in which they or their families have an interest.

• Without approval, faculty members may not sit on a review committee (such as a Food and Drug Administration panel) judging a technology in which they or their families have an interest.

• Without approval, faculty members may not serve as a managing executive for a profit-making biomedical company, nor may they have a financial interest in a business that competes with services provided by the university or its hospitals.

• Without approval, faculty may not make clinical referrals to a business in which they or their families have an interest.

• Faculty members must disclose to the public (not just to an administrator) their financial interest in a subject which they discuss in a research publication, a formal presentation, or an expert commentary, and they must do so "simultaneously" as they speak or publish.

rewritten to include an extensive series of questions about the financial involvement of the investigators and their students. This information must be cleared by the human subjects review comittee, and the patients must see it, too.

Because of the diversity among individual schools, Korn argues that it makes no sense to issue blanket prohibitions for the entire country, as the NIH guidelines attempted to do. Korn himself advocates using the system of Institutional Review Boards that watch over research on human subjects to do a similar job for conflicts of interest. In this approach, each institution would have to assure the government that it had put an effective system into place, subject to spot auditing by some federal supervisor like the NIH. He thinks this would allow for the greatest local autonomy while maintaining high standards. Stanford works under "the philosophy that people are generally decent and behave well," says Korn. "You don't have to tie them up in a bunch of minutiae." Although employees at state universities must work under very strict prohibitions, "rules like those would be anathema on this campus." In general, Korn thinks national policy should avoid detail and give broad, philosophic direction.

This is precisely the aim of two major reports issued this spring by university leaders—one drafted by a panel Korn chaired for the Association of Academic Health Centers (AAHC), and the other by a group at the Association of American Medical Colleges (AAMC), headed by Michael Jackson of the George Washington University School of Medicine.

Korn's AAHC report, issued on 22 February [1990], traces the boom in academic-industrial collaborations since the 1970s, now encouraged by federal law, and it notes that there are several areas of growing concern. For example, it says the possibilities for conflicts are "legion" in spin-off companies started by faculty members, because the founders live in both the academic and profit-making worlds and control resources and young people's careers in both.

An academic's chief loyalty, both reports say, must always be to the university, but they remain a bit vague in the measures they would use to reinforce that loyalty.

One point on which all experts seem to agree is the need to disclose potential conflicts in advance. A set of guidelines issued recently by the AAMC says universities ought to develop procedures for full disclosure of financial and professional interests not only for use by the school but to inform "the interested public." In addition, the AAMC paper says, institutions should review researchers' personal holdings, including those of the immediate family, at least once a year. Questionable cases should be passed up the chain of command to the university president or, better, to a standing committee. Those at odds with the rules "must be handled expeditiously and conclusively," the AAMC panel believes, and "all decisions must be documented."

The AAHC goes further, saying that "significant" financial or other relationships, if they raise a potential conflict of interest, should be "fully and accurately disclosed in all speeches, writing, advertising, public communications, or collegial discussions" of sponsored research.

These guidelines are new, but others like them have been in existence at major universities for some time—and "honored in the breach," according to C. Kristina Gunsalus, associate vice chancellor for research at the University of Illinois at Champaign. The way to make principles work, Gunsalus says, is to develop a reporting system that will win faculty cooperation and actually do the tedious job of screening and reading the disclosure forms. You must look for trouble, as she does, because "it is extremely difficult for the most honest and upright of scholars to acknowledge their own conflicts for what they are."

Representative Weiss says that while he "applauds" the AAMC and AAHC for developing conflict-of-interest guidelines, they do not go far enough. He favors "strong minimum standards for all research institutions." Unless everyone plays by the same rules, Weiss says, "universities that make serious efforts to minimize conflicts of interest could be at a disadvantage in recruiting scientists who enjoy lucrative financial relationships with the private sector."

The consensus among those who actively manage faculty conflicts is that one must begin with written forms. They are "the only thing that everybody agrees is absolutely essential," says John Lombardi, the former provost at Johns Hopkins, now president of the University of Florida at Gainesville. "If you actually disclose and write down the relationships you have, the conflict of interest is much easier to discern."

Lombardi finds that 95% of the cases turn out to be fine. But "5% are very difficult because they skirt the borders of a conflict of interest. Then you have to do what rule-makers don't like to do: you have to exercise judgment."

Both Gunsalus and Lombardi say that when the system works well, it encourages the faculty to venture out into the commercial world, because the responsibility for error—if something goes wrong—falls squarely on the official who gave approval and not on the individual researcher.

III. BUSINESS ETHICS

It could be argued there is no such thing as business ethics because there is nothing special about applying general ethical standards to business situations. This may be true in the abstract, but in fact there is good reason to focus on business ethics as a special area within the broad field of ethics because specific business situations can raise ethical questions that go well beyond the mandates of professional codes and general ethical maxims.

In both private and public employment, foresters frequently handle large transactions in timber or land. Their technical work in estimating timber volumes or appraising land values can affect business decisions by others, including public agencies. Their recommendations for cutting and regeneration practices can affect forest owners' revenues, costs, and harvest timing. These activities can create many opportunities for ethical lapses if thoughtful vigilance is not exercised. Not uncommonly, junior people with sound technical training but limited business experience are entrusted with this work. Lacking adequate preparation, they are in greater danger of making ethical mistakes or of detecting troublesome situations much later than they should.

The field of business ethics is not a new one, but it has received increased attention in business schools, business publications, and the media in general. There is now a *Journal of Business Ethics*, and there are many organizations and consultants assisting corporations with their ethical policies. Most major corporations now have formal ethical codes and programs. If you are an employee or shareholder of a large corporation, you might want to obtain a copy of its formal ethical code and consider it in light of what you learn from these readings.

Public servants are no less in need of business ethics training than independent professionals or employees of private firms. Some individuals are so full of their sense of acting for a larger public good that they are willing to overlook what they perceive to be minor "white lies," deceptions, or other actions that they would criticize if taken by someone else. Indeed, public sector foresters, because they may be less familiar with the details of business practice and custom, may need to pay special attention to these issues. Public forest land managers are involved in large timber sales and need to be able to conduct themselves with a sense of clarity and confidence on ethical matters because they are handling other people's land and money.

The readings in this section come from a wide variety of writers, most of them from outside the forestry profession. The experiences and ideas discussed in these selections significantly enrich our exploration of ethics by widening our field of view.

The consequences of individual business decisions are wide-ranging. For example:

1. As an individual, dishonesty or its appearance injures your reputation.

2. Injury to your reputation will have later repercussions in your business.

3. Willingness to engage in or tolerate questionable practices will ultimately injure your own personal growth and self-esteem.

4. If you are willing to contribute to an atmosphere of lowered standards, you can expect to be victimized by others.

5. Unethical acts will injure firms or individuals with whom you do business, whether they discover this or not.

6. Your employer's reputation, and possibly his or her business, will suffer sooner or later if unethical practices are tolerated. Consequences could include litigation, civil or criminal penalties, and disbarment from public contracts.

7. In the long run, tolerance of shady practices injures the foundations of our entire business system and, indeed, of our democratic government because corruption spreads quickly and sooner or later infects the political system.

After many years of absence, a sawmill owner began attending church regularly. One day, as they shook hands on the way outside, the minister said, "Glad to see you attending church again, Mr. Jones. I understand your brother is working with you at the mill again. Will we be seeing him at church again soon?" Jones replied, "Hope not, Reverend ... someone's got to buy the logs!" This story always gets a laugh from a forestry audience, but it also gives cause for reflection. Surely the reputation of log buyers can be redeemed!

Additional Reading

Andrews, Kenneth R. *Ethics in Practice: Managing the Moral Corporation.* Cambridge: Harvard Business School, 1989.

Barinaga, Marcia. Confusion on the cutting edge. *Science* 257(1992):616–691.

For a brief, easy, but useful and stimulating introduction, see Blanchard, Kenneth; and Peale, Norman Vincent. *The Power of Ethical Management.* New York: Fawcett Crest, 1988.

Bok, Sissela. *Secrets: On the Ethics of Concealment and Revelation.* New York: Vintage Books, 1984. Contains an excellent chapter on whistle blowing.

Donaldson, Thomas; and Werham, Patricia. *Ethical Issues in Business: A Philosophical Approach.* Englewood Cliffs, NJ: Prentice-Hall Publishers, 1979.

Harvard Business School. *The Business of Ethics and the Ethics of Business.* Boston: Harvard Business School Publishing Division, no date. This volume is a compendium of *Harvard Business Review* articles. They are well-written and practically focused, though oriented more toward the higher level executive issues and less to down-to-earth matters.

Nader, Ralph; Petkas, Peter J.; and Blackwell, Kate. *Whistle Blowing: The Report of the Conference on Professional Responsibility.* New York: Grossman, 1972. Includes excellent case studies.

Stark, A. What's the matter with business ethics? *Harvard Business Review* May–June 1993. Argues that business ethics is too often taught in terms of abstractions.

Taylor, John F. A. The ethical foundations of the market. In *The Masks of Society: An Inquiry into the Covenants of Civilization.* New York: Appleton-Century-Crofts, 1966.

U.S. General Accounting Office. *Whistleblower Protection: Agencies' Implementation of the Statutes Has Been Mixed.* Washington, D.C.: GAO/GGD-93-66, 1993.

Journalist Peter Marks offers an excellent overview of the contending views of teaching ethics in business schools. On the one hand, some students and teachers are skeptical that ethics can actually be taught or that the students most needing ethical instruction will bother to learn. On the other, committed ethicists and instructors are working hard to equip students with better tools for ethical reasoning at the beginning of their business careers.

The author is a writer with *Newsday,* a Long Island newspaper.

CHAPTER SEVENTEEN

Ethics and the Bottom Line

PETER MARKS

There they were, those wild and crazy MBA students from Wharton, cavorting onstage with fake arrows through their ears, making fun of the nerdy accounting majors and the hyperactive finance jocks and sending up virtually every sacred ritual of business school.

In a skit called "I Spit on Your Résumé," the students let loose on the tortures of the job interview. In "Attack of the Fundraisers," they skewered corporate philanthropy. In a takeoff on "Let's Make a Deal," a student playing a housewife from Butte, Mont. was plucked from the audience to mount a leveraged buyout of a giant cosmetics conglomerate.

It was the road show of the "Wharton Follies," an annual highlight of student life at the University of Pennsylvania's Wharton School and a chance for the MBAs-in-training to thumb their noses at their pressure-cooker world. On an arctic night in February, hundreds of Wharton alumni paid $25 each to sit in an auditorium in midtown Manhattan and watch as about 40 students from the Philadelphia school put on a three-hour show that lampooned everything from insider trading to b-school romances to The Wall Street Journal.

You didn't know about the lighter side of insider trading? In a parody called "Jailhouse Stock," an Elvis impersonator gyrated in front of a prison set and sang about his role in a scam to sell oil-company stock short.

"Appellate judge sent me to a federal jail/Traded inside, and, baby, I got nailed," Elvis lamented.

If virtually every fact of the world of business was fair game—though the students said they had been asked by school officials not to single out for public ribbing such high-profile Wharton alums as

This article reprinted by permission from *Newsday*, May 1990, pp. 26–32.

junk-bond impresario Michael Milken—they saved some of their most scathing satire for an issue that has become more and more topical on the campuses of America's business schools as the Wall Street insider-trading scandals unfold: Can American businesspeople be trusted?

In a skit near the conclusion of the show, a young woman playing the role of class valedictorian at graduation stepped away from her podium, grabbed a microphone and, to the music of Michael Jackson's "Beat It," explained how she made it to the head of the class:

> I cheated, cheated,
> Cheated, cheated,
> And maybe I sound conceited,
> You're probably angry,
> I'm overjoyed,
> I work on Wall Street,
> You're unemployed.

The Wharton alumni, many of whom had come to the "Follies" in taxis and limos from offices in the New York brokerage houses, consulting firms and Fortune 500 companies that recruit heavily at prestigious business schools such as Wharton, did not have to have the joke explained. The image of the Wall Street high-flyer has taken a beating in recent years, with newspapers and network news shows carrying story after story about the shady financial dealings—and the resulting court cases—involving some of the best-known figures on the Street.

The skit revealed how deeply the travails of Milken and Ivan Boesky and Dennis Levine—millionaire financial wizards who have been indicted or convicted for Wall Street dealings—have penetrated the psyches of the coming generation of business leaders. But it also revealed the insecurity some of them feel about how easily corruption can occur. "I want a good job/Who cares if it's right?" the valedictorian sang. "To hell with ethics/I sleep at night."

These days, the nation's premier business schools are under increasing pressure to ensure that their students do understand the difference between right and wrong—or at the very least, can recognize ethical dilemmas in a complex business world and be able to figure out for themselves how to respond. It is not that anyone seriously thinks of Harvard or Wharton or Columbia or Northwestern as a school for scandal, but there is a growing sense among some academicians, ethicists and business leaders that ethics should be as vital a component of a Master's in business administration as courses in statistics or management.

"Ethics should be a part of every business-making decision and procedure," says W. Michael Hoffman, a philosophy professor and head of the Center for Business Ethics at Bentley College in Waltham, Mass. "It should be integrated throughout all courses in the business curriculum."

Or, as Kirby Warren, the acting dean of Columbia University's business school, puts it: "There's a little bit of thief in all of us—and a lot of good in most of us. I would like to inject ethical issues into as many courses as I can."

It may never draw as many students as international finance or stir up as much fear and loathing as managerial accounting, but one thing seems certain: ethics is in.

At Harvard Business School, administrators are trying to figure out how to spend a $30 million gift from alumnus John S. R. Shad, former chairman of the Securities and Exchange Commission and then head of the recently bankrupt investment firm Drexel Burnham Lambert, who wants the money used for a broad-based curriculum on ethics. In a five-year, $5 million program, Arthur Andersen & Co., the mammoth accounting firm, is training business professors from across the country at a center in St. Charles, Ill., who will be including segments on ethics in their courses. And one of the hottest issues at Wharton last year had to do with ethics: factions within the student body fought over a revamping of the Wharton honor code.

Ethics has even hit the lecture circuit. Students in the Columbia Business School ethics club, MORES, have initiated a lecture series by white-collar felons. The first speaker was Dennis Levine, the former Wall Street investment banker who was jailed after making millons trading on illegally obtained information about impending corporate takeovers. Levine's Columbia appearance was a small dose of real life in the rather esoteric discipline of ethics, a subject that has not had an easy time finding its niche in the pragmatic world of business. It is, in fact, a b-school rarity: There is no number-crunching, no economic theory, no need even for the trusty old HP 12C, the calculator of choice among MBAs-to-be. Dostoyevsky appears on the syllabuses of ethics seminars almost as often as Milton Friedman. One Columbia business professor says "Crime and Punishment" is required reading for some of his management classes because the novel's antihero, Raskolnikov, is "someone who had to rationalize" terrible deeds— though his problems were a lot more serious than expense-account padding.

The concern about ethics, of course, is not limited to the world of business—or business schools, for that matter. People in many professions, from doctors and scientists to politicians and journalists, confront ethical problems and sometimes have been known to have lapses in judgment. In fact, business school officials say it was a political scandal, Watergate, that in the early 1970s provided the initial impetus for the creation of a number of ethics courses at some leading business schools.

But even with the heightened interest in ethics, most MBAs today do not take full-blown ethics classes. If the vast majority of students do not consider it wholly irrelevant—the derisive adjective of choice for the subject is "touchy-feely"—they say it may be better suited to a segment on the "MacNeil-Lehrer Newshour" than a seminar squeezed into their schedules between organization theory and labor-market analysis.

"A lot of people here say, 'I'm here to get a job,' " says Greg Fukutomi, who got his MBA at Columbia last year and is now a candidate for a doctorate in the management of organizations. "They say, 'I'm a good person; I'm not going to screw up.' I don't think any of the guys who did insider trading thought they'd screw up. They probably thought, 'This is all part of making a deal.' "

Even business-school leaders are unsure how much ethics is appropriate. Donald Jacobs, dean of the Kellogg Graduate School of Management at Northwestern University—Kellogg was ranked the country's best overall business school in 1988 by a highly regarded survey in Business Week magazine—at first spurned an offer from a wealthy alumnus recently to donate a large sum for an endowed chair in ethics.

"I've always believed that just teaching ethics as ethics is not our role because we have no expertise," Jacobs said in an interview in the school's sleek, $15 million conference center overlooking Lake Michigan. "What we do know are the functional areas—we have real experts in finance, we have real experts in marketing."

At the urging of students and some faculty, however, Jacobs subsequently modified his position. The dean now says he decided to accept a $1 million gift from private donors to endow a chair in ethics because he came to the conclusion that Kellogg could use an ethics "champion" who will encourage the teaching of ethics throughout the school.

"We're really very excited about the whole business," Jacobs says, adding that the search to fill the chair is continuing.

Like most major two-year MBA programs, Northwestern's Kellogg school offers a handful of ethics courses as electives and encourages its faculty to address ethical issues in lectures and seminars. The accrediting agency for business schools, the American

Assembly of Collegiate Schools of Business, mandates that schools include in their core curriculum some teaching of ethics, but how schools interpret the requirement varies greatly.

And still, some students at places such as Columbia, Wharton and Northwestern, vying for post-graduation jobs in banking and marketing and consulting that command starting salaries of up to $80,000 a year, express a hunger for some moral framework, some code of behavior for themselves and their peers—a yearning for what they might call a deeper sense of the "macro" picture. Many of the MBAs have already been out in the working world and returned to business school to boost or change their careers—and have been a bit overwhelmed by the repeated assaults on their sense of right and wrong in the working world.

"It's a constant temptation," says a Wharton student who has worked as an options trader on the Chicago Board of Trade. The futures exchanges are now the target of a federal probe into allegations of profit-skimming by traders. The Wharton student, who asked that his name not be used, says that his boss has been "a role model for ethics," but others he knew were a lot less savory. "You wondered how they woke up with themselves every morning," he says.

Dan Braude, a second-year student at Columbia Business School, says some of his friends snickered when he revealed he was taking an elective called Managing the Socially Responsible Corporation. "They disdain it either because they think they'll never use it or because they think they'll never need it," he says. "They laughed and said, 'Ooh, a touchy-feely course.'"

At Wharton, where the captains of industry are often invited to speak to students about their companies and how they made it, the most popular speakers of late have not been officials from Touche Ross or McKinsey & Co. or Bankers Trust—which among them hired 62 of the 292 members of the Class of 1988—but Ben Cohen and Jerry Greenfield, the iconoclastic entrepreneurs who founded Ben & Jerry's ice cream.

A number of students say that the ice-cream makers showed them a side of business they really didn't know much about: Ben and Jerry told them about their guidelines specifying that the highest-paid company employees can earn no more than five times as much as the lowest-paid. They also eschewed business suits—no small act of defiance to a schoolful of would-be executives and one that communicated to the students that a businessperson need not sacrifice individuality to be successful.

"You could really feel good about what they did," says Randy Burkert, a finance major from Philadelphia. "You find out you don't have to subject yourself to being a corporate clone," adds Jeffrey Sakaguchi, a combined finance and management major from Seattle.

Many students say that business school is good at black-and-white issues and less good in the gray areas. And ethics often falls into that gray area. The term can apply to a variety of courses and topics that speak to issues of right and wrong, of values and morality. In business school, the topic often gets a practical spin: Is it wrong to post-date an invoice if everyone else is doing it? Some professors say ethics entails matters of social responsibility as well: Do companies have an obligation to the community that goes beyond the bottom line?

"We are trying to teach some reasoning, which is something that students can subscribe to more than to 'You can do this, and you can't do that,'" says Donald Haider, associate professor of public management at Kellogg.

On a business-school campus, ethical dilemmas don't always crop up in predictable ways. To give a class a better feel for labor negotiations, Columbia Business School professor James Kuhn a couple of years ago divided students into teams for a simulation game, giving time to each to plan strategy before bargaining began. One group took to the game with such a vengeance that it got a tape recorder and hid the microphone in the room where another group was holding a strategy meeting.

Kuhn was a bit stunned by the ethical—and legal—lapse. " 'Do you realize what you've done?' I said to them. 'You've bugged the room!' They were a bit embarrassed."

But some in the business world wonder how much the notion of right and wrong can be taught at the graduate level. "If an MBA candidate doesn't know the difference between honesty and crime, between lying and telling the truth, then business school, in all probability, will not produce a born-again convert," IBM Chairman John Akers wrote recently in a journal published by the Massachusetts Institute of Technology.

"I don't think someone can teach you the rules of ethics, about where you draw the line," says Anne Clarke, who has just completed her first year as a marketing major at Northwestern. "The best you can do is learn from other people's examples."

"We're not here to make these people good," argues Wharton's Joanne Ciulla, senior fellow in the legal studies and management departments who teaches courses on ethics. "We're here to show them how to manage the environment so that people can *be* good."

It is noon on a Monday early in Wharton's spring semester and the corridors of Steinberg/Dietrich Hall—which derives half of its name from millionaire financier Saul Steinberg, a Wharton alumnus who recently donated $15 million to his alma mater—are teeming with accounting and finance majors on their way to class.

Displayed prominently on a brick wall on the main floor are photos of a dozen titans of commerce and jurisprudence. They are Wharton alumni for whom students have voted a spot on their "Wall of Fame" for "outstanding achievement and contribution to society." Among the group: CBS founder William Paley, Class of 1922; Supreme Court Justice William Brennan, '28; CBS President Laurence Tisch, '43; Steinberg, '59; billionaire Donald Trump, '68, and Michael Milken, '79 (whose inclusion has led some students to rechristen it the "Wall of Shame.")

In a classroom not far from Trump's serene visage, Ciulla's Business Ethics course is under way. About 50 students, one in a suit, the rest in jeans, fill the seats for a wide-ranging discussion of the day's ethical thicket, a case involving the H. J. Heinz Co. of ketchup fame. During the mid-1970s, company officials learned that some mid-level managers had doctored invoices in 1974 so that income would look higher in 1975—making it easier for them to meet their 1975 goals and earn bonuses from the company's incentive plan. The issues at hand: Why did it happen? What did it mean? And what should the company have done about it?

Like many professors at Wharton and other top MBA programs, Ciulla uses the case-study method, a teaching technique that was pioneered by the Harvard Business School, to engage the class on a thorny subject. The case, which was published by Harvard, had been assigned reading. Within minutes, the students are debating everything from the value of the company's code of ethics to the recent statements about ethics by the Bush administration to the ethical problems facing Drexel Burnham Lambert.

"OK, let's see what they did," Ciulla says, trying to bring the discussion back to the Heinz case. "They predated some invoices, and they postdated some invoices. But midlevel managers tell me, 'We do this all the time.' Does this do any harm to the company?"

The students aren't sure. It seems an abstract dilemma for many of them and some are not quite certain where ethics fits in.

"The incentive plan was screwed up. But that's because the ethics plan was so weak," says a young man in a gray sweater.

"It seems like everyone knew what was going on," adds a young woman. "It's kind of like it had been approved, it was OK. Upper management has to take responsibility too."

Ciulla steers the class through the case, pushing students to relate the managers' actions to their own notion of fairness—and how the company might remedy the problems. "The question of who gets caught—this does raise really serious questions of justice," she says, asking students to look at the issue from the vantage point of the company's top brass and decide what to do to the offending worker. "You have to do something dramatic; you have to take a stand."

"Torture him," a student calls out. It is in classes like these that the dissecting of corporate ethics takes place at the top schools. The cases they study range from how the Ford Motor Co. responded after a number of rear-end explosions involving the Pinto model (required reading: "Can a Corporation Have a Conscience?") to crises of fiscal conscience faced by accountants ("Is Business Bluffing Ethical?"). There is, in fact, a vast array of cases for virtually every ethical issue from the advertising practices of the tobacco industry to Johnson & Johnson's handling of the poisoned-Tylenol-capsules scare.

The cases can be fictitious or based on real situations and often present complex dilemmas that are laid out in pages and pages of material.

In the case of Nestlé Alimentana S.A., a subsidiary of the Nestlé Co., students are asked to assess whether the company was irresponsible in marketing baby formula in the Third World and not adequately explaining the nutritional superiority of breast-feeding. Issues in the case range from what should be important to the company besides financial return to its stockholders to whether the company in some way was damaged by the incident.

In the case of Campbell Soup, classes explore a situation in which a company becomes the target of a boycott by farm labor organizers, who have criticized what they see as the company's harsh efforts in replacing migrant workers in tomato harvesting with mechanical harvesters. The case deals with the question of the social responsibility of a major corporation when dealing with a disadvantaged segment of the community.

And in the case of "the Constant Charger," a fictitious MBA student is flown from Chicago to New York for interviews with two firms and submits the identical hotel bill to each of the companies for reimbursement. The question: Knowing this, would you hire the student? Northwestern's Haider says many students answer that they would: "My experience with that case over 10 years is that half the students see nothing wrong with double compensation."

Those who teach the subject are often on the lookout for new ways to present the issues. Carol Einiger, an investment banker at the First Boston Corp. who taught a well-received Columbia seminar on the ethics of investment banking, arranged for her class to screen the film "Wall Street." Boris Yavitz, the director of Columbia's Management Institute and a professor of public policy, invites to his class executives from some of the companies under study in the case histories. To give her class an idea of how people in other professions view ethics, Ciulla had her students meet with a Philadelphia lawyer who represents a number of reputed mobsters.

During the question-and-answer session with attorney Nicola Nastasi, held one evening in Penn's Annenberg Center, across the way from Steinberg/Dietrich Hall, the students took the offensive, pressing the lawyer to concede that defending people like Nicodemo "Littly Nicky" Scarfo, reputed head of the Mob in Philadelphia, would make him question the ethics of his role.

"Would you consider it a hollow victory if you got a guy off who you knew was guilty?" one student wanted to know. "I feel quite good when I'm successful," Nastasi replied. "A just result is what under the law is right."

"Seems like you have to put your ethics on hold," another voice said.

" 'Ethical' to me means right and wrong. 'Moral' to me means good and bad," Nastasi said. "There is no moral code dealing with lawyers."

For some students, exposure to ethics sometimes reinforces a belief that they can make a difference. "A course like that raises your expectations," Sakaguchi says. "When we come here," Doreen Lee, a marketing major, adds with the certainty of a believer, "we're committing ourselves to a broader good."

In an interview in her small, spare office, Ciulla, a philosophy professor by training, relaxes in her desk chair as she describes the challenge of making ethics relevant to someone who may never have studied Mill or Kant. "Part of the trick," she says, "is how much philosophy you can teach without losing the business student. . . . Some people think ethics is what you hear when you go to church. But it requires a kind of precision of thinking."

"Ethics is more like a practical skill," says David Schmidt, director of the Trinity Center for Ethics and Corporate Policy, an offshoot of Manhattan's Trinity Church that serves as a consultant to universities and corporations. "And the way you get better at a practical skill is to use it."

In the 2nd floor of Uris Hall, the building that serves as the nerve center for Columbia's MBA students and faculty, a dozen young men and women in suits of slate gray and navy blue are seated outside the doors to a dozen tiny, airless rooms. Each is holding a folder or briefcase in his or her lap. They sit in silence, waiting to be summoned to be grilled about their goals and skills. A painter trying to capture the tableau might call it "Ambition in Repose."

These are the interview rooms at Columbia Business School, an area where, quite literally, careers are made. On this day in March, recruiters from a dozen firms will hold a series of half-hour interviews, seeing as many as 15 students in the course of the day. Some are looking for summer hires among the first-year students, others for permanent employees. They're among hundreds of companies that look over the graduating crop at Columbia each year.

The Industrial Bank of Japan is shopping for an assistant loan officer. CBS wants a senior financial analyst. Eli Lilly International, a pharmaceutical manufacturer, is looking for a candidate with a background in marketing and finance, and Chemical Bank is seeking someone for its consumer banking division.

It is all part of the culture of careerism, an environment that some have criticized as having more in common with trade schools than medical and law schools, a place where the job hunt is a blood sport. On a wall near the Shearson-Lehman-Hutton student lounge, a poster announces the hours tailors from Hickey-Freeman will be on campus to measure students for half-price business suits. For a modest fee, Columbia even videotapes students to help them develop poise in interviews.

In the bustling Uris Deli, a self-service cafeteria on the main floor of the business-school building, the job-seekers can be heard comparing notes as they wolf down blueberry muffins and coffee between morning classes.

"How did the interview go yesterday?" a man in a camel's-hair coat asks a woman carrying a leather briefcase.

"It wasn't bad at all," she replies. "Actually, it was a little annoying. She was 25 years old—a real human-resources type. And she didn't know anything." The intense scramble for jobs among business-school students has fueled the image of the MBA as a yupped-up climber obsessed with money and petrified at the prospect of getting a "ding"—business schoolese for a job rejection. Students say that the competition can get so cutthroat that some of their classmates go to interviews for jobs they have no intention of ever taking, in an effort to hoard job offers. ("Networking slobs/Driving turbo Saabs," is how the "Wharton Follies" described these MBAs.)

"It's a time when people get very nervous," Dan Braude says. "It's difficult to concentrate both on schoolwork and looking for a job."

The emphasis on money has always posed ethical problems for business schools. Two years ago the conflict burst into public view when corporate raider Asher Edelman offered $100,000 to students in his class at Columbia who could point Edelman to a company that he would take over. The school soon quashed the idea, but for a long time after, the episode was the talk of the school.

"That Asher Edelman incident really started the discussion of whether we really deserve the large salaries we get," Braude says. "We have only four terms here," says Chris White, a 38-year-old finance major who graduated last month. "If you have a course that is going to pay you $100,000 and four others that don't, you're going to write off those four other courses. That has to distort the purpose."

White chaired the school's ethics club one semester last year, during which he invited Dennis Levine to speak to about 90 Columbia students. During his 20-minute talk, according to several who attended, Levine—whose testimony was a crucial factor in the conviction of Ivan Boesky on insider-trading charges—spoke of his legal problems and his regrets and the devastating effect the case has had on his life. "I think Dennis was very effective in bringing out the long-term consequences on your family and kids," White says. "I had kids come up to me who said it made them think what it might be like telling this to a child who was not born when they committed a crime."

On the whole, however, some professors wonder whether the students who choose ethics courses are the ones who would most benefit from them. "The ones who have ethics or are ethical in their behavior may not be the ones who need it," says Northwestern's Donald Haider. "So you're kind of preaching to the converted."

And even for those who take the classes, there is a feeling in the b-schools that in the end, a good upbringing is going to count a lot more in the business world than an A in ethics. "When you cut right through it," Columbia's Kirby Warren says, "so many of these ethical dilemmas are not ethical dilemmas if you grow up in a halfway decent family. You know what's right and wrong."

Still, the business indicators all point to an upswing in the teaching of ethics. At their training center in Illinois, Arthur Andersen officials—who hire 5,000 graduates a year from the nation's colleges—have provided their ethics training to 400 professors from 90 universities and hope to eventually reach 2,000 instructors.

"I want to influence hundreds of thousands of kids," says Donald Baker, the Arthur Andersen executive who developed the program.

But it remains to be seen how much ethics students can take. As the valedictorian in the "Wharton Follies" described the new interest in ethics, "They say that people follow it—don't make me choke/Ethics are for those academic folk."

Mr. Boulding served for many years on the economics faculty of the University of Michigan. He is an innovator who has long been committed to understanding the business world in its social context by considering the interplay among value systems, culture, and business. He suggests that a capitalist society depends fundamentally on "a minimum of honesty," which systems of constraint and punishment cannot replace. Boulding shows us why ethical conduct in business practice matters a good deal more than might first appear—it is not merely a private matter.

CHAPTER EIGHTEEN

Ethics and Business:
An Economist's View

K. E. BOULDING

It would be presumptuous for a mere economist to attempt to set forward a complete theory of ethics. However, one cannot talk about this subject at all without outlining a tentative ethical theory. It is not necessary to assume that a single set of ethical principles will guide conduct in all spheres of social life and in all the roles which an individual may play. Nevertheless, if we are to talk about the ethical principles of a part of a society and of an aspect of human behavior, we must be able to see these in the framework of a larger ethical system. Even though I can claim only amateur status as a moral philosopher, I feel it is necessary to say something about ethical principles in general before I can begin to apply them to a business society.

The first principle of my ethical theory is that all individual human behavior of any kind is guided by a value system; that is, by some system of preferences. In this sense everybody has a personal ethical system. No one could live, move, or act without one. We can distinguish between what might be called the "real" personal ethic, which might be deduced from a person's actual behavior, and the "verbal" ethic, which would be derived from his statements. We find it a common—indeed, almost a universal—phenomenon that a person will give lip service to one set of ethical principles, but that in his behavior he will follow another set of values. Without a set of values of some kind, however, his behavior is inexplicable. Even if his behavior is random and irrational it is presumably because he sets a high value on randomness and irrationality. This is a view of human behavior which is derived mainly from economic theory. The

This article was presented at a seminar sponsored by the College of Business Administration, Pennsylvania State University, State College, PA, on 19 March 1962. Reprinted by permission from Boulding's *Beyond Economics* (Ann Arbor: University of Michigan Press), pp. 227–236.

economist thinks of behavior in terms of choice. He envisions the individual at any moment faced with a number of alternative images of possible futures, out of which one is selected. The fact that it is selected is a demonstration that it has a higher value for the individual than any of the alternative choices. For the economist, a value system is simply a system of rank order. We look at the field of possible futures and we rank them first, second, third, and so on, which is what the economist means by a utility or welfare function. And having performed this complicated and arduous act, we then behave in such a way that the future which we have ranked first has, we believe, the best chance of being realized. In these terms, of course, even the worst of men has an ethic of his own. If a person prefers to be cruel, mean, and treacherous, he is expressing a value system which ranks these types of behavior high.

It is clear that merely to say that everyone behaves according to a personal ethic or value system does not solve the problem of ethical theory, the major perplexity of which is to develop a rule of choice among possible personal value systems. The individual is faced not only with images of the world which are ranked according to a particular value system, he is faced with a number of different value systems according to which the world may be ranked. Just as there is a problem of choice among alternative futures, so there is a problem of choice among ways of choosing these alternative futures. It is assumed that out of all possible ways of choosing—that is, out of all possible value systems—only one is "right" or "best." This is the ethical value system.

By way of illustration, let us imagine an individual who is at a point of time where he faces three possible futures: A, B, and C. Let us suppose, to simplify matters, that he visualizes the effects of his actions on himself and on others in terms of a single variable which we will call "riches." If he chooses A, he will be a little richer, and other people will be richer too. If he chooses B, he will be greatly richer, and nobody else will be worse off. If he chooses C, he will be a little poorer, but other people will be a lot poorer. Which he chooses, of course, depends on his value system; that is, on his personal ethic. If he has an altruistic personal ethic, in which he enjoys the riches of others as if they were his own, he will probably choose A. If he has a selfish personal ethic, in which he places a high value on his own riches but is indifferent to the condition of others, he is likely to choose B. If he has a malevolent personal ethic, in which he takes a positive satisfaction in the misfortunes of others, he is likely to choose C. Thus, any choice is possible depending on the personal ethic of the individual making the choice.

The ethical problem, however, is the problem of what my personal ethic shall be. Some may wish to argue that this choice is not really open to the individual, that his personal ethic is simply a result of his life experience and of the various punishing or rewarding consequences of various acts. If we admit the possibility of choice at all, however, it seems to me that we must admit the possibility of choice among value systems as well as among possible futures; a determinism which excludes the ethical problem also excludes the possibility of any rational behavior at all.

My third principle is that no *a priori* proof is possible in any proposition of ethical theory; that is, we cannot arrive at a rule of choice which will always give us the best personal ethic by a process involving pure logic, without reference to the world of experience. This does not mean, however, that ethical problems are in principle insoluble. The problem here is essentially one of limiting the field of choice among personal ethical systems. We may not by pure reasoning alone be able to limit the choice to a single system. This does not mean, however, that no limitation of this choice is possible. In the first place some limitation of choice can be made by a *reductio ad absurdum* argument. Suppose we had a society in which the prevailing personal ethic involved killing all children at birth. It is clear that a society of this kind would not persist beyond a generation and its prevailing personal ethic would die with it. Thus, even though we may not wish to set up survival value as an absolute standard for the choice of personal ethical systems, it is clear that survival value strongly limits the choices which may in practice be made. The history of personal value systems can be regarded as something like an evolutionary process in which there is constant mutation as charismatic individuals are able to impose a personal value system on a society or are able to initiate a subculture within a society. In all societies groups are constantly arising which proclaim new ethical standards and which seek to obtain adherents to these standards. Each mutation, however, encounters the selective process, and in the biological world some mutations have survival value and some do not. Some may have survival value in the short run but not in the long. For example, the Shakers developed a personal value system which excluded sexual intercourse. What ever its merits (and these were, no doubt, for some individuals considerable, or the Shakers would never have come into being), the system proved fatal to the subculture in the long run.

I am not suggesting that survival is the only test of validity. I am suggesting that it narrows the field of choice. It does not necessarily narrow the field to a single position. Within cultures that have survival value there are still better or worse cultures by other criteria. A culture, for instance, may have survival value and yet be extremely disagreeable for the individuals to live in. On the other hand, if it is agreeable but does not have survival value, we must exclude it from our system of possible choices simply because it excludes itself. Thus, even though an ultimate and final answer to the question as to whether any particular value system is "right" may not be possible, the question as to its rightness is meaningful simply because it is possible, by taking careful thought and by knowing more about the world, to limit the field of choice. The more we limit the field of choice, however, the harder it becomes to resolve the arguments about how to limit it further. Some, for instance, may not wish to exclude those value systems which lead to unhappiness if by excluding them we also exclude certain aspects of nobility and creativity. The solution which seems to be working itself out is one in which we have a number of different cultures, each embodying a different ethical principle. These differences, however, may correspond to deep differences in human nature, which may even have a genetic base. Within a complex society there is room for many such subcultures and many ethical systems, ranging from the Amish to the Zoroastrians. It is one of the great virtues of the division of labor, as Durkheim pointed out, that it permits a diversity of subcultures and therefore a diversity of ethical systems within the framework of a larger society; it is not necessary to impose a single ethical system on the whole society.

My fourth proposition is that corresponding to every culture or subculture there is an ethical system which both creates it and is created by it. In other words, any ethical system is embodied in a social system of which it is an essential part. Changes in ethical systems inevitably produce changes in the social system, and changes in the social system likewise react upon the ethical system. Sometimes a change in the ethical system is embodied in explicit form, in the shape, for instance, of a Bible, a Koran, or a Book of Mormon, around which a subculture is then built. Sometimes the ethical system exists in an almost unconscious set of rules of behavior and norms of conduct (as in the development of mercantile capitalism) which never become embodied in sacred writings or achieve any charismatic power, yet which profoundly affect human conduct and are transmitted from generation to generation. The relationships here are complex in the extreme. Where an overt ethical system, such as the Christian ethic, contains elements which are inappropriate to the social system in which it is embedded, it may remain the overt system because the covert system by which people really act is different. On the other hand, the overt system exercises a constant pressure on the society in which it is recognized. The dynamic character of a society, in fact, often depends on there being a certain tension between its overt and its covert ethical systems. A culture in which the overt and covert ethic coincide and in which there is no hypocrisy is likely to be deplorably stable. Where there is hypocrisy, there is a force within the culture itself making for change. A good example of this is the constant pressure in the American society towards a better integration of the Negro community: a pressure which is imposed upon us by the fact that our overt ideals and our actual behavior do not correspond. Because the overt ideals have a sacred character about them, they are not easy to change. Therefore, they exercise a constant pressure on the society to bring its practice closer to its professions. Where there are no overt ideals, a gap between ideals and practice is easily closed by changing the ideals. This is why a society with impossible ideals is likely to be highly dynamic, whereas a society with possible ideals is likely to stagnate.

I am arguing then that we must reject the type of ethical relativism which says that all ethical answers are equally valid, just as we must reject the cultural relativism which refuses to raise the question about the value of a culture. Even though we cannot regard ethical norms as absolute in the sense that they are independent of the cultural milieu, the ethical problem in any culture is a meaningful problem because it is usually fruitful to raise the question as to whether better solutions than the one currently in vogue are possible. Once a society ceases to raise this kind of question, once it ceases to examine itself in ethical terms, that society is very likely doomed to stagnation and to eventual decay.

It is now high time to get past the preliminaries and begin to apply these principles to the ethics of a business society. We should notice first that no society is a pure business society. No society, that is, has ever organized itself around the institutions of exchange alone. Every society has a government which organizes the threat system and every society has integrative institutions in the family, the church, the school, the club, and so on. Furthermore, business institutions themselves, such as the corporation, the bank, the organized commodity or security exchanges, and the labor union, are not organized solely by exchange even though we can properly regard exchange—that is, buying and selling and the transformations of inputs and outputs that go on in the process of production—as being the major organizer of the system. There remain, however, coercive or threat system elements in the threat of cutting off exchange, the withdrawal of custom, quitting, firing, and so on. There may be internal threat systems in the shape of industrial discipline, and there are also extensive integrative systems in the attempt to build morale, loyalty, the corporate image, systems of authority and instruction, and so on. Nevertheless, we shall not go far wrong if we suppose that anything which undermines the institutions of exchange and the organizing power of exchange undermines the business system.

Again the first thing to note is that there are certain individual value systems which undermine the institutions of exchange and are, therefore, extremely threatening to a business system. An exchange system, for instance, cannot flourish in the absence of a minimum of simple honesty because an exchange system is an exchange of promises, and honesty is the fulfillment of promises. If we extend the concept of honesty a little further into the fulfillment of role expectations, we see this also is essential for the successful operation of a system based on exchange. This is why the institutions of capitalism cannot operate successfully in the total absence of what might be called the puritan virtues. Thus, if capitalism is to work successfully, there must be defenses in the society against dishonesty. These defenses may lie in part in the threat system; that is, in the system of law and police. But I suspect that a good part of the burden must be carried by the integrative system in the internalization of these moral standards in the individual. This is done in the home, in the church, and in the school, but it is also done, of course, by the example of those around the individual and especially the example of his peer group. For this reason, the building of honesty into a culture in which it does not already exist is a difficult matter, for dishonesty tends to perpetuate itself through the teaching process which it develops. Here again there is a constant struggle between the overt and the covert elements in the value system; however, if this results in a collapse of the overt system—in a general lapse into cynicism and the overt acceptance of a dishonest covert system—a society is doomed.

The second set of problems relates to the political images, value systems, and institutions which provide the framework for a market-oriented economy. Because exchange is not sufficient to organize society by itself, even in a society which is organized mainly by exchange, there must be a minimum governmental framework. The success of the market economy of the United States, for instance, can be attributed in large part to the fact that there has been no hesitation in using the instruments of government for economic purposes where this has seemed to be desirable. An exchange economy, for instance, may pass over into a degree of monopolistic organization which undermines it. Hence, we have antitrust and related legislation, which may not always carry out its intent but which profoundly affects the structure of the economy and the norms of business behavior. As it has been well said, the ghost of Senator Sherman sits at every board of directors' table in America. Similarly, because a pure market economy is susceptible to meaningless fluctuations of inflation or of unemployment, there must be a cybernetic machinery in government to stabilize the general level of the economy. Political images and value systems which are hostile to this necessary governmental framework, even though they may be derived from a moral commitment to the exchange economy, are in fact inimical to it.

The third ethical problem of a business society arises out of the fact that the institutions of exchange in themselves do not develop enough of the integrative system. Exchange is a highly abstract relationship. It is something, indeed, which can be done just as well by a machine as by a person in many cases, as the rise of vending machines indicates. It is precisely because exchange is an abstract relationship that it is capable of creating organizations on such a large scale. A large organization would be impossible if every member had to have a rich personal relationship with every other. This, indeed, is the rock on which all utopian communities have foundered. At some points in the society, however, there must be personal relationships which are richer than those of the exchange relationship if the society and its activities are to have meaning and significance for the individuals. The poets, preachers, and philosophers have, on the whole, taken a rather dim view of trade ("all is seared with trade," says Gerard Manley Hopkins, "bleared, smeared with toil"). In spite of the success of a business society in increasing productivity and in providing for human wants, it has a tendency to undermine itself because of its inability to generate affection. It is a positive-sum game in which everybody benefits, but in which the game itself is apparently so lacking in emotional effect that it does not produce loyalty, love, and self-sacrifice. Very few people have ever died for a Federal Reserve Bank, and nobody suggests that they should. Consequently when the institutions of a business society come under attack from those who are emotionally committed to another way of organizing economic life, they often fail to generate in their own supporters the same degree of emotional commitment. The situation is, indeed, even worse than I have suggested, for when business institutions, such as corporations, attempt to develop emotional commitment and try to develop welfare capitalism, baseball teams, company songs, and the school spirit, the overall effect is likely to be slightly ridiculous, like that of a good solid workhorse putting on wings and trying to set up in business as a Pegasus.

Thus, we face this dilemma: if a business society is to survive it must develop an integrative system and integrative institutions, but the peculiar institutions of a business society (such as markets, corporations, banks, and so on), because they are essentially instrumental in character, are not capable of developing a powerful integrative system in themselves. If market institutions are to survive, therefore, they must be supplemented by a matrix of integrative institutions, such as the family, the church, the school, and the nation, which develop individual value systems based on love, self-sacrifice, identification with goals outside the person, and altruism.

Textbook author Vincent Barry's reflections on the use of official position, conflict of interest in financial investments, bribery, and gifts and entertainment are applicable, as we have often seen, in both the public and private sectors. Every forester has probably observed persons in authority engaging in variations of the unethical practices cited in this article. By acting in such ways, these leaders are implicitly setting the ethical standards that will be used by their peers and subordinates. At the time this article was written, Mr. Barry, editor of an early text on business ethics, served on the faculty at Bakersfield College, CA.

CHAPTER NINETEEN

Moral Issues in Business

VINCENT BARRY

The Need for Business Ethics

Although numerous contemporary realities point to the need to investigate business ethics, for simplicity we will concentrate on three: the ethical content of business decisions, the human dimension in business, and the current moral quandary that business faces.

We can obtain a good understanding of the profuse ethical content of business in the United States today by organizing our discussion around two factors. One factor is that the part business plays in influencing the social system has recently become more apparent. The second factor is related. In the last fifteen years society's expectations of what business can and should do to help solve social problems have increased proportionately with society's understanding of those problems.

Business's Role in the Social System

Business is not an independent, self-sustaining mechanism. On the contrary, it's related to the world around it, and what's around it relates to business. We call this interdependent network of relationships observable in society the social system. *Specifically, a social system is a combination of interrelated people and their institutions which operate as a whole.*

Because business functions within the social system, it is affected by concerns and values outside narrow self-interests. Social interests and assessments of worth inevitably enter the work place, often introducing ethical content into business decisions.

For example, in recent years the public has become increasingly impatient with the extent of economic hardship, such as hard-core poverty and unemployment, especially when it applies to traditionally disadvantaged groups. Similarly, average Americans are less tolerant of continued racial and sexual discrimination in the work place. Society has become concerned that everyone be treated fairly. In a word, justice and equality have become important social values and have infiltrated the business world perhaps as never before. This moral awareness has heightened public concern about the *right* thing to do about providing equal job opportunities, distributing wealth, restoring ravaged cities, and so on.

As part of the social system, business inevitably feels the press of society's concerns and values: justice, equality, truth, honesty, respect for life. Frequently values clash with one another. Thus we want everyone to have an equal employment opportunity, but we don't want to give anyone preferred and unfair consideration. We want to be protected from the hazards of a product, but we want the freedom to use it if we wish. We want our air cleaned up, but we brook no restrictions on the size of the car we will drive. Predictably, these clashes show up in business. They demand choices, often moral decisions. The result? Business decisions are taking on moral overtones. This phenomenon is being intensified by a second factor: society's expectations of business.

Social Expectations

There's no question that for the better part of United States history business has had one primary objective: the maximization of profit. In the past, business viewed its social charge as little more than providing goods and services as efficiently as possible. This left it free to focus on the economic realities of costs and prices. Its main responsibility was to produce products for a profit and at a fair price. True, business did have other social responsibilities, such as to provide jobs or to refrain from restraint of trade, but these were more economic than social. In brief, according to this classical view, business acted responsibly when it made efficient use of its resources in producing goods and services at prices consumers were willing to pay.

Furthermore, the range of societal expectations seems to be ever widening. Society is quickly coming to expect business to help solve problems which it is only indirectly involved with. Thus, at various times, business has been called on to employ its resources and expertise to solve community problems, to set up and maintain health-care facilities, to underwrite artistic endeavors, to subsidize educational enterprises, and to provide day-care centers for children of working mothers.

In short, numerous interests are pressuring business to address itself to a wide range of social problems and needs. Whether or not this is advisable, justifiable, or even possible doesn't concern us here. What does is the fact that society's expectation that business behave "morally"—that is, in a socially responsible way—has never been greater. As a result, business decisions which once raised only economic concerns now ignite moral debate. Lacking a firm grounding in business ethics, many business people find it difficult to participate meaningfully in such an exchange. In fact, they frequently view it as a preamble to the dismantling of the free enterprise system.

Undoubtedly the implications of business's influential role in the social system and society's rising expectations of social responsibility have put tremendous pressure on business people to determine their ethical whereabouts. This pressure increases the burden on an already morally troubled and unsettled group who long have silently suffered from the war between their private values and their business lives. Their plight deserves attention; it alone seems enough to justify the study of business ethics.

The Individual in Business

One needn't observe the business scene very long before noting the double bind business people frequently experience because of the dual roles they sometimes must play. The first role is that of the private individual: a decent and ethical human being who readily admits the need for moral principles on a personal and social level. The second role is that of a business person: a human being who rarely exhibits any of the marked moral sensitivity of the private individual. Not only do these roles or "personalities" share little moral ground, the second one can often brutalize the first when value conflicts arise in a business context.

As a dramatic example of this, consider the sentiments of Dan Drew, church builder and founder of Drew Theological Seminary, who shared these thoughts with business people of the nineteenth century.

> Sentiment is all right up in the part of the city where your home is. But downtown, no. Down there the dog that snaps the quickest gets the bone. Friendship is very nice for a Sunday afternoon when you're sitting around the dinner table with your relations, talking about the sermon that morning. But nine o'clock Monday morning, notions should be brushed aside like cobwebs from a machine. I never took any stock in a man who mixed up business with anything else. He can go into other things outside of business hours but when he's in the office, he ought not to have a relation in the world—and least of all a poor relation.[1]

It seems that many business people live by this code. "Downtown" the dominant business-person personality represses the values that the private-individual personality lives by at home. When conflicts arise, the "decent" personality is sacrificed on the altar of expedience with a prayerful "That's business."

But don't assume that business people wouldn't prefer things to be different. On the contrary, a basic assumption of this author is that they would prefer to do what is right rather than what's wrong. But how? Personal codes of ethics don't seem to apply to specific business decisions, or at least business people can't see the application. They need specific direction on making moral decisions in a business context.

True, some people would probably disagree. Rather than an immersion in business ethics, they'd suggest that business people realize the wisdom of Dan Drew's observations. Personal values are and must be kept separate and distinct from business lives. If business people could just accept and practice this, the dissenters claim, they would eliminate much internal conflict in trying to reconcile personal values and business demands. Because many business people seemingly practice this view—though rarely expressing it as forcefully as Drew did—it bears examination. If the view is sound, then there would be little need to study business ethics. But if the view is unsound, then we should investigate business ethics, if only to facilitate moral decision making for the people in it. There are a number of objectionable points about the view as described. First, to argue that morality has no place in business is to take a constricted and narrow view of business. It's to think of business only as an economic enterprise. But we've already seen that business functions as part of a social system. As such it is characterized by a network of human relationships that are secured and preserved by rules, standards, and principles that regulate conduct between individuals. To deny this, as the aforementioned view does, is to deny the nature of business.

Conflicts of Interest

You don't have to work for a firm for very long, either at a management or nonmanagement level, before realizing that your interests often collide with the interests of the company. You want to dress one way, the firm wants you to dress another way; you'd prefer to show up for work at noon, the firm expects you to be present at 8 A.M.; you'd like to receive $50,000 a year for your labor, the firm gives you a fraction of that figure. Whatever the value in question, attitudes toward it can differ. Thus the reward, autonomy, and self-fulfillment that workers seek aren't always compatible with the worker productivity that the firm desires.

Sometimes this clash of values can take a serious form as in what is officially termed in business a conflict of interest. *In business a conflict of interest arises when employees, at any level, have an interest in or are parties to an interest in a transaction substantial enough that it does or reasonably might affect their independent judgment in acts for the firm.*[2] Looked at another way, the firm has the right to expect employees to use independent judgment on its behalf. Conflicts of interest arise when employees jeopardize this independent judgment.

The primary source of the firm's right to expect independent judgment on its behalf and the concomitant employee responsibility to respect that right is the work contract. When one is hired, one agrees to discharge contractual obligations in exchange for pay. Thus one does specific work, puts in prescribed hours, expends energy, all in return for remuneration, usually in the form of money. Implicit in any work contract is that employees will not use the firm for personal advantage. This doesn't mean that individuals shouldn't seek to benefit from being employed with a firm, but that in discharging their contractual obligations, employees will not subordinate the welfare of the firm to personal gain.

Financial Investments

Conflicts of interest can be present when employees have financial investments in suppliers, customers, or distributors with whom their firms do business. For example, Fred Walters, purchasing agent for Trans-Con Trucking owns stock with Timberline Paper. When ordering office supplies, Fred buys exclusively through a Timberline affiliate, even though he could get the identical supplies cheaper from another supplier. This is an actual conflict of interest, but, again, even if Walters never advantages himself in this way, he is potentially conflicted.

How much of a financial investment compromises one's independent judgment in acts for the firm? The question defies a precise answer. Ordinarily it's acceptable to hold a small percentage of stock in a publicly owned supplier that's listed on a public stock exchange. Some firms often state what percentage of outstanding stock in a company employees may own. This varies from 1 to 10 percent.

Use of Official Position

A serious area of conflict of interest involves the use of one's official position for personal gain. Cases in this area can range from using subordinates for nonbusiness-related work to using one's important position with an influential firm to enhance one's own financial leverage and holdings.

Although not apparent in the Lance case, many abuses of official position arise from capitalizing on privileged information. Capitalizing on such information is called insider dealings.

Insider Dealings

Insider dealings refers to the ability of key employees to profit from knowledge or information that has not yet become public. Thus, Marie Kellogg, vice-president of Target Investments, learns at a board meeting that the firm is interested in developing a large parcel of land in a remote corner of the city. In the weeks following, Kellogg purchases a number of acres of the land inexpensively. Sure enough, some months later Target's forecast is borne out. And just as predictably Kellogg makes a tidy profit.

Participants in deals like these ordinarily defend their actions by claiming that they didn't injure anyone. It's true that trading by insiders on the basis of nonpublic information seldom directly injures anyone. But moral concerns arise from indirect injury. As one author puts it, "What causes injury or loss to outsiders is not what the insiders knew or did, rather it is what they themselves [the outsiders] did not know. It is their own lack of knowledge which exposes them to risk of loss or denies them an oportunity to make a profit."[3] Consider, for example, the Texas Gulf Sulphur stock case.

In 1963 test drilling by Texas Gas Company indicated a rich ore body near Timmins, Ontario. In a press release of April 12, 1964, insiders at Texas Gulf attempted to play down the potential worth of the Timins property by describing it as "a prospect." But on April 16 a second press release termed the Timmins property "a major discovery." In the interim, insider investors made a handsome personal profit through stock purchases. At the same time, stockholders who unloaded stock based upon the first press release lost money. Others who might have bought the stock lost out on a chance to make a profit.

In 1965 the Security and Exchange Commission (SEC) charged that a group of insiders, including Texas Gulf directors, officers, and employees, violated the disclosure section of the Securities and Exchange Act of 1934 by purchasing stock in the company while withholding information about the rich ore strike the company had made. The courts upheld the charge, finding that the first press release was "misleading to the reasonable investor using due care."[4] As a result, the courts not only ordered the insiders to pay into a special court-administered account all profits they made by trading on the inside information but also ordered them to repay profits made by outsiders whom they had tipped. The courts then used this account to compensate persons who had lost money by selling their Texas Gulf Sulphur stock on the basis of the first press release. We mention these facts to illustrate how indirect injury can result from insider dealings and also to note the risk that insiders run in capitalizing on privileged information, both of which facts bear on the morality of such trading.

In addition, insider dealings clearly raise moral questions not easily resolved. When can employees buy and sell securities in their own companies? How much information must they disclose to stock-holders about the firm's plans, outlooks, and prospects? When must this information be disclosed? Also, if people in business are to operate from a cultivated sense of moral accountability, it's important for them to understand who is considered to be an insider. In general, an *insider* would be anyone who has access to privileged information. In practice, determining precisely who this is isn't always easy. On the one hand, corporate executives, directors, officers, and other key employees surely are insiders. But what about outsiders whom a company temporarily employs, such as a contractor doing work for a firm? It seems reasonable to say that such outsiders in their temporary roles bear the same responsibilities as insiders do with respect to privileged information.

Related to but distinct from information that's privileged is information that's secret or classified. How employees use such pro-prietary data can raise important moral concerns.

Proprietary Data

Proprietary data is classified or secret information which employees are exposed to in discharging their obligations to the firm. Moral questions involving the disposition of proprietary data sometimes arise when an employee leaves a firm. A classic case involved Donald W. Wohlgemuth, who was a worker in the spacesuit department of B. F. Goodrich Company in Akron, Ohio. Eventually Wohlgemuth became the general manager of the spacesuit division and was privy to Goodrich's highly classified spacesuit technology for the Apollo flights. Shortly thereafter Wohlgemuth, feeling his salary insuffi-cient, joined Goodrich's competitor, International Latex Corpora-tion in Dover, Delaware. His new position was manager of engineering for the industrial area that included making spacesuits in competition with Goodrich. As you might expect, Goodrich protested by seeking an order restraining Wohlgemuth from working with Latex. In 1963 the Court of Appeals of Ohio denied Goodrich's request for an injunction, respecting Wohlgemuth's right to choose his employer. But it did provide an injunction restraining Wohlgemuth from revealing Goodrich's trade secrets.[5]

Cases like Wohlgemuth's are fundamentally different from those involving insider dealings, for they pit a firm's right to protect its secrets against an employee's right to seek employment wherever he or she chooses. This doesn't mean that moral problems don't arise in proprietary data cases or that they're easily resolved. After all, one can always ask whether Wohlgemuth was moral in what he did, the legality of his action notwithstanding. Such a question seems especially appropriate when one learns that Wohlgemuth, when asked by Goodrich management whether he thought his action was moral, replied, "Loyalty and ethics have their price and International Latex has paid this price."[6]

Clearly, then, considerably more moral groundbreaking is needed in the areas of insider dealings and use of proprietary data. But one thing seems clear: both areas raise serious ethical questions. Similar questions surround practices involving bribes and kickbacks.

Bribes and Kickbacks

A bribe is a consideration for the performance of an act that's inconsistent with the work contract or the nature of the work one has been hired to perform. The consideration can be money, gifts, entertainment, or preferential treatment.

Bribes constitute another area where conflicts of interest arise. To illustrate, in exchange for a "sympathetic reading of the books," a firm gives a state auditor a trip to Hawaii. Or a sporting goods company provides the child of one of their retailers with free summer camp in exchange for preferred display, location, and space for their products. In both instances, individuals have taken considerations inconsistent with the work contract or the nature of the work they contracted to perform.

Bribery sometimes takes the form of *kickbacks, a practice which involves a percentage payment to a person able to influence or control a source of income.* Thus, Alice Farnsworth, sales representative for Sisyphus Book Co., offers a book-selection committee member a percentage of the handsome commission she stands to make if a Sisyphus civics text is adopted. In a word, the committee member receives a kickback for the preferred consideration.

There are a number of noteworthy observations to make that bear on the morality of bribes and kickbacks. First, since they are ordinarily illegal, they raise the same general moral issues as any act of law breaking does. This observation applies to bribes and kickbacks that American companies and their overseas subsidiaries might give to foreign officials. Such acts were declared illegal in 1977, following the admission that nearly 400 American companies had made payments to foreign officials and political leaders amounting to about $300 million. Note that this legislation does not outlaw so-called "grease payments," payments designed to prod foreign officials into carrying out policies set by their governments. Nevertheless, such payments raise moral questions about cooperation in extortion. Moreover, questions of illegality arise even when a firm gives gratuities without asking for or expecting favors in return. An important case was Gulf's pleading guilty in 1977 to giving trips to an Internal Revenue Service auditor who was in charge of the company's tax audit. According to government sources, Gulf's plea marked one of the first times a major U.S. corporation acknowledged that gifts to public officials are illegal even when no favors are asked or expected.

In addition to the legal ramifications of bribes and kickbacks, there's the morally relevant fact that such practices can easily injure the firm. Employees who take bribes can conveniently overlook the interests of the firm in pursuit of their own gain. But even when bribes appear to hurt no one, moral concerns arise. In the case of the sporting goods company that wanted special treatment let's suppose that the product is a good, fairly priced one. People buy it and are satisfied with it. The manufacturer is happy, the retailer is happy, the consumer is happy. In brief, it appears that everyone's interests are served. So what's wrong, if anything? Perhaps nothing. But before reaching a conclusion, note that bribery always involves cooperation with others who are seeking to gain preferential treatment through an act that undermines the free market system. True, such an economic system might be inherently immoral. In that case, an act that undermines it is not necessarily wrong; it may be right. We will not debate this thesis here. . . . For now, we wish to note that bribes tend to subvert the free enterprise system, because they function to give advantage in a way that is not directly or indirectly product-related. Bribes attack the foundation of the free market system, which is competition.

As suggested by our examples, gifts are often given as bribes. But even when they're not intended or received as bribes, gifts and entertainment often involve conflicts of interest.

Gifts and Entertainment

In determining the morality of gift giving and gift receiving in a business situation and in deciding whether a conflict of interest exists, a number of factors must be considered.

1. *What is the value of the gift?* Is the gift of nominal value, or is it substantial enough to influence a business decision? Undoubtedly definitions of *nominal* and *substantial* are open to interpretation and are often influenced by situational and cultural variables. Nevertheless, many firms consider a gift of ten dollars or less given infrequently—perhaps once a year—a

nominal gift. Anything larger or more frequent would constitute a substantial gift. Although this standard is arbitrary and
inappropriate in some cases, it does indicate that a rather
inexpensive gift might be construed as substantial.

2. *What is the purpose of the gift?* Dick Randall, a department store
manager, accepts small gifts, like pocket calculators, from an
electronics firm. He insists that the transactions are harmless
and that he doesn't intend to give the firm any preferred treatment in terms of advertising displays in the store. So long as
the gift is not intended or received as a bribe and it remains
nominal, there doesn't appear to be any actual conflict of interest in such cases. But relative to the purpose of the gift, it would
be important to ascertain the electronics firm's intention. Is it to
influence how Randall lays out displays? Does Randall himself
expect it as a palm-greasing device before he'll ensure that the
firm receives equal promotional treatment? If so, extortion
may be involved. Important to this question of purpose is a
consideration of whether the gift is directly tied to an accepted
business practice. For example, appointments books, calen-
dars, or pens and pencils with the donor's name clearly
imprinted on them serve to advertise a firm. Trips to Hawaii
rarely serve this purpose.

3. *What are the circumstances under which the gift was given?* A gift
given during the holiday season, or for a store opening, or to
signal other special events is circumstantially different from
one unattached to any special event. Whether the gift was
given openly or secretly should also be considered. A gift with
the donor's name embossed on it usually constitutes an open
gift, whereas one known only to the donor and the recipient
would not.

4. *What is the position and sensitivity to influence of the person
receiving the gift?* Is the person in a position to affect materially a
business decision on behalf of the gift giver? In other words,
could the recipient's opinion, influence, or decision of itself
result in preferential treatment for the donor? Another important point is whether the recipients have made it abundantly
clear to the donors that they don't intend to allow the gift to
influence their action one way or the other.

5. *What is the accepted business practice in the area?* Is this the general
way of conducting this kind of business? Monetary gifts and
tips are standard practice in numerous service industries.
Their purpose is not only to reward good service but to ensure
it again. But it's not customary to tip the head of the produce

department in a supermarket so that the person will put aside the best of the crop for you. Clearly, where gratuities are an integral part of customary business practice they are far less likely to pose conflict of interest questions.

6. *What is the company's policy?* Many firms explicitly forbid the practice of giving and receiving gifts in order to minimize even the suspicion that a conflict may exist. Where such a policy exists, the giving or receiving of a gift would constitute a conflict of interest.

7. *What is the law?* This consideration is implicit in all facets of conflicts of interest. Some laws, for example, forbid all gift giving and receiving among employees and firms connected with government contracts. Again, where the gift transaction violates a law, a conflict of interests is always present.

Notes

1. Quoted in Robert Bartels, ed., *Ethics in Business* (Columbus: Ohio State University Press, 1963), p. 35.

2. Davis and Blomstrom, p. 182.

3. John A. C. Hetherington, "Corporate Social Responsibility, Stockholders, and the Law." *Journal of Contemporary Business.* Winter, 1973, p. 51.

4. "Texas Gulf Ruled to Lack Diligence in Minerals Case." *Wall Street Journal* (Midwest Edition), February 9, 1970, p. 1.

5. This case is treated in "Trade Secrets: What Price Loyalty?" (Ref. 6)

6. Michael S. Baram, "Trade Secrets: What Price Loyalty?" *Harvard Business Review.* November–December, 1968, p. 67.

Forest management regularly involves valuing timber and land for purchase or for sale. In public agencies, formal written appraisals are required as a basis for negotiating real estate prices. Foresters also regularly obtain appraisals from other professionals and perform appraisals for clients. Thinking about the ethics of real estate appraisal is therefore helpful in examining the ethics of forestry.

In this article, John Kokus, real estate professor at the American University in Washington, D.C., summarizes principal ethical canons of the appraisal profession.

CHAPTER TWENTY

Ethics for the Real Estate Appraiser

JOHN KOKUS, JR.

We all know in general terms what ethics is about, yet few of us have a specific definition. For example, Thomas Watson, former president of IBM, once said: "If you reach for a star you will never get a star but neither will you get a handful of mud." Setting high standards may not produce blameless character but it does help.

The confusion surrounding "ethics" is understandable. Ethics means different things to different people. Its rules are moral, rather than legal. Ethics activity is voluntary; legal activity is mandatory. Gross ethical violation may become illegal. However, laws only set forth minimum standards of conduct.

It is not enough that an act stay within the limits of the law. Laws do not control the conscience. Ethics precedes the law; in fact, as they gain support, ethical ideas eventually become law. For example, affirmative action programs in fair-housing legislation speak to the ethics as "thou shall," in the same way the law speaks to unethical acts as "thou shall not." Ethical conduct is often well above the legal minimum required.

Ethics appears also to deal with subjective or judgmental issues. Freedom of action can result in abusive practices. At such times the law steps in to legislate mandatory state laws, or professional associations draft voluntary guidelines, all of this against a backdrop of human and social tradition. Through the years human activity, especially in business, has resulted in both ethical and unethical practices.

Ethics is also influenced by culture and time. During the Middle Ages, charging interest on loans was condemned as usury. As commercial society evolved, capital came to be seen as a necessary, nonevil agent of productive endeavor. The concept of interest was

This article reprinted by permission from *The Appraisal Journal*, October 1983, pp. 540–545. Copyright © 1983 by *The Appraisal Journal*, 430 N. Michigan Ave., Chicago, IL, 60611.

explained as rent charged for the use of money. Yet some usury laws remain, even though ethics has adapted over the course of time.

The Code of Ethics of the National Association of Realtors

Dating from 1913, the Code includes the following: The term Realtor® has come to connote competency, fairness, and high integrity resulting from adherence to a lofty ideal of moral conduct in business relations. No inducement of profit and no instruction from clients can ever justify departure from this ideal. In the interpretation of his obligation, a Realtor® can take no safer guide than that which has been handed down through the centuries, embodied in The Golden Rule, "Whatsoever ye would that men should do to you, do ye even so to them."

To be free, and to retain freedom, is to act responsibly. One of these responsibilities is found in the very first paragraph of the Preamble to the *Realtor Code of Ethics*. The founding fathers of real estate—that early group of real estate professionals—also manifested their wisdom with the choice of these words, "the highest and best use of the land."

The U.S. Constitution provides for and encourages the right of private ownership of land, subject to the government's powers to regulate and control the use of that land, to tax it, to condemn and pay fair market value for the land for higher public purpose, and to acquire land by escheat. Throughout all of these private and public processes the highest and best use of real property valuation is paramount. The value of land as a productive resource is specifically acknowledged.

There are very grave responsibilities to acknowledge and value these gifts of the soil and man's labor thereon. When land is valued at its highest and best use, all parties benefit. Individual property owners receive the financial benefits of ownership, the various levels of government receive the tax revenues of productive land use, future generations receive the gift of land reservations, and our system of enterprise prospers and moves forward, giving merit to its concept of free institutions and personal liberties. This sentiment, and its signifi-cance, is heralded in the first sentence of the Preamble of the Code of Professional Ethics and Standards of Professional Conduct of the American Institute of Real Estate Appraisers: "Real estate is one of the basic sources of wealth and its proper use is essential to the economic well-being of any society." The appraisal profession thereby assumes a unique and profound trust.

Regulation No. 10 of the American Institute of Real Estate Appraisers embodies both the ethical code and the professional conduct standards of the appraisal profession affiliated with the National Association of Realtors. This group of members who are both Realtors and appraisers have voluntarily adopted two sets of Codes of Ethics. Regulation 10, the appraisal code of ethics, was first adopted in 1958. The Preamble to Regulation 10 clearly states that, rather than the "negative" threat of disciplinary action, "it is a personal desire for individual excellence coupled with a personal desire for the respect of your peers and the respect and confidence of the society which you serve that provides the most effective incentive to a true professional." While ethics can be regulated by any profession, its highest expression will always be found in the voluntary commitment and dedication by the individual. It is the philosophy of human conduct, with emphasis on right and wrong as moral questions.

Moreover, all aspects of ethics can never be specifically defined for any profession. Codes of ethics and state laws and regulations have frequently been amended to reflect necessary changes in any professional business as it relates to society. An understanding of a wide variety of practices is ever necessary if unfavorable consequences that tarnish the public image of appraisers are to be avoided.

Interpreting and Applying Ethical Decisions

Ethics, in general, is relatively clear and straightforward, personally and socially desirable; however, becoming specific about ethics and arriving at decisions for complex problems is difficult as two illustrations show.

The first is the classic case of lifeboat ethics. An ocean-going vessel sank in the high seas with 24 survivors clutching onto a single lifeboat capable of holding a maximum of 12 persons. The captain perceived his options as either 24 certain deaths or 12 possible survivors. He chose the lesser evil and ordered 12 of the 24 lifeboat occupants ejected into the sea. Against all odds, the lifeboat reached land with nine persons still alive.

Tried in British Admiralty Court, the captain was found guilty, but was given a lenient and suspended sentence for his crime against humanity. Evidently the court regarded his action as clearly and always illegal, but under the trying life-and-death circumstances, the only ethical action possible. Agreement was easy; ethically, the issue was to save lives. Disagreement arose over the specifics of how to accomplish the task, in this case sacrificing lives to save lives.

A difficult case of national ethics involved the 52 Americans held hostage in Iran. The ethical issue was whether to honor and sanctify individual human life or to uphold national sovereignty and international law. To enforce the law and reprimand the offenders would be to jeopardize individual human life. All Americans agreed that the nation needed to act with honor and courage. Some believed that the ethical interpretation of courageous action was to act forcefully and immediately, while others argued that the correct ethical action was negotiation and patience.

The United States was applauded for its use of delicate restraint and condemned for its failure to exercise a powerful response. In normal everyday living, we are not usually confronted with these decisions. However, we are confronted with those that arise in our lines of work and our professions.

Appraisal Ethics

Examination and study of the Code of Professional Ethics reveals that the explanatory comments are stated in positive language. Of the eight canons, four are stated in positive mandate, while four are admonitory in their warnings.

Positive Canons

Canon 2. A member or candidate of the Institute must assist the Institute in carrying out its responsibilities to the users of appraisal services and to the public.

Canon 3. When performing a real estate appraisal assignment, a member or candidate of the Institute must perform such appraisal assignment without advocacy for the client's interests or the accommodation of his or her own interests.

Canon 7. In arriving at an analysis, opinion or conclusion concerning real estate, a member or candidate of the Institute must use his or her best efforts to act competently and comply with the Institute's standards of professional practice relating to competency.

Canon 8. In communicating an analysis, opinion or conclusion concerning real estate, a member or candidate of the Institute must comply with the Institute's standards of professional practice relating to written and oral appraisal reports.

Admonitory Canons

Canon 1. A member or candidate of the Institute must refrain from mis-conduct that is detrimental to the real estate appraisal profession.

Canon 4. A member or candidate of the Institute must not violate the confidential nature of the appraiser-client relationship by improperly disclosing the confidential portions of a real estate appraisal report.

Canon 5. In securing real estate appraisal assignments and in promoting a real estate appraisal practice, a member or candidate of the Institute must refrain from conduct which is deceptive, misleading or otherwise contrary to the public interest.

Canon 6. A member or candidate of the Institute who has specific knowledge of the requirements of the Institute's standards of professional practice must not deliberately or recklessly fail to observe such requirements.

The canons are clear and straight forward in general terms, and there is obvious acceptance as well as perhaps unanimous agreement on their general usefulness. They are uncontroversial principles. It is interpretations that become polemical and disputed.

Members of the Appraisal Institute agree to treat each other fairly, but may disagree over the interpretation of fairness. In assessing a candidate's qualifications for admission, fairness and frankness are essential guidelines. To what extent, however, are personally unobserved or secondhand stories about a person admissible? At what point does information about someone become hearsay? The charge to member appraisers is "to report to the proper Institute official or committee *all* [emphasis added] favorable and unfavorable *information* [emphasis added] they possess relating to the character or other qualifications of an applicant."

Character

Character is a more difficult judgment. What is morally acceptable to one person may be offensive and disreputable to another. Whose standard of character do we accept? If we accept Webster's definition of character as "moral excellence," how big a slip is allowed before an individual loses the reputation of character? To what extent is forgiveness allowed if one has slipped in the past? Or, can subsequent good deeds and a righteous lifestyle countermand previous character defect? We strive for objectivity but these are subjective questions.

Diligence and Service

The goal of an appraiser is to perform diligently and objectively. As a result, more challenging and responsible assignments will presumably be offered. As one's industriousness intensifies, there is less time to do other things, including service on appraisal committees. Yet that is what the code of ethics requires. This ethically vexatious situation actually intensifies over time as the appraiser becomes better qualified and more competent. As one becomes better qualified to serve, one finds less time to serve. There lies the moral dilemma.

Diligence

A related problem arises when work is overbooked. All of us would like to have assignments back to back or on hold. The concomitant danger is that work can become hurried and sloppy, or that an expansion of a present contract requested by the client and agreed to by the appraiser can also result in a work backup. The ethical question of whether to accept new work or expansion of current work increases proportionately as the work load increases. There are no easy answers; the application of care, skill, and integrity requires exactly the right amount of diligence to each assignment.

Qualifications

As work load increases, a real estate appraisal assignment may be referred to another appraiser. This does not mean to make referrals to cronies but to those "best qualified." The only questions are, how does one know who is best qualified? Is there one acknowledged expert? Is the appraiser the one who has done the most work in this field, the one everyone else talks about, the one who has gone to court the most, or the one most requested to do a particular type of appraisal?

Limits of a Member's Experience

A related issue reads as follows: "A Member who has been granted the MAI designation should not abuse that designation by using it to suggest inherent competence to perform services which in actuality exceed the limits of the member's experience and training." The public perceives the MAI designation to be the leading one, and believes that an appraiser has earned the MAI because he or she can handle any appraisal assignment. Explanatory Comment, Canon 7, "hedges" the earned competence of the MAI appraiser who, by training, education, and assignment, is expected to perform with diligence, objectivity, and competence.

However, insofar as an appraiser's conscience precludes accepting an assignment that exceeds the limits of the member's experience and training, is it not as morally wrong or unethical to limit the personal growth that comes from undertaking and surmounting successively greater challenges? It would be difficult to perceive personal growth and societal advancement in any other way. The MAI designation is an earned recognition, not a bestowed grant. We respect the MAI's signal achievement and competence.

It is further stated that "the interests of the public and the profession require that every member properly advise clients and potential clients of *the actual range of the member's services and experience"* [emphasis added]. Someone had to be first in the performance of an appraisal assignment. Was that individual unethical in exceeding his or her grasp at the time? What exactly are the limits of our experience, training, and services? Perhaps it is as unethical to perform by historical ways and past practice as it is to challenge and risk future performance. Would it be as morally wrong to underachieve as it would be to overreach, given the risk to professional acclaim? To fail is not to lose professionalism; not to try is.

The more one examines any professional code of ethics, the more difficult decisions become. This is the price of reasoned thought.

Thomas Dorsey, a senior vice president at American Savings of Florida, approaches the ethics of appraisal from a different perspective, that of a banker and expert in construction and real estate. His observations on ethics expand on the previous selections and suggest the importance of public regulation of appraisers—a topic just beginning to affect forestry in a significant way. In his brief essay, he raises, but does not elaborate upon, the age-old conflict all professionals encounter sooner or later: "The appraiser cannot act contrary to the interests or requirements of the client. At the same time, the appraiser cannot perform in a manner which conflicts with the dictates of conscience, with ethics, and as evidence of both, with professional standards." Since appraisers are often expected to be advocates for a client's monetary interests by supporting estimates of values consistent with the client's interests, they test their ethical and professional standards every day.

CHAPTER TWENTY-ONE

Ethics, Appraisal Standards, and Client Relationships

THOMAS A. DORSEY

Labor to keep alive in your breast that little spark of celestial fire—conscience.[1]   GEORGE WASHINGTON

Ethics, appraisal standards, and client relationships: different yet inseparable. Ethics and standards are the keystones of profession. Appraisals, in order to be defensible and respectable as unbiased opinions of value, require a strong sense of ethics throughout. The appraiser must subscribe to a demanding set of professional standards. The appraiser's client relies on a reputation for ethical and professional behavior in determining the qualifications of the appraiser and the value of the appraisal report. A healthy appraiser-client relationship, therefore, includes a common set of ethical values.

This idea of an interrelationship between ethics, standards, and client relationships is not a concept unique to the appraisal profession. Other professions embrace standards which imply similar interrelationship and interdependence. Noted academic and business leaders likewise reinforce this interpretation of a professional's responsibilities: Ethics should not be distinguished as separate from standards, and neither can be considered as independent of client relationships.

More often than not, ethics and standards are in the background. The appraiser and the client are concerned with matters such as time and expense. Somehow, the needs of the appraiser and of the client must be met in a manner which precludes constraints which hinder the appraiser's ability to perform and undermine the client's ability to procure a report which in and of itself will have value. What are ethics, what are appraisal standards, and what do the two together have to do

This article reprinted by permission from *Real Estate Appraiser and Analyst,* Fall/Winter 1987, pp. 22–25.

with client relationships? This is an attempt to answer these questions and reach a conclusion as to their importance to real estate appraisers today.

Ethics

Webster defines ethics as "standards of conduct and moral judgment," or as "the system or code of morals of a particular . . . group, profession."[2] Simply put, it's what lets us know the difference between right and wrong. Twenty years ago, and for our purposes here, Sanders Kahn put it more succinctly: "Ethics is Profession."[3]

Ethics is also a code of conduct, which, in order to be effective, demands a strong degree of leadership and commitment. We rely on professional organizations to provide the leadership, and we view membership in a reputable organization as evidence of a sense of ethics. Of course, appearances are often deceiving. Irrespective of the rule that a book should not be judged by its cover, more often than not it is. A presumption of competence is made when an organization refers to itself as professional, and when an individual relies on membership in that organization to certify or attest to his or her competency. That is why ethics requires an implementation and control section, beginning with standards.

In defining ethical behavior, professionals take the first step in determining what standards should be considered as professional. Ethics in this light might be considered as a set of parameters. Christensen, Andrews, and Bower acknowledge this role in their text, *Business Policy:* "Policy for ethical and moral personal behavior, once the level of integrity has been decided, is not complicated by a wide range of choice."[4]

Standards

In 1970, L. W. Ellwood introduced his now famous *Ellwood Tables for Real Estate Appraising and Financing* as follows:

> I believe experience can teach lessons which may lead to sound judgment. I believe sound judgment is vital in selecting the critical factors for appraisal. But, I also believe the bright 17 year old high school student in elementary astronomy can do a better job estimating the distance to the moon than the old man of the mountains who has looked at the moon for 80 years. So, I find it difficult to accept the notion that dependable valuation of real estate is nothing more than experience and judgment.

I would not give a red cent for an appraisal by the 'expert' who beats his breast and shouts: 'I don't have to give reasons. I've had 40 years experience in this business. And this property is worth so much because I say so.'

After all, value is expressed as a number. And, no man lives who, through experience, has all numbers so filed in the convolutions of his brain that he can be relied upon to choose the right one without explicable analysis and calculation.[5]

Experience alone does not make a professional, and so we have standards—standards of behavior, and standards of performance.

. . . Most professional groups, appraisers and otherwise, have rules and regulations to guide and govern their membership, and to inform the public as to the nature and ethical character of the particular organization. Appraisers have somewhat of a unique challenge, and perhaps an opportunity, brought about by the lack of unity in the appraisal profession. Estimates of the number of persons representing themselves as appraisers of real estate range to 250,000 and higher, and the number of groups touting themselves as "professional" increase without end (by one count, 40 in the U.S. alone). Yet there are at most eleven groups which can truly be considered as national associations of professionals, and membership in these barely exceeds 70,000. When the list is narrowed further to include only professionally designated appraisers, those holding the full endorsement of an individual organization, the number falls sharply. The two groups arguably the most widely recognized, the Society of Real Estate Appraisers (SREA) and the American Institute of Real Estate Appraisers (AIREA), together have 13,000 designated members (overlap caused by dually designated members is ignored here), just five percent of the 250,000 "appraisers" nationwide. When we speak of standards, therefore, we most often find ourselves speaking of standards for a particular group, and not standards for a profession.

There are some problems inherent in the independent course which the professional groups have to date followed. Not least of these is the alphabet soup of designations which acts to confuse the users of appraisal services. What the appraisal profession needs is a "United States of Appraisers." It is currently approaching a "Federated States" status, in what may be either a viable alternate to this ultimate goal, or only a step along the way. This is occurring through the embryonic Uniform Appraisal Standards Board, an effort by leading appraisal groups to set standards for an entire industry . . . standards because, as headlined in a recent issue of *Realtor News*, "Wanting to follow our Code of Ethics isn't quite enough."

A basic tenet of the relationship between government and the governed is that whenever a void in the private sector is perceived, regulators step in to protect the public interest. The R-41 memoranda are but one confirmation of this premise.

A Regulated Industry, A Regulated Profession

On September 11, 1986, the Federal Home Loan Bank Board (FHLBB) Office of Examinations and Supervisions circulated a memorandum entitled "R-41c, Appraisal Policies and Practices of Insured Institutions." Although the FHLBB's requirements do not affect every appraiser and every appraisal, they are significant enough that they have given birth to new conferences and seminars designed specifically to address this single policy statement.

We tend to forget what is taught in introductory appraisal courses: real estate is a regulated industry. Its instruments of conveyance are taxed and recorded in the public records. The real estate itself is taxed. Transactions are more often than not made possible by financing which is provided by federally regulated institutions. Government agencies, Freddie Mac and Fannie Mae, are increasingly active as players in the mortgage market. In this environment, can appraisers remain free from regulation?

R-41c set forth management and content requirements, as they relate to the nature and use of the appraisal report. It does not seek to be all-encompassing or to replace "standards" of other organizations. It does emphasize particular concerns of the FHLBB, and between the lines, one might read frustration in the FHLBB's inability to refer to or to endorse one set of standards and one appraisal organization. This sentiment is shared by others, and leaves the profession open to criticism. One conclusion of a recent Congressional study, for instance, was that "appraiser ineptitude, negligence and misconduct are widespread." Indeed, in referring to the many who hold themselves out as appraisers, the report stated "it's a shaky profession that threatens the safety and soundness of the nation's financial system."[6]

The appraisal profession itself recognizes the problem. Recent articles published in *The Real Estate Appraiser and Analyst* have included "The Appraisal Profession: Problems? Solutions?" and "The Need for Regulation of Appraisers."[7]

The opportunity present is to design a set of standards which will meet the needs of the public and the profession. Satisfaction of the public's needs requires that the standard have "teeth," or some means of enforcement. Satisfaction of the profession's needs might be simply stated as recently presented by the SREA and the AIREA to members of Congress:

There is a need for regulation of appraisers, primarily to address the issue of the 90 percent that are unsupervised and unattached to a professional organization. This regulation must include:

1. Standards of professional practice: Must be at least as strong as we have now.

2. Appraisal certification board: Identifies qualifications that are needed and administers testing.

3. Appraisal standards board: Develops the standards for the profession, and reviews, adds to, and deletes from the standards.

4. Enforcement: Is a critical part of the system.

5. Continuing education: Keeps appraisers current.[8]

This action by leading appraisal groups represents a more global approach in the effort to define standards and to implement procedures to educate and police an entire industry. It is an endorsement of a Uniform Appraisal Standards Board, and at the same time a recognition of the industry's failure heretofore to support a profession which can speak for all appraisers with a single voice. The result of current efforts may be licensing or certification, or both, or neither. Almost certainly, however, a major cornerstone in the building of a profession will be in place.

Standards are important and necessary, because they both state that the profession is committed to ethical behavior, and define how that behavior is to be evidenced and evaluated. Implicit in any definition of professional standards is the requirement that adherence be mandatory. That brings with it the need for disciplinary procedures. If standards are to be effective, they must have teeth, and they must be applied to an entire industry and not merely to the several leading groups within that industry. Once agreed standards have been defined and are in place, all who would hold themselves as appraisers or as users of appraisal services should be properly educated and informed.

General Henry M. Robert wrote "Where there is no law, but every man does what is right in his own eyes, there is the least of real liberty."[9] Without order, without uniformity of standards throughout the profession, we cannot and do not serve either our own interests or those of the public.

The Client

Enter the client. The client identifies the property to be appraised, and we as appraisers solve the appraisal problem. That the client *employs* the appraiser in no way diminishes the appraiser's obligation to render an independent and unbiased opinion. In fact, that circumstance only acts to underscore the appraiser's duty.

It is, after all, the client to whom the appraiser owes the fiduciary relationship. As an example, the State of Florida instructs its real estate licensees, "The rule requiring good faith toward a principal prohibits all dealing that is not open and above board . . ."[10] This is significant not only as it relates to ethics, but because "the breach or betrayal of a trust invited or accepted is fraud."[11] Another source is perhaps more direct: "The agent owes a duty of undivided loyalty to his principal. He must not put himself in a position where his individual interests conflict with his duties to his principal."[12]

In the purest of views, the client employs the appraiser for legitimate purposes or ends and does not seek to direct the value. There are exceptions. Just as not all who hold themselves as appraisers are professional, not all who find themselves in the role of client behave in a completely ethical manner. Appraisers who value their own reputations avoid clients without ethics. The question is, what leads the ethical client to the ethical appraiser?

Earlier, we said that ethics are evidenced by professional standards. Implied therein, clients should be comforted by the knowledge that the appraiser subscribes to a demanding set of such standards. But more than likely, the client's attitude toward standards approaches, "Who cares?". The client seeks a reputable, sound, and in some cases, salable, opinion. The client is consciously aware of the need for ethics in appraising, and to a much lesser degree, aware that professional standards are important. The sophisticated and more regulated client places importance on standards. And even this client arrives first for ethics.

The well-informed client may value membership in one "professional" organization over the other. That client will also place value on referrals from other known professionals and from peers. The less-informed are faced with an alphabet soup, an array of possibilities, and conflicting recommendations. This is the challenge of those segments of the industry who truly are the appraisal profession.

Peter F. Drucker wrote the following in his popular work, *Management:* "The professional has to have autonomy. He cannot be controlled, supervised, or directed by the client. He has to be private in that his knowledge and his judgment have to be entrusted with the decision."[13]

Trust. That is a key. Trust earned and trust given. And reputation is a result. The professional's first responsibility was laid out some 2,500 years ago. The Hippocratic oath stated in part "above all, not knowingly to do harm."[14] This is what the client expects and deserves.

The appraiser is duty-bound to disclose all that is significant, and to summarize findings and conclusions in a clearly developed, concise and logical report. That report should not only avoid concealment of material information, it must not raise any unanswered questions. In other words, the complete report will stand on its own, as a professional statement or argument. One purpose of the report is to communicate, and that is certainly in keeping with the client's requirement. Another is to communicate all that is both relevant and known to the appraiser. The appraiser's ethical restraints, apart from standards, dictate that only realistic assumptions and limiting conditions be made. These same ethics dictate that a strong sense of what is right and wrong governs the appraiser's judgment in identifying potential conflicts: between appraiser and client, buyer and seller, the property and its environment, user and appraiser, or wherever an objective reader of the report might surmise or suspect even an appearance of impropriety. Because that appearance, if ignored and if not disclosed and satisfactorily explained, will call into question the integrity of the appraiser and the credibility of the report.

The appraiser cannot act contrary to the interests or requirements of the client. At the same time, the appraiser cannot perform in a manner which conflicts with the dictates of conscience, with ethics, and as evidence of both, with professional standards.

The client wants an appraisal which is useful, and that is simply one completed by an appraiser of good repute, in a manner which conforms to the reporting requirements of the client or the ultimate user of the appraisal report, *and* to the standards of the profession.

But the appraiser is more than an appraiser. He or she ofttimes fills the role of office manager and business developer. A cheerleader for the profession. A promoter of an individual business enterprise. An arbiter in the realm of office politics. Shakespeare wrote in *As You Like It*, "one man in his time plays many parts." Performance and conduct outside the office and even outside of appraising will reflect upon reputation and ultimately upon the public's and the client's perception of the appraiser's ethics. The appraiser's code of ethics is therefore more than a workplace necessity, it must define the appraiser's behavior in personal as well as professional pursuits.

Today is the Tomorrow We Worried About Yesterday

Appraisers are moving swiftly to the forefront of the real estate industry. The need for a profession with ethics and standards is more than recognized, it is required. There is no turning back. Perhaps the hint of regulation was our "Rubicon." Certainly, it has become an endorsement of the values (not estimates of value) and standards which are common to those most widely accepted as professionals.

Those who pursue a career in politics forsake a degree of privacy, blur the distinction between what is public and what would be private. To a certain extent, a commitment to a profession, to the appraisal profession, limits acceptable behavior to that set forth in a particular set of ethics. Those same ethics lead professionals to associate with others of like character, to form professional societies, associations and institutes. These of themselves lead to common standards to better identify and measure the abilities of those who would call themselves professionals. This is hardly a self-serving endeavor. For it provides a direct benefit to the public, to the client. Clients require professional services, and it is we who bear the burden of defining what is professional, of building a profession, and of informing the public and our client of our success in so doing. In the final analysis, it is the appraisal profession itself which leads clients to qualified appraisers. Thanks is due to those who have worked throughout the years to build the reputation of the profession, along with the reputation of certain associations which endorse appraisers.

Today is the tomorrow we worried and dreamed and talked about yesterday. We have opportunity and promise, but these go hand-in-hand with responsibility and a duty to protect the public trust. Appraisers—professionals—must lead and work together to build that single self-regulating profession which will survive any degree of outside scrutiny, and continue to serve the needs of the public, our client.

Notes

1. *Quotations from Our Presidents* (Mount Vernon, New York: The Peter Pauper Press, 1969).

2. *Webster's New Collegiate Dictionary* (Springfield, Massachusetts: G. & C. Merriam Company, 1973).

3. Sanders A. Kahn, *Down to Earth* (Chicago, Illinois: Society of Real Estate Appraisers, 1985).

4. C. Roland Christensen, DCS, Kenneth R. Andrews, Ph.D., and Joseph L. Bower, DBA, *Business Policy*, Fourth Edition (Homewood, Illinois: Richard D. Irwin, Inc., 1978).

5. L. W. Ellwood, MAI, *Ellwood Tables*, Third Edition (Chicago, Illinois: American Institute of Real Estate Appraisers, 1970).

6. Congressman Barnard's Report to the U.S. Congress. 1986.

7. Lloyd D. Hanford, Jr., "The Appraisal Profession: Problems? Solutions?," and Barry A. Diskin, Patrick Maroney and Frank A. Vickory, "The Need for Regulation of Appraisers," *The Real Estate Appraiser and Analyst*, Summer 1986, pp. 13–18; pp. 19–27.

8. Recommendations for Uniform Appraisal Standards, 1986.

9. Sarah Corbin Robert, *Robert's Rules of Order*, Newly Revised (Glenview, Illinois: Scott, Foresman and Company, 1970).

10. *Florida Real Estate Commission Handbook*, State of Florida, 1985.

11. Ibid.

12. Harold F. Lusk, SJD, Charles M. Hewitt, JD., DBA, John D. Donnell, JD., DBA, A. James Barnes JD., *Business Law*, Second Edition, UCC Edition (Homewood, Illinois: Richard D. Irwin, Inc., 1970).

13. Peter F. Drucker, *Management—Tasks, Responsibilities, Practices* (New York, New York: Harper & Row, 1974).

14. Ibid.

In this brief article, Andy Harrison, an editor for the respected trade publication *Pulp and Paper*, reports on a discussion at an industry meeting concerning the abuse of each other's technical secrets in situations involving the purchase of high-technology equipment. In these cases there are many opportunities for either the purchaser or seller to steal the other party's ideas and innovations for their own benefit. Only through the application of new knowledge and the diffusion of that knowledge, however, will productivity, product quality, and product yields advance in society's overall interest.

CHAPTER TWENTY-TWO

Where Do We Draw the Line on Ethics?

ANDY HARRISON

The House Ethics Committee has been created to seek out unethical practices in our government. However, the function of this committee seems to raise more questions than it answers. For example, will it become a real deterrent to unethical practices or degenerate into a witch hunt for certain members of Congress? Can people even agree on what is and is not ethical? And, perhaps most important, since ethics legislation is actually a form of behavioral censorship, an age-old question must be asked: "Who will censor the censors?"

The paper industry encounters ethics problems of sorts every time a mill "deals" with a supplier, or vice versa. At a recent meeting of the local Ohio TAPPI section in Middletown, this subject was discussed at length, indicating a growing concern over mill/supplier relationships. This sensitive topic seemingly would not produce much give-and-take open discussion, but participation was surprisingly high and lively.

Howard Crosson of Betz Paperchem opened the session by presenting the supplier's view. Paul Hoelderle of Fletcher Paper Co. followed with the mill perspective. Afterwards, almost all attendees talked freely about their experiences and thoughts on "proper code of conduct."

One mill representative said, "We bought an air knife system from this vendor, and after installing the equipment, we added our own designs to improve the operation of the system. After discovering these improvements, the vendor then incorporated them into its air knife design for future sales."

Is it ethically "okay" for a vendor to take an idea from the mill that involves the vendor's product? On one hand, the mill is providing the supplier with design technology that most likely will be sold to other

---

This article reprinted by permission from *Pulp & Paper*, May 1989, p. 7.

mills (the competition). On the other hand, the vendor is supplying new technology to the industry by spreading it to others. Nobody wants to give away any exclusive technology, yet how would the paper industry be as advanced as it is today without the learning process that goes on between mills and suppliers? Suppliers do learn from mills and vice versa. When this happens, everyone wins. However, there always seems to be that nagging doubt, "Is the vendor taking advantage of me?"

Vendors can ask the same question about mills, however. "We spent many manhours designing and proposing the right system for this mill. When it came down to a purchase, however, the mill used our design and bought from another vendor."

Suppliers sell their products by offering every possible edge to the customer. They will offer the best product on the market, the best service, the most reliable product, the best price, the best design, and/or the best knowledge. If one supplier has all of these items, life becomes very simple. But fortunately, the free-market system tends to make things more complicated. Usually, a half-dozen vendors will bid on a project, and they are all relatively close. What is going to sway the mill?

The mill is always hungry for knowledge. But superior knowledge on the product or process doesn't necessarily sell the product. The supplier still has to give away some of this knowledge to attract the mill. The question is, Does the mill have the right to use information or ideas obtained from one vendor on a product it buys from another vendor? All may be fair in love and war, but animosities build and a mill/supplier relationship deteriorates.

The most important aspect of the mill/supplier relationship is communication. Open communication must be established and maintained so that the needs of each are realized and not just assumed. If the mill and vendor can build trust in each other, many potential problems can be avoided.

New technology and information are always in high demand. Much of the industry's research is done by suppliers, and many mills rely heavily on that research. Both mill and supplier prosper by such a working relationship. Keeping new information secret can be a competitive edge to the individual mill or supplier for the short term but not to the paper industry in the long term.

Ethical behavior is inevitably monitored by market forces. If, for example, a vendor tells one mill what the mill down the street is doing, this might be a tip that the vendor is acting unethically. Generally, that vendor won't get much business. In an industry that communicates as well as the paper industry supposedly does, unethical behavior will be noted and, for the most part, not tolerated.

The business aspects of professional forestry provide practitioners with many practical day-to-day questions of honesty and trust, especially in the area of conflicts of interest. Conflicts of interest are situations in which there is a real or potential conflict between, for example, the financial interests of a professional and the interests of the professional's employer or client, or those of another client. Potential conflicts of interest are easy to see in the most extreme cases, yet, in reality, as the author notes, "all too often there is a gray area in which a forester must make a decision."

Ted Stuart, a consulting forester in Virginia, contributed this piece to *The Consultant* years ago. Stuart is a leader on the issue of forestry ethics within the Association of Consulting Foresters.

CHAPTER TWENTY-THREE

Conflict of Interest:
A Question of Ethics

EDWARD STUART, Jr.

It is gratifying to note that at long last more consideration is being given to ethical behavior in the forestry profession. This subject has been discussed at various forestry seminars, articles have been published, and there seems to be some concentrated effort to require forestry schools to include the subject in their curriculum. All too often in the past the subject of ethics is given secondary importance and very little time is ever allocated for a full discussion of the issue. When it has been discussed, it is generally on a philosophical plane rather than on the hard practical aspects of day-to-day problems. Despite the improvements, we still have a long way to go.

Today the credibility of professional foresters is being challenged in Congress, in the courts, in the press, and even in the classrooms. A measure of credibility, in addition to professional competency, with a doubting public, is the professional behavior of the forester as guided by a strong ethical code, conscientiously adhered to and strongly enforced.

One of the most favorable signs recently was the acceptance of the Society of American Foresters (in conjunction with the SAF/ACF Liaison Committee) to look into and support the concept of required ethical courses in our forestry schools. All too often forestry educators have down-graded the need of a required course or seminar. Their excuses would be a crowded curriculum, that unethical behavior by foresters was unknown, and the subject of ethics could easily be woven in with their present courses.

Unfortunately, there have been numerous cases of unethical conduct brought to the author's attention over the years and in practically all cases it has been because of lack of understanding of our code. Also, there have been many inquiries on certain situations as to

This article reprinted by permission from *The Consultant*, October 1981, pp. 85–87.

how to act in order not to violate any Canon. All of which substantiates the need for proper guidance and instruction in our schools.

Very few forestry schools today offer instruction or guidance on ethical behavior. A few schools, such as the University of Maine, do hold periodic seminars. The importance of good professional behavior is very apparent in all other professions and some of these are requiring courses in ethics as a prerequisite to graduation. It is easy for lecturers to moralize and discuss the philosophy of ethics. This does not help the forester when he is confronted with actual day-to-day problems. What is needed are case study courses that allow the student to make his decisions based on the facts of the case with the Code of Ethics in front of him. Not only does this type of instruction apply to forestry students, but it should apply periodically to all practicing foresters.

Let us examine some situations which a forester can be expected to encounter which seem to be causing some uncertainty. The number one problem is conflict of interest. This is very common and covers many activities. Foresters must know what constitutes a conflict of interest—or even the appearance of a conflict of interest. Let me cite one or two examples that seem to be occurring.

One of the most flagrant examples of unethical behavior recently reported involved a widow who owned a large tract of timber in a distant state. She was informed by a timber brokerage firm, staffed by foresters, that her timber was heavily infested with bark beetles and that in order to salvage something she should sell immediately. They, of course, would be glad to sell the timber for her. This elderly lady had the good sense to retain a consulting forester who resided in the areas of the timber and the report came back that there was no infestation and the timber did not require harvesting. The consultant also obtained the verification of the local State Forester. This case certainly was a conflict of interest, as well as being deliberate misinformation.

A management forester, employed by a lumber company to assist the private landowner, marks a block of timber for harvesting, estimates the volume, and determines (in his opinion) the market price—he submits this to the landowner. His employer then purchases the timber based on his forester's appraisal. The timber was not exposed to the open market. There is no question that a practice of this type is a conflict of interest or, at the least, an appearance of conflict—regardless of how unbiased the forester was in determining the timber value.

Many of the large wood-using firms in different parts of the country have entered into the Landowners Assistance Program (LAP). Many believe that because of the financial expense they have in assisting the individual landowner over the years, they should have the right to purchase the wood when it matures, either because of an option, right of first refusal, etc., and then based on their appraisal. If they would allow the timber to be exposed to the open market or the timber appraised by an unbiased third party, the appearance of a conflict of interest could be avoided.

[The accompanying] illustration is an advertisement placed in a local newspaper by a local lumber company. It is left to our readers' opinion whether this is a good or poor business practice and whether or not any unethical behavior is involved.

In some cases a consulting forester will accept an assignment to make a management plan for a landowner which, of course, includes a harvesting plan. He accepts this assignment with payment of his fee to be made at time of first timber sale. At first glance, there seems to be no conflict of interest. However, there is an appearance of a conflict of interest as it can be easily considered that the forester updated his cutting plan to get his fee at an early date.

Other questions concerning conflict of interest are: Can a forester appraise a tract of timber and base his fee contingent upon a percentage of his findings? Can a forester purchase timberland for himself when a portion of this time is involved in advising potential buyers? Can a publicly employed forester accept Federal forestry financial assistance when he is responsible for recommending allocation of Federal funding in his area? Under what circumstances can a forester cruise the same tract of land for two different clients? When can he accept compensation from two parties for the same service? What about moonlighting? Is it a conflict of interest? When?

Solutions to the above appear to be simple, but, unfortunately, that is not the case. Discussion of specific cases has often demonstrated a wide divergence of opinion. The Society of American Foresters and the Association of Consulting Foresters codes cannot begin to offer guidelines for this one canon. There are just too many situations that could occur. All too often there is a gray area in which a forester must make a decision. Basic compulsory instruction in our forestry schools and post graduate seminars such as the Practicing Foresters Institute will assist the forester in making the right decision.

Although this article has confined itself to only one canon of our Code, the same considerations can be applied to all of the other canons of both ACF and SAF Codes.

The importance of foresters understanding their code and conducting themselves accordingly cannot be overstressed. Professionals will be suspect unless they can demonstrate professional competence combined with personal integrity.

SELLING YOUR TIMBER?

Should you use a private consulting forester or an industrial forester? Both exist to serve you, the landowner. It is up to you to decide. Some factors to consider are these. Request information from the foresters about their credentials and experience. Ask for references from people with whom the forester has dealt in the past. Visit a completed woodlot and inspect the work of the forester. Ask what their services consist of and what the fee is. Determine for yourself which forester will net you the most dollars for your woodlot. Here is a case example.

A private consulting forester professionally marks your woodlot and puts the timber out to bid. This forester estimated you had 100,000 board feet and the highest bid was $100/mbf. What will you net in this transaction?

$$
\begin{array}{r}
100 \text{ mbf at } \$100/\text{mbf} = \$10,000 \\
\text{less consultant's fee (average 10\%)} \quad \underline{1,000} \\
\$9,000
\end{array}
$$

In this example, the buyer, whether logger or sawmill, has purchased the rights to cut all trees marked by the consulting forester and estimated to be 100,000 board feet. Actually the logger will usually cut 10–30% more than the consultant estimates, since it is only an estimate and if the consultant wants to stay in business he had better see that the purchaser receives what he has contracted to purchase. Therefore, the consultant's estimate is generally, conservative. There can also be volume removal variations due to logger skill, mill log-specifications, etc. all of which can affect the board footage realized from a woodlot. The above method assures the landowner of a guaranteed amount of money for his timber before any timber is cut. This is a commonly used method of selling timber.

An alternative to the above is to sell the same timber instead to an industrial forester. This forester with equal skill and experience may also estimate the woodlot to contain 100,000 board feet of timber. However, the contract can specify that the landowner be paid for all the lumber cut, not just the 100,000 feet estimated. If we assume a minimum 10% overrun in our example case, resulting in an actual total cut of 110,000 bd. ft., what will be your net income from this timber sale?

$$100,000 \text{ ft. at } \$100 = \$10,000$$
$$10,000 \text{ ft. overrun at } \$100 = 1,000$$
$$\underline{\text{Industrial forester fee} \qquad 0}$$
$$\text{Landowner's net } \$11,000$$
$$\text{or } \$2,000 \text{ more}$$

than what a consulting forester could put in your pocket on the same transaction. Sellers beware, before you sign any work agreement papers with any forester, check him out, and check out the bottom line.

(Company name, address, phone number)

Harvard Business School emeritus professor Kenneth Andrews was editor of the *Harvard Business Review* from 1979 to 1985. In this article, he contrasts the view that ethics is a matter of individual conscience and upbringing with his own view that organizations have duties to foster, support, and cultivate high ethical standards in their employees. Andrews wants to know why there is so much falsification and concealment of important information in business, most of it being done by "good" people. He observes that three qualities are needed for ethical independence: "competence to recognize ethical issues," "self-confidence to seek out alternative points of view" and decide what is right, and the "tough mindedness" to make decisions when lacking all the facts or the time to deliberate on them. Andrews believes that these qualities can be cultivated in people and that management has a positive responsibility to do so. He especially emphasizes the need for trust, both within and between organizations.

After reading this essay, review the essay by Kermit Johnson, who also writes about the pressures on individuals in large organizations.

CHAPTER TWENTY-FOUR

Ethics in Practice

KENNETH R. ANDREWS

As the 1990s overtake us, public interest in ethics is at a historic high. While the press calls attention to blatant derelictions on Wall Street, in the defense industry, and in the Pentagon, and to questionable activities in the White House, in the attorney general's office, and in Congress, observers wonder whether our society is sicker than usual. Probably not. The standards applied to corporate behavior have risen over time, and that has raised the average rectitude of businesspersons and politicians both. It has been a long time since we could say with Mark Twain that we have the best Senate money can buy or agree with muckrakers like Upton Sinclair that our large companies are the fiefdoms of robber barons. But illegal and unethical behavior persists, even as efforts to expose it often succeed in making its rewards short-lived.

Why is business ethics a problem that snares not just a few mature criminals or crooks in the making but a host of apparently good people who lead exemplary lives while concealing information about dangerous products or systematically falsifying costs? My observation suggests that the problem of corporate ethics has three aspects: the development of the executive as a moral person; the influence of the corporation as a moral environment; and the actions needed to map a high road to economic and ethical performance—and to mount guardrails to keep corporate wayfarers on track.

Sometimes it is said that wrongdoing in business is an individual failure: a person of the proper moral fiber, properly brought up, simply would not cheat. Because of poor selection, a few bad apples are bound to appear in any big barrel. But these corporate misfits can subsequently be scooped out. Chief executive officers, we used to

---

think, have a right to rely on the character of individual employees without being distracted from business objectives. Moral character is shaped by family, church, and education long before an individual joins a company to make a living.

In an ideal world, we might end here. In the real world, moral development is an unsolved problem at home, at school, at church—and at work. Two-career families, television, and the virtual disappearance of the dinner table as a forum for discussing moral issues have clearly outmoded instruction in basic principles at Mother's knee—if that fabled tutorial was ever as effective as folklore would have it. We cannot expect our battered school systems to take over the moral role of the family. Even religion is less help than it once might have been when membership in a distinct community promoted—or coerced—conventional moral behavior. Society's increasing secularization, the profusion of sects, the conservative church's divergence from new lifestyles, pervasive distrust of the religious right—all these mean that we cannot depend on uniform religious instruction to armor business recruits against temptation.

Nor does higher education take up the slack, even in disciplines in which moral indoctrination once flourished. Great literature can be a self-evident source of ethical instruction, for it informs the mind and heart together about the complexities of moral choice. Emotionally engaged with fictional or historic characters who must choose between death and dishonor, integrity and personal advancement, power and responsibility, self and others, we expand our own moral imaginations as well. Yet professors of literature rarely offer guidance in ethical interpretation, preferring instead to stress technical, aesthetic, or historical analysis.

Moral philosophy, which is the proper academic home for ethical instruction, is even more remote, with few professors choosing to teach applied ethics. When you add to that the discipline's studied disengagement from the world of practical affairs, it is not surprising that most students (or managers) find little in the subject to attract them.

What does attract students—in large numbers—is economics, with its theory of human behavior that relates all motivation to personal pleasure, satisfaction, and self-interest. And since self-interest is more easily served than not by muscling aside the self-interest of others, the Darwinian implications of conventional economic theory are essentially immoral. Competition produces and requires the will to win. Careerism focuses attention on advantage. Immature individuals of all ages are prey to the moral flabbiness that William James said attends exclusive service to the bitch goddess Success.

Spurred in part by recent notorious examples of such flabbiness, many business schools are making determined efforts to reintroduce ethics in elective and required courses. But even if these efforts were further along than they are, boards of directors and senior managers would be unwise to assume that recruits could enter the corporate environment without need for additional education. The role of any school is to prepare its graduates for a lifetime of learning from experience that will go better and faster than it would have done without formal education. No matter how much colleges and business schools expand their investment in moral instruction, most education in business ethics (as in all other aspects of business acumen) will occur in the organizations in which people spend their lives.

Making ethical decisions is easy when the facts are clear and the choices black and white. But it is a different story when the situation is clouded by ambiguity, incomplete information, multiple points of view, and conflicting responsibilities. In such situations—which managers experience all the time—ethical decisions depend on both the decision-making process itself and on the experience, intelligence, and integrity of the decision maker.

Responsible moral judgment cannot be transferred to decision makers ready-made. Developing it in business turns out to be partly an administrative process involving: recognition of a decision's ethical implications; discussion to expose different points of view; and testing the tentative decision's adequacy in balancing self-interest and consideration of others, its import for future policy, and its consonance with the company's traditional values. But after all this, if a clear consensus has not emerged, then the executive in charge must decide, drawing on his or her intuition and conviction. This being so, the caliber of the decision maker is decisive—especially when an immediate decision must arise from instinct rather than from discussion.

This existential resolution requires the would-be moral individual to be the final authority in a situation where conflicting ethical principles are joined. It does not rule out prior consultation with others or recognition that, in a hierarchical organization, you might be overruled.

Ethical decisions therefore require of individuals three qualities that can be identified and developed. The first is competence to recognize ethical issues and to think through the consequences of alternative resolutions. The second is self-confidence to seek out different points of view and then to decide what is right at a given time and place, in a particular set of relationships and circumstances. The third is what William James called tough-mindedness, which in management is the willingness to make decisions when all that needs to be known cannot be known and when the questions that press for answers have no established and incontrovertible solutions.

Unfortunately, moral individuals in the modern corporation are too often on their own. But these individuals cannot be expected to remain autonomous, no matter how well endowed they are, without positive organized support. The stubborn persistence of ethical problems obscures the simplicity of the solution—once the leaders of a company decide to do something about their ethical standards. Ethical dereliction, sleaziness, or inertia is not merely an individual failure but a management problem as well.

When they first come to work, individuals whose moral judgment may ultimately determine their company's ethical character enter a community whose values will influence their own. The economic function of the corporation is necessarily one of those values. But if it is the only value, ethical inquiry cannot flourish. If management believes that the invisible hand of the market adequately moderates the injury done by the pursuit of self-interest, ethical policy can be dismissed as irrelevant. And if what people see (while they are hearing about maximizing shareholder wealth) are managers dedicated to their own survival and compensation, they will naturally be more concerned about rewards than about fairness.

For the individual, the impact of the need to succeed is doubtless more direct than the influence of neoclassical economic theory. But just as the corporation itself is saddled with the need to establish competitive advantage over time (after reinvestment of what could otherwise be the immediate profit by which the financial community and many shareholders judge its performance), aspiring managers will also be influenced by the way they are judged. A highly moral and humane chief executive can preside over an amoral organization because the incentive system focuses attention on short-term quantifiable results.

Under pressures to get ahead, the individual (of whose native integrity we are hopeful) is tempted to pursue advancement at the expense of others, to cut corners, to seek to win at all cost, to make things seem better than they are—to take advantage, in sum, of a myopic evaluation of performance. People will do what they are rewarded for doing. The quantifiable results of managerial activity are always much more visible than the quality and future consequences of the means by which they are attained.

By contrast, when the corporation is defined as a socioeconomic institution with responsibilities to other constituencies (employees, customers, and communities, for example), policy can be established to regulate the single-minded pursuit of maximum immediate profit. The leaders of such a company speak of social responsibility, promulgate ethical policy, and make their personal values available for emulation by their juniors. They are respectful of neoclassical economic theory, but find it only partially useful as a management guide.

As the corporation grows beyond its leader's daily direct influence, the ethical consequences of size and geographical deployment come into play. Control and enforcement of all policy becomes more difficult, but this is especially true with regard to policy established for corporate ethics. Layers of responsibility bring communication problems. The possibility of penalty engenders a lack of candor. Distance from headquarters complicates the evaluation of performance, driving it to numbers. When operations are dispersed among different cultures and countries in which corruption assumes exotic guises, a consensus about moral values is hard to achieve and maintain.

Moreover, decentralization in and of itself has ethical consequences, not least because it absolutely requires trust and latitude for error. The inability to monitor the performance of executives assigned to tasks their superiors cannot know in detail results inexorably in delegation. Corporate leaders are accustomed to relying on the business acumen of profit-center managers, whose results the leaders watch with a practiced eye. Those concerned with maintaining their companies' ethical standards are just as dependent on the judgment and moral character of the managers to whom authority is delegated. Beyond keeping your fingers crossed, what can you do?

Fortunately for the future of the corporation, this microcosm of society can be, within limits, what its leadership and membership make it. The corporation is an organization in which people influence one another to establish accepted values and ways of doing things. It is not a democracy, but to be fully effective, the authority of its leaders must be supported by their followers. Its leadership has more power than elected officials do to choose who will join or remain in the association. Its members expect direction to be proposed even as they threaten resistance to change. Careless or lazy managements let their organizations drift, continuing their economic performance along lines previously established and leaving their ethics to chance. Resolute managements find they can surmount the problems I have dwelt on—once they have separated these problems from their camouflage.

It is possible to carve out of our pluralistic, multicultured society a coherent community with a strategy that defines both its economic purposes and the standards of competence, quality, and humanity that govern its activities. The character of a corporation may well be more malleable than an individual's. Certainly its culture can be shaped. Intractable persons can be replaced or retired. Those committed to the company's goals can generate formal and informal sanctions to constrain and alienate those who are not.

Shaping such a community begins with the personal influence of the chief executive and that of the managers who are heads of business units, staff departments, or any other suborganizations to which authority is delegated. The determination of explicit ethical policy comes next, followed by the same management procedures that are used to execute any body of policy in effective organizations.

The way the chief executive exercises moral judgment is universally acknowledged to be more influential than written policy. The CEO who orders the immediate recall of a product, at the cost of millions of dollars in sales because of a quality defect affecting a limited number of untraceable shipments, sends one kind of message. The executive who suppresses information about a product's actual or potential ill effects or, knowingly or not, condones overcharging, sends another.

Policy is implicit in behavior. The ethical aspects of product quality, personnel, advertising, and marketing decisions are immediately plain. CEOs say much more than they know in the most casual contacts with those who watch their every move. Pretense is futile. "Do not *say* things," Emerson once wrote. "What you *are* stands over you the while, and thunders so that I can not hear what you say to the contrary." It follows that "if you would not be known to do anything, never do it."

The modest person might respond to this attribution of transparency with a "who, me?" Self-confident sophisticates will refuse to consider themselves so easily read. Almost all executives underestimate their power and do not recognize deference in others. The import of this, of course, is that a CEO should be conscious of how the position amplifies his or her most casual judgments, jokes, and silences. But an even more important implication—given that people cannot hide their character—is that the selection of a chief executive (indeed of any aspirant to management responsibility) should include an explicit estimate of his or her character. If you ask how to do that, Emerson would reply, "Just look."

Once a company's leaders have decided that its ethical intentions and performance will be managed, rather than left untended in the corrosive environment of unprincipled competition, they must determine their corporate policy and make it explicit much as they do in other areas. The need for written policy is especially urgent in companies without a strong tradition to draw on or where a new era must be lauched—after a public scandal, say, or an internal investigation of questionable behavior. Codes of ethics are now commonplace. But in and of themselves they are not effective, and this is especially true when they are so broadly stated that they can be dismissed as merely cosmetic.

Internal policies specifically addressed to points of industry, company, and functional vulnerability make compliance easier to audit and training easier to conduct. Where particular practices are of major concern—price fixing, for example, or bribery of government officials or procurement—compliance can be made a condition of employment and certified annually by employees' signatures. Still, the most pervasive problems cannot be foreseen, nor can the proper procedures be so spelled out in advance as to tell the person on the line what to do. Unreasonably repressive rules undermine trust, which remains indispensable.

What executives can do is advance awareness of the kinds of problems that are foreseeable. Since policy cannot be effective unless it is understood, some companies use corporate training sessions to discuss the problems of applying their ethical standards. In difficult situations, judgment in making the leap from general policy statements to situationally specific action can be informed by discussion. Such discussion, if carefully conducted, can reveal the inadequacy or ambiguity of present policy, new areas in which the company must take a unified stand, and new ways to support individuals in making the right decisions.

As in all policy formulation and implementation, the deportment of the CEO, the development of relevant policy—and training in its meaning and application—are not enough. In companies determined to sustain or raise ethical standards, management expands the information system to illuminate pressure points—the rate of manufacturing defects, product returns and warranty claims, special instances of quality shortfalls, results of competitive benchmarking inquiries—whatever makes good sense in the special circumstances of the company.

Because trust is indispensable, ethical aspirations must be supported by information that serves not only to inform but also to control. Control need not be so much coercive as customary, representing not suspicion but a normal interest in the quality of operations. Experienced executives do not substitute trust for the awareness that policy is often distorted in practice. Ample information, like full visibility, is a powerful deterrent.

This is why purposely ethical organizations expand the traditional sphere of external and internal audits (which is wherever fraud may occur) to include compliance with corporate ethical standards. Even more important, such organizations pay attention to every kind of obstacle that limits performance and to problems needing ventilation so that help can be provided.

To obtain information that is deeply guarded to avoid penalty, internal auditors—long since taught not to prowl about as police or detectives—must be people with enough management experience to be sensitive to the manager's need for economically viable decisions. For example, they should have imagination enough to envision ethical outcomes from bread-and-butter profit and pricing decisions, equal opportunity and payoff dilemmas, or downsizing crunches. Establishing an audit and control climate that takes as a given an open exchange of information between the company's operating levels and policy-setting levels is not difficult—once, that is, the need to do so is recognized and persons of adequate experience and respect are assigned to the work.

But no matter how much empathy audit teams exhibit, discipline ultimately requires action. The secretary who steals petty cash, the successful salesman who falsifies his expense account, the accountant and her boss who alter cost records, and, more problematically, the chronically sleazy operator who never does anything actually illegal—all must be dealt with cleanly, with minimum attention to allegedly extenuating circumstances. It is true that hasty punishment may be unjust and absolve superiors improperly of their secondary responsibility for wrongdoing. But long delay or waffling in the effort to be humane obscures the message the organization requires whenever violations occur. Trying to conceal a major lapse or safeguarding the names of people who have been fired is kind to the offender but blunts the salutary impact of disclosure.

For the executive, the administration of discipline incurs one ethical dilemma after another: How do you weigh consideration for the offending individual, for example, and how do you weigh the future of the organization? A company dramatizes its uncompromising adherence to lawful and ethical behavior when it severs employees who commit offenses that were classified in advance as unforgivable. When such a decision is fair, the grapevine makes its equity clear even when more formal publicity is inappropriate. Tough decisions should not be postponed simply because they are painful. The steady support of corporate integrity is never without emotional cost.

In a large, decentralized organization, consistently ethical performance requires difficult decisions from not only the current CEO but also a succession of chief executives. Here the board of directors enters the scene. The board has the opportunity to provide for a succession of CEOs whose personal values and characters are consistently adequate for sustaining and developing established traditions for ethical conduct. Once in place, chief executives must rely on two resources for getting done what they cannot do personally: the character of their associates and the influence of policy and the measures that are taken to make policy effective.

An adequate corporate strategy must include noneconomic goals. An economic strategy is the optimal match of a company's product and market opportunities with its resources and distinctive competence. (That both are continually changing is of course true.) But economic strategy is humanized and made attainable by deciding what kind of organization the company will be—its character, the values it espouses, its relationships to customers, employees, communities, and shareholders. The personal values and ethical aspirations of the company's leaders, though probably not specifically stated, are implicit in all strategic decisions. They show through the choices management makes and reveal themselves as the company goes about its business. That is why this communication should be deliberate and purposeful rather than random.

Although codes of ethics, ethical policy for specific vulnerabilities, and disciplined enforcement are important, they do not contain in themselves the final emotional power of commitment. Commitment to quality objectives—among them compliance with law and high ethical standards—is an organizational achievement. It is inspired by pride more than by the profit that rightful pride produces. Once the scope of strategic decisions is thus enlarged, their ethical component is no longer at odds with a decision right for many reasons.

As former editor of [*Harvard Business Review*], I am acutely aware of how difficult it is to persuade businesspeople to write or speak about corporate ethics. I am not comfortable doing so myself. To generalize the ethical aspects of a business decision, leaving behind the concrete particulars that make it real, is too often to sermonize, to simplify, or to rationalize away the plain fact that many instances of competing ethical claims have no satisfactory solution. But we also hear little public comment from business leaders of integrity when incontestable breaches of conduct are made known—and silence suggests to cynics an absence of concern.

The impediments to explicit discussion of ethics in business are many, beginning with the chief executive's keen awareness that someday he or she may be betrayed by someone in his or her own organization. Moral exhortation and oral piety are offensive, especially when attended by hypocrisy or real vulnerability to criticism. Any successful or energetic individual will sometime encounter questions about his or her methods and motives, for even well-intentioned behavior may be judged unethical from some point of view. The need for cooperation among people with different beliefs diminishes discussion of religion and related ethical issues. That persons with management responsibility must find the principles to resolve conflicting ethical claims in their own minds and hearts is an unwelcome discovery. Most of us keep quiet about it.

In summary, my ideas are quite simple. Perhaps the most important is that management's total loyalty to the maximization of profit is the principal obstacle to achieving higher standards of ethical practice. Defining the purpose of the corporation as exclusively economic is a deadly oversimplification, which allows overemphasis on self-interest at the expense of consideration of others.

The practice of management requires a prolonged play of judgment. Executives must find in their own will, experience, and intelligence the principles they apply in balancing conflicting claims. Wise men and women will submit their views to others, for open discussion of problems reveals unsuspected ethical dimensions and develops alternative viewpoints that should be taken into account. Ultimately, however, executives must make a decision, relying on their own judgment to settle infinitely debatable issues. Inquiry into character should therefore be part of all executive selection—as well as all executive development within the corporation.

And so it goes. That much and that little. The encouraging outcome is that promulgating and institutionalizing ethical policy are not so difficult as, for example, escaping the compulsion of greed. Once undertaken, the process can be as straightforward as the articulation and implementation of policy in any sphere. Any company has the opportunity to develop a unique corporate strategy summarizing its chief purposes and policies. That strategy can encompass not only the economic role it will play in national and international markets but also the kind of company it will be as a human organization. It will embrace as well, though perhaps not publicly, the nature and scope of the leadership to which the company is to be entrusted.

To be implemented successfully over time, any strategy must command the creativity, energy, and desire of the company's members. Strategic decisions that are economically or ethically unsound will not long sustain such commitment.

Journalist John Case writes for *INC.* magazine, a publication about fast-growing companies. In this article, he presents the results of an *INC.* survey in which readers responded to the ethical issues raised by a hypothetical, although common, business situation. The cases are followed by a number of real-life business conflicts experienced by survey respondents and the results of these conflicts. Think of how you would answer the questions raised by these cases.

CHAPTER TWENTY-FIVE

Honest Business

JOHN CASE

The discussion here at INC. that morning got a little heated—which is why we're inviting you to test your ethical judgments against those of other businesspeople all around the country.

The connection? Let me explain. Several months ago, eight of this magazine's editors gathered around the big oak conference table for a critical assessment of the previous month's issue. When the discussion turned to an article called "Supply-Side Financing" (February 1989)—well, that's when a few voices were raised.

"It's just wrong," said one editor. "Bad business and lousy ethics. How can we appear to be condoning it?"

"It" was the story of Dennis Chang, president and chief executive of Jasmine Technologies Inc., in San Francisco. Strapped for cash in his company's early days, Chang had paid his suppliers as little as he possibly could. But what really galled the protesting editor was Chang's pledge of personal guarantees well beyond his net worth. "The first guarantee devalues all subsequent guarantees," wrote author Ellyn E. Spragins, "but your suppliers don't know that."

The debate raged, the idealists among the editors charging Chang with unethical behavior, and the realists accusing the idealists of terminal naïveté. ("If nobody stretched their creditors, nobody would ever get a business started.") By May a chorus of letter writers had joined in. "What happened to the days when an individual's word meant something?" asked one. "You've provided an outline for larceny," charged another. Author Spragins was unapologetic: "Supplier finance is no more inherently unethical than any other business practice," she retorted. Chang himself argued in a later letter that he had built personal relationships with his vendors, getting credit "openly and honestly." Of course, there were risks—"but isn't

---

that the nature of being in business?"

Somehow I doubt we'll ever resolve this dispute to everyone's satisfaction. But it got us thinking: despite all the hoopla over business ethics, no one has paid much attention to the choices entrepreneurs must make every day. If you leave one company to start another, is it OK to take a co-worker with you? What about luring away a customer? When you're borrowing money, how much do you have to tell your lender? And—remembering Dennis Chang—what are your obligations to your suppliers?

As they say on late-night TV, wait! Don't answer yet.

Instead, take this little test.... A few months ago we sent this same questionnaire to a sampling of INC. readers. Several hundred responded, not only with checkmarks but with strong opinions and stories of similar situations they had lived through....

Ready? Here's the background:

You've been involved with computers all your working life. Five years ago Rob Firman asked you to join his new data-processing company; today you're a sales manager in Firman's 20-person shop. But you really want to start your own data-processing business, and for the past few months you've spent your evenings doing spreadsheets, developing customer lists, and otherwise mapping out the new venture. Firman knows nothing about your plans. You'll tell him as soon as you're ready to give notice.

In the meantime, you continue to put in a hard day's work. But now and then a difficult situation comes up....

The Key Employee

It's 5:30, and there's a knock on your office door. Quickly you rehearse your pitch one last time, even as you're calling for Lowrey to come in. The programmer enters and sits down, grinning.

"Did you hear?" he asks. "Seems the boss wants me to put on a tie and begin acting like an executive."

You congratulate him, but the doubt once more flickers across your mind. The start-up you're planning needs Lowrey. He's the quickest technical guy you've ever met. He has an incredible knack for making clients feel at ease. And more than once he's told you he'd like to be involved in a new venture someday.

If it weren't for Firman, you'd offer him a job right now.

Face it, you think, Firman put you where you are today. He not only hired you and trained you, he gave you more responsibility than you were ready for, confident that you'd grow into it. And you have— so quickly that now you're chafing under his sometimes-arrogant authority. When you leave, Firman may feel betrayed. But you steel yourself against the remorse. That's business. Firman is a big boy.

But what about Lowrey? Just last week Firman told you his hopes for the young man. He talked of broadening Lowrey's responsibilities, of eventually bringing him into management. Just the guy I need, he said, to help me take this company past $2 million.

You look again at Lowrey, sitting in your office, smiling to himself. You know he's essential to Firman's plans. But you also know where his long-term hopes lie, and that he'll take a job with your start-up if you offer one. You even think that Firman, if he weren't personally involved, would encourage you. "A man has to look out for himself," he likes to say.

What Do You Do?

- Offer Lowrey a job.
  He's the one who'll be making the choice, after all. Firman can take care of himself.

- Quit thinking about the start-up
  until you leave your job. You shouldn't be doing it on Firman's payroll.

- Bite your tongue.
  It's OK to keep planning the start-up, but don't mention it to Lowrey until you've left.

The Customer

Lisa Meggett sips her coffee, leaving her words hanging. You bite off some bread, trying to chew slowly and think fast.

Meggett is managing partner of a consulting firm, one of Firman's customers, and you've been handling the account. Today you've asked her to lunch, ostensibly to discuss her longer-term data-processing needs. In fact, you want a chance to talk up your new company. You figure you can offer her the same services she's getting from Firman, only for 30% less. And now she has left an opening you could drive a minicomputer through.

"I'm afraid I have bad news," she had said. "I hope you won't take it personally, because I've always enjoyed working with you. But computers are so cheap now, we're planning on bringing our data processing in-house. We should save 20% over what we're paying you. Unless you can match that I'll have to close our account."

Give me a break, you think—not only can I beat it, I can show her exactly where her figures are wrong. She hasn't counted the cost of hiring, training, and keeping busy the new people she'll have to bring in. She hasn't counted overhead. The sales pitch is a no-brainer, and Meggett's firm could be a big customer for your planned new company. But that's not what's troubling you.

You were just leaving for lunch today when Firman came in. He was edgy. "Listen," he told you, "don't take any bull from these people. I play golf with one of the partners over there, and he's been going on about buying this new computer. I bet you hear all about their big in-house plans and how much money they'll save.

"My attitude is, let 'em go. They'll find out it's not so easy, and they'll come crawling back to us. This isn't tiddledywinks we're playing here. If that's their plan, say so long and wish them well."

Now, back at the restaurant, Meggett looks up, awaiting your response. You can take Firman's approach even though (you can't help thinking) it amounts to shafting the customer. Or you can make your own pitch, offering Meggett the best deal she'll get and maybe landing your first account.

What Do You Do?

- Do as Firman says.
  You're on his payroll, after all.

- Make your pitch.
  Serving the customer better is what business is all about.

- Have an open-ended discussion.
  Make your pitch after you've left your job.

And now . . .
You've made the big move, leaving Firman and starting up a brand-new business. But you find that the dilemmas are still with you. . . .

The Banker

Driving back to the office—*your* office, you think proudly, with your five-month-old company finally moved in—you review the events of the past few hours.

What a meeting! The sales pitch you had so carefully prepared was scarcely necessary—it was as if the customer had long since decided to sign the contract. The amazing part was, he also wanted you to take on all his affiliated offices. Now, as you drive, the wheels are spinning in your mind. You'll need several new programmers and a couple of support people. You'll need new machines and more space.

The tricky part, you realize, will be handling Norton at the bank.

So far, Norton has been everything an entrepreneur could ask for in a banker. Conservative, sure—he's a banker, isn't he?—but nevertheless helpful. He arranged a modest line of credit secured by some of your hardware, and you are already into him for $25,000 or so. What's keeping him happy, of course, is your hefty cash balance and your prompt payment record. This move will change all that. You'll have to drain your cash, hurrying up your receivables and stretching your payables from now until summer if possible. You'll have to utilize Norton's line to the fullest.

Braking at a light, you realize your excitement is tinged with frustration. If you're straight with Norton—if you tell him everything you're planning—he'll tell you you can't handle that much growth, that you're crazy to be taking on so much new work so soon. He'll tell you that your debt will be too high and your cash too low. He might even say he's going to have to cut back your credit line—exactly what you can't afford.

On the other hand, you think: it's not his company. His loan is secured by the equipment. And there's no reason you have to tell him everything you're doing; you won't even be audited for another seven months, and by then you should be making money. You are, after all, in business to grow—and you've got one big opportunity to do just that.

What Do You Do?

- Come clean to Norton.
  Try to convince him you can handle the growth. If you can't, don't accept the new job.

- Keep quiet.
  Tell Norton as little as possible. Take the new job, manage the cash carefully.

- Compromise.
  Tell Norton little but admit to yourself he has a point. Take the new job but look hard for more investment capital.

The Supplier

That's him now, you think, as you look out your office window and see Dick Wilson getting out of his car. You know Wilson a little; you've even played golf with him a couple of times. So it was natural to call him. In any event, he represents the best office-furnishings company in town, and you suddenly have several new offices to outfit. He enters your office, sticking out a meaty hand.

"Thanks for coming out, Dick," you say. "Here—let's take a little stroll and you can see the space we're talking about."

Walking through the vacant rooms, you know you've got him where you want him. You did business with Wilson's company at the beginning, paying them promptly with your start-up capital. When they run their credit check they'll learn you have plenty of cash left and a good payment record. No reason for them not to cooperate on this one—ship the furniture and equipment out, bill you 30 days, not even pick up the phone until 45 or 60 days.

Granted, they don't know what you're planning. They don't know that the rooms Wilson's measuring will soon be occupied by several expensive programmers and several expensive machines. Or that the cash drain on your company will be severe. In the end, you calculate, it'll be roughly six months before Wilson's company gets all its money. You'll send them just enough to show your good faith—and to keep the moving trucks away from the door.

Now, talking with Wilson back in your office, you consider telling him the whole situation. Would it scotch the deal? Maybe, and then you'd be left with a bunch of unfurnished offices and the word out that you were short of cash. On the other hand, who's hurt if you say nothing? Wilson's company will get its money eventually, and you'll be ready for the new business you're taking on. Even in the worst situation, all they need to do is come get their furniture.

What Do You Do?

- Tell Wilson
  all your plans. If he backs out, find another supplier.

- Keep mum.
  Credit is his company's problem, not yours.

- Postpone.
  Don't contract for more furnishings until you can afford them, even if that means postponing the planned expansion.

The Key Employee

What the Respondents Said

12% Offer Lowrey a job.
 6% Quit thinking about the start-up.
82% Bite your tongue.

In Their Own Words . . .

- "Make the offer! Employers must compete for employees as well as for customers."

- "Making a low-key offer to Lowrey isn't 'stealing' him. But let Firman know your plans."

- "I offered the job while I was still employed. It was a mistake. The person didn't go with me, and it could have ended up in a costly lawsuit."

- "Not offering the job is unfair to Lowrey. But be honest with Firman. He was once involved in a start-up—remind him of that."

- "Until your plans are definite you're risking your career—and Lowrey's. It's not a done deal until you're out."

- "I encounter this as 'Firman.' I encourage workers with entrepreneurial leanings to discuss things with me. Maybe we can help them get started so both of us benefit."

- "Quit thinking about it! Give your best to one thing at a time."

- "I've been on Firman's side of the issue and didn't like what my colleague did."

- "Congratulate Lowrey, but hint that someday you want to have your own company. Don't say when."

In Real Life

On the spot: Andy Friesch, founder, Heartland Adhesives & Coatings Inc., Germantown, Wis.

Friesch was a top sales producer for a $100-million-plus company in the industrial adhesives industry. But he felt that his opportunities were limited there, the company was "underutilizing what I was best at," and anyway he had dreams of going out on his own. Two of his co-workers, one in marketing and one in the technical end, seemed like prime candidates to help him get started.

What he decided: In Friesch's mind, there was no question, "I asked them to join me. Some people think I'm a rebel on this, but so long as you don't violate a noncompete agreement you have to look out for yourself first. I said, look, I'm not twisting your arm; if you want to stay here I respect that. But I have a vision and I want to follow it.

"I don't want to give the imipression I'm not a loyal person. But when it comes to corporate America, individuals have to do what's in their own best interest. Corporations look out for themselves first and employees second."

The upshot: No one came. "They were willing to talk about going off with me, but when it came right down to it they didn't want to make the sacrifice." Were there hard feelings when he left? "Not from my peers. My superiors—I don't think they like what I've done, or what it says about their company."

The Customer

What the Respondents Said

12% Do as Firman says.
22% Make your own pitch.
66% Have an open-ended discussion.

In Their Own Words . . .

- "Do as Firman says. But set up an unofficial meeting and make your pitch."

- "Tell Meggett here's what Firman says—but then tell her you're starting up and can save her both headaches and money."

- "Make your pitch. Firman has lost the concept of customer service."

- "If Firman won't serve the customer, I certainly will."

- "When I realized my boss really didn't care, I went back to the customer and made my pitch without the boss's knowledge. At that point it was none of his business."

- "Don't try it. One of our salesmen left and took some of our accounts. We sued him and won."

- "If you're still taking Firman's money you have to do your best for him."

- "While on the payroll you must represent the firm or tell the boss why not."

- "Regardless of Firman's directions, he'll really be upset if *you* lose the account."

- "In a similar situation, I quit almost immediately and landed the account."

In Real Life

On the spot: John G. McCurdy, founder of Sunny States Seafood, Oxford, Miss.

Working for another Mississippi seafood distributor, McCurdy was furious with his boss. On paper they were partners, though McCurdy's 8% interest paled beside the other man's 92%. But it seemed to McCurdy that he was doing all the work while the other fellow took home the money—and ordered him around besides. So McCurdy decided to go out on his own. Hearing of his decision, several customers asked to go with him; some even called him up after he had established his own company, asking to do business.

What he decided: "I wanted to do anything I could to close that company and put him on welfare. But taking accounts just wasn't the right thing to do. I'm only 24 years old, I've got another 40 years left in business, and if I do somebody like that, somebody's going to do me like that. I don't live by the Golden Rule all the time, but that time I did."

The upshot: McCurdy is succeeding, and has recently opened up a franchised store in Tupelo. His former employer: pfffft. "He got into debt and had to sell out to a big catfish company."

The Banker

What the Respondents Said

44% Come clean to Norton.
40% Keep quiet, tell Norton little.
26% Compromise, say little but look for investment.

In Their Own Words . . .

- "Banks don't like surprises. Norton's an entrepreneur's dream—don't blow it!"

- "Convince Norton you can handle it. If he disagrees, take it to another bank for evaluation."

- "I'd keep quiet. Growth is my business, banking my banker's."

- "Bankers haven't the foggiest idea how to run a company. Tell them only what they require."

- "Without my bank I'm out of business. So I don't do anything to test the relationship."

- "I'd ask Norton to come to lunch and show him this wonder-ful opportunity, but tell him when I run the numbers they look thin. Get him excited, give him the facts, ask for his assistance."

- "Could you get the job and phase in the affiliates on a dated schedule?"

- "Work with the new client to get some up-front money."

- "I spent 10 years as a banker on Wall Street. No surprises to your banker!"

- "Bankers are never very knowledgeable or competent."

In Real Life

On the spot: W. Mark Baty Jr., Accredited Business Services, Cleveland

Some years ago, Baty was running his own metals recycling busi-ness and was approached by a big customer. To land the account he needed a load lugger—a specialized truck with a hydraulic lift on the back—costing some $60,000. Approaching the bank for a loan, he knew his overall financial situation was a little shaky: he was in debt to his parents and in-laws, stretched out on his credit cards, and finagling his receivables and payables. "I'd try to collect my receiv-ables in 10 days if I could, and I'd put off my payables for 30 or 60 days, whatever the traffic would bear." But he also knew he could get the account if he had the equipment.

What he decided: "You don't lie to a bank—but I certainly don't believe in offering them more information than they ask for. You don't tell them the loans your parents made to you, the money you've used from credit cards for your business. Playing with your accounts receivable and payable, you definitely don't tell your banker that. You're doing a little bit of creative financing, you're using everything you can to keep your business going."

The upshot: He got the loan, paid it off, and the banker was never the wiser. "You're not lying to anybody, and you're not falsifying records. It's something that's done every day in business."

The Supplier

What the Respondents Said

54% Tell Wilson.
11% Keep mum.
24% Postpone.

In Their Own Words . . .

- "Tell him. Then set a payment schedule agreeable to all."

- "Tell him. You can always find eager vendors willing to grant terms. It isn't worth your ill health fending off collectors and repo men."

- "Keep quiet. Once you start to do somethng, it's ridiculous to vacillate."

- "We kept quiet. We kept one strong credit reference and finessed a number of purchases off that reference."

- "I got several bids in a similar situation, making it clear that a delayed payment schedule was an important decision-making criterion."

- "Postpone. Too many start-ups overextend their expenses."

- "Buy used furniture. If you have to have all the best trappings at this point, you probably should have stayed where you were."

- "I stretched out payments once, and I regret it. Be up-front and negotiate terms, then stay within them."

- "I bought door panels and bridged boxes to make temporary desks."

- "We got the equipment and later went into Chapter 7. Because I guaranteed the payment, the equipment company got everything they were owed."

In Real Life

On the spot: Kurt Listug, cofounder of Taylor Guitars, Santee, Calif.

Listug was only 21 when he and his partner founded their guitar manufacturing company; the partner was all of 19. "We had no idea what we were getting ourselves into." But they plowed ahead, ordering supplies and materials even though they weren't quite sure how they would pay the vendors. Among the orders: $1,200 worth of rosewood, purchased from C. F. Martin & Co., a venerable and considerably larger guitar manufacturer.

The $1,200 didn't get paid. And didn't get paid. "I remember being so broke," recalls Listug. "We only owed them $1,200—how could we not have it? But we didn't, for a long time." Martin finally gave up, sending the bill to a collection agency. But Listug's company still couldn't pay it, and the creditor too threw in the towel.

What he decided: "Years later, after they had written it off, we started to catch up. We made a list of our bills and added that one to it. Finally we had the money and we paid them—out of the blue. They were shocked by the whole thing."

The upshot: Listug's company now has 35 employees—and a deal with a supplier who gets his wood in India. "We're not buying from Martin now. But we've bought rosewood pretty regularly from them over the years."

IV. ENVIRONMENTAL ETHICS

Environmental ethics pose the single most difficult problem for forest managers. Our desires to improve the forest, secure its future productivity, and husband all of its values and resources are often challenged by the stark realities of costs, political forces, or an absence of authority to do what is best for the land. Indeed, in many instances there is serious scientific uncertainty about what actions are best for the land.

As a number of authors have noted, most recently Roderick Nash, environmental ethics represents an extension of ethics to broader realms beyond individual people. The concept of human responsibilities toward nature has raised religious and ethical discussion in all ages, but never more so than today, as our ability to affect the very climate of the planet is becoming clear to all.

The conflicts in land management have been deepened by our growing knowledge of the value of unmanaged ancient forests, whether they stand in the Amazon, Sarawak, Southwest Oregon, Coastal Alaska, or in the tiny patches left in Wisconsin or the Southern Appalachians. These conflicts, between those who want to preserve the resources as they are and those who want to utilize them, often include litigation, high-powered lobbying, and intense publicity campaigns that draw wide media attention. Forest managers are finding themselves caught in the middle, forced to set priorities for resources that are already overstretched. In extreme cases, the survival of human communities is pitted against the survival of plant and animal species, putting forest land managers in a powerful but very difficult position.

The forestry profession in its early years was a powerful, prophetic voice for conservation, for wise use, and for democratic values in resource ownership, control, and use. As the profession has matured, it has turned more toward the many specialized technical and administrative tasks involved in implementing the vision of forest improvement for the future. There is nothing wrong with this. Prophets don't always make good administrators, and the creation of a large and generally effective community of managers, scientists, and technicians who care for the nation's public and private forests is a significant social accomplishment.

In recent decades, however, the traditional approach to wise land use has been challenged by the growing public demand for an environmental agenda in land management policy. At the same time, much of the public has lost interest in the issues surrounding the supply of raw materials and in the traditional economic concerns of rural areas everywhere. These conflicts and strains have created new personal and ethical tensions in the forestry profession.

The role of an environmental ethic in forestry has evolved considerably since Aldo Leopold's groundbreaking 1933 *Journal of Forestry* article. At this writing, a Land Ethic Canon has just been added to the Society of American Forester's Code of Ethics. The issues surrounding this proposed canon are reviewed in James Coufal's essay reprinted here.

Environmental ethics is about stewardship. In our differing roles as managers, planners, scientists, landowners, government advisers, and decisionmakers, we all face the task of integrating a land ethic into our ways of thinking about choices for the land and for the people who depend on it. As is true of professional and business ethics, there is a gulf between the general environmental ethical mandates and their application in daily life. This gulf must be filled by reflective and serious dialogue and by the prudent judgments of foresters and their associates, supervisors, and clients. This group of readings can help by providing ideas for discussion and encouraging creative solutions to difficult environmental choices.

Additional Reading

Abbey, Edward. *Desert Solitaire: A Season in the Wilderness.* New York: Simon and Schuster, 1972. A classic of environmental and nature writing; see especially the sections Polemic: industrial tourism and the national parks and Episodes and visions.

Albanese, Catharine L. *Nature Religion in America: From the Algonkian Indians to the New Age.* Chicago: University of Chicago Press, 1990.

Craig, R. Land ethic canon proposal: a report from the Task Force. *Journal of Forestry* 90(1992):40–41.

Errington, Paul. *A Question of Values.* Ames: Iowa State University Press, 1987. Excellent, thoughtful essays by a pioneering wildlife ecologist. See especially the essays, A question of values, and An appreciation of Aldo Leopold.

Fayter, Paul. Senses of the natural world: recent works in the philosophy and history of science. *Forest History* 34(1990):85–91. A review of 7 books about environmental ethics.

Finley, James C.; and Jones, Stephen B., Eds. Practicing Stewardship and Living a Land Ethic. Proceedings, 1991 Conference on Pennsylvania State Forest Resources Issues. College Station, PA: Pennsylvania State University School of Forest Resources, 1991.

Lansky, M. *Beyond the Beauty Strip: Saving What's Left of our Forests.* Gardiner, Maine: Tilbury House Publishers, 1992. A severe critique of industrial forest management and policies in the Northeast. An example of recent challenges to current forest practices.

Leopold, Aldo. *A Sand County Almanac.* New York: Oxford University Press, 1966. Also available in paperback.

Nash, James A. *Loving Nature: Ecological Integrity and Christian Responsibility.* Nashville, TN: Abingdon Press, 1991. An excellent and readable statement of a religious grounding for environmental ethics. See especially ch. 3, The ecological complaint against Christianity, and ch. 8, Political directions for ecological integrity.

Nash, Roderick Frazier. *The Rights of Nature: A History of Environmental Ethics.* Madison: University of Wisconsin Press, 1989. A wide-ranging and readable history, destined to be a classic.

Passmore, John. *Man's Responsibility for Nature.* New York: Charles Scribner's Sons, 1974.

Rolston, Holmes, III. *Environmental Ethics.* Philadelphia: Temple University Press, 1989.

_____ . *Philosophy Gone Wild: Essays in Environmental Ethics.* Buffalo, NY: Prometheus Books, 1986. See especially ch. 1, Is there an environmental ethic?; ch. 8, Just environmental business; and ch. 9, Valuing wildlands.

Rolston, Holmes, III; and Coufal, James. A forest ethic and multivalue forest management. *Journal of Forestry* 89(1991):35–40.

Sale, Kirkpatrick. *The Conquest of Paradise: Christopher Columbus and the Columbian Legacy.* New York: Plume, 1990. An informative account of Spanish and English first encounter with the New World and its people, from a Green perspective.

Smith, D. I. Forest management and the theology of nature. *American Forests* 99(1992):13–16. Comments on the issues in the setting of current debates over forest management in Idaho.

Stegner, Wallace. *The American West as Living Space.* Ann Arbor: University of Michigan Press, 1988.

Tanner, Thomas, ed. *Aldo Leopold: The Man and His Legacy.* Ankeny, Iowa: Soil Science Society of America, 1987.

Worster, Donald. *Under Western Skies: Nature and History in the American West.* New York: Oxford University Press, 1992.

You can find the famous Lynn White essay Fritsch refers to in "The historical roots of our ecologic crisis." *Science* 155(1967):1203–1207. See also the extensive discussion of White's influence in Nash, *The Rights of Nature.*

The *Journal of Forestry* (May 1991) carried many letters and articles commenting on the proposed Land Ethic Canon, which has been adopted. Read these articles and letters and then see what you think of the canon, now shown as Canon 1, reprinted in the first section of this book.

Conservation became a new force in both federal and state governments in the early 20th century as resource experts and politicians struggled to find ways to stabilize rural economies, hold soils in place, and put the unemployed to work. Aldo Leopold was an early leader in game management and was far ahead of his time in the field of conservation. His essay, "The Conservation Ethic" was first published in the *Journal of Forestry*. Leopold's insights into nature, his extraordinary eloquence, and his understanding of society have made his works conservation classics of the first order. This essay is better known as it appeared in *A Sand County Almanac* in 1949, published after Leopold's untimely death. It is printed here in the form in which it appeared in the *Journal of Forestry* in 1933 (reprinted June 1989) following its presentation to the Southwest Division of the American Association for the Advancement of Science.

CHAPTER TWENTY-SIX

The Conservation Ethic

ALDO LEOPOLD

When god-like Odysseus returned from the wars in Troy, he hanged all on one rope some dozen slave-girls of his household whom he suspected of misbehavior during his absence.

This hanging involved no question or propriety, much less of justice. The girls were property. The disposal of property was then, as now, a matter of expediency, not of right and wrong.

Criteria of right and wrong were not lacking from Odysseus' Greece: witness the fidelity of his wife through the long years before at last his black-prowed galleys clove the wine-dark seas for home. The ethical structure of that day covered wives, but had not yet been extended to human chattels. During the three thousand years which have since elapsed, ethical criteria have been extended to many fields of conduct, with corresponding shrinkages in those judged by expediency only.

This extension of ethics, so far studied only by philosophers, is actually a process in ecological evolution. Its sequences may be described in biological as well as philosophical terms. An ethic, biologically, is a limitation on freedom of action in the struggle for existence. An ethic, philosophically, is a differentiation of social from anti-social conduct. These are two definitions of one thing. The thing has its origin in the tendency of interdependent individuals or societies to evolve modes of coöperation. The biologist calls these symbioses. Man elaborated certain advanced symbioses called politics and economics. Like their simpler biological antecedents, they enable individuals or groups to exploit each other in an orderly way. Their first yardstick was expediency.

The late Dr. Leopold was a longtime professor of wildlife management at the University of Wisconsin. This article was presented to the Southwest Division of the American Association for the Advancement of Science in Las Cruces, NM, on 1 May 1933. Reprinted from *Journal of Forestry,* June 1989, pp. 26–28, 41–45, by permission of the Society of American Foresters.

The complexity of coöperative mechanisms increased with population density, and with the efficiency of tools. It was simpler, for example, to define the anti-social uses of sticks and stones in the days of the mastodons than of bullets and billboards in the age of motors.

At a certain stage of complexity, the human community found expediency-yardsticks no longer sufficient. One by one it has evolved and superimposed upon them a set of ethical yardsticks. The first ethics dealt with the relationship between individuals. The Mosaic Decalogue is an example. Later accretions dealt with the relationship between the individual and society. Christianity tries to integrate the individual to society, Democracy to integrate social organizations to the individual.

There is as yet no ethic dealing with man's relationship to land and to the non-human animals and plants which grow upon it. Land, like Odysseus' slave-girls, is still property. The land-relation is still strictly economic, entailing privileges but not obligations.

The extension of ethics to this third element in human environment is, if we read evolution correctly, an ecological possibility. It is the third step in a sequence. The first two have already been taken. Civilized man exhibits in his own mind evidence that the third is needed. For example, his sense of right and wrong may be aroused quite as strongly by the desecration of a nearby woodlot as by a famine in China, a near-pogrom in Germany, or the murder of the slave-girls in ancient Greece. Individual thinkers since the days of Ezekiel and Isaiah have asserted that the despoliation of land is not only inexpedient but wrong. Society, however, has not yet affirmed their belief. I regard the present conservation movement as the embryo of such an affirmation. I here discuss why this is, or should be, so.

Some scientists will dismiss this matter forthwith, on the ground that ecology has no relation to right and wrong. To such I reply that science, if not philosophy, should by now have made us cautious about dismissals. An ethic may be regarded as a mode of guidance for meeting ecological situations so new or intricate, or involving such deferred reactions, that the path of social expediency is not discernible to the average individual. Animal instincts are just this. Ethics are possibly a kind of advanced social instinct in-the-making.

Whatever the merits of this analogy, no ecologist can deny that our land-relation involves penalties and rewards which the individual does not see, and needs modes of guidance which do not yet exist. Call these what you will, science cannot escape its part in forming them.

Ecology—Its Role in History

A harmonious relation to land is more intricate, and of more consequence to civilization, than the historians of its progress seem to realize. Civilization is not, as they often assume, the enslavement of a stable and constant earth. It is a state of *mutual and interdependent coöperation* between human animals, other animals, plants, and soils, which may be disrupted at any moment by the failure of any of them. Land-despoliation has evicted nations, and can on occasion do it again. As long as six virgin continents awaited the plow, this was perhaps no tragic matter—eviction from one piece of soil could be recouped by despoiling another. But there are now wars and rumors of wars which foretell the impending saturation of the earth's best soils and climates. It thus becomes a matter of some importance, at least to ourselves, that our dominion, once gained, be self-perpetuating rather than self-destructive.

This instability of our land-relation calls for example. I will sketch a single aspect of it: the plant succession as a factor in history.

In the years following the Revolution, three groups were contending for control of the Mississippi valley: the native Indians, the French and English traders, and American settlers. Historians wonder what would have happened if the English at Detroit had thrown a little more weight into the Indian side of those tipsy scales which decided the outcome of the Colonial migration into the cane-lands of Kentucky. Yet who ever wondered why the cane-lands, when subjected to the particular mixture of forces represented by the cow, plow, fire, and axe of the pioneer, became bluegrass? What if the plant succession inherent in this "dark and bloody ground" had, under the impact of these forces, given us some worthless sedge, shrub, or weed? Would Boone and Kenton have held out? Would there have been any overflow into Ohio? Any Louisiana Purchase? Any transcontinental union of new states? Any Civil War? Any machine age? Any depression? The subsequent drama of American history, here and elsewhere, hung in large degree on the reaction of particular soils to the impact of particular forces exerted by a particular kind and degree of human occupation. No statesman-biologist selected those forces, nor foresaw their effects. That chain of events which in the Fourth of July we call our National Destiny hung on a "fortuitous concourse of elements," the interplay of which we now dimly decipher *by hindsight only.*

Contrast Kentucky with what hindsight tells us about the Southwest. The impact of occupancy here brought no bluegrass, nor other plant fitted to withstand the bumps and buffetings of misuse. Most of these soils, when grazed, reverted through a successive series of more and more worthless grasses, shrubs, and weeds to a condition of unstable equilibrium. Each recession of plant types bred erosion; each increment to erosion bred a further recession of plants. The result today is a progressive and mutual deterioration, not only of plants and soils, but of the animal community subsisting thereon. The early settlers did not expect this, on the cienegas of central New Mexico some even cut artificial gullies to hasten it. So subtle has been its progress that few people know anything about it. It is not discussed at polite tea-tables or go-getting luncheon clubs, but only in the arid halls of science.

All civilizations seem to have been conditioned upon whether the plant succession, under the impact of occupancy, gave a stable and habitable assortment of vegetative types, or an unstable and uninhabitable assortment. The swampy forests of Caesar's Gaul were utterly changed by human use—for the better. Moses' land of milk and honey was utterly changed—for the worse. Both changes are the unpremeditated resultant of the impact between ecological and economic forces. We now decipher these reactions retrospectively. What could possibly be more important than to foresee and control them?

We of the machine age admire ourselves for our mechanical ingenuity; we harness cars to the solar energy impounded in carboniferous forests; we fly in mechanical birds; we make the ether carry our words or even our pictures. But are these not in one sense mere parlor tricks compared with our utter ineptitude in keeping land fit to live upon? Our engineering has attained the pearly gates of a near-millennium, but our applied biology still lives in nomad's tents of the stone age. If our system of land-use happens to be self-perpetuating, we stay. If it happens to be self-destructive we move, like Abraham, to pastures anew.

Do I overdraw this paradox? I think not. Consider the transcontinental airmail which plies the skyways of the Southwest—a symbol of its final conquest. What does it see? A score of mountain valleys which were green gems of fertility when first described by Coronado, Espejo, Pattie, Abert, Sitgreaves, and Couzens. What are they now? Sandbars, wastes of cobbles and burroweed, a path for torrents. Rivers, which Pattie says were clear, now muddy sewers for the wasting fertility of an empire. A "Public Domain," once a velvet carpet of rich buffalo-grass and grama, now an illimitable waste of rattlesnake-bush and tumbleweed, too impoverished to be accepted as a gift by the states within which it lies. Why? Because the ecology of this Southwest happened to be set on a hair-trigger. Because cows eat brush when the grass is gone, and thus postpone the penalties of overutilization. Because certain grasses, when grazed too closely to bear seed-stalks, are weakened and give way to inferior grasses, and these to inferior shrubs, and these to weeds, and these to naked earth. Because rain which spatters upon vegetated soils stays clear and sinks, while rain which splatters upon devegetated soil seals its interstices with colloidal mud and hence must run away as floods, cutting the heart out of country as it goes. Are these phenomena any more difficult to foresee than the paths of stars which science deciphers without the error of a single second? Which is the more important to the permanence and welfare of civilization?

I do not here berate the astronomer for his precocity, but rather the ecologist for his lack of it. The days of his cloistered sequestration are over:

> Whether you will or not,
> You are a king, Tristram, for you are one
> Of the time-tested few that leave the world,
> When they are gone, not the same place it was.
> Mark what you leave.

Unforeseen ecological reactions not only make or break history in a few exceptional enterprises—they condition, circumscribe, delimit, and warp all enterprises, both economic and cultural, that pertain to land. In the cornbelt, after grazing and plowing out all the cover in the interests of "clean farming," we grew tearful about wild-life, and spent several decades passing laws for its restoration. We were like Canute commanding the tide. Only recently has research made it clear that the implements for restoration lie not in the legislature, but in the farmer's toolshed. Barbed wire and brains are doing what laws alone failed to do.

In other instances we take credit for shaking down apples which were, in all probability, ecological windfalls. In the Lake States and the Northeast lumbering, pulping, and fire accidentally created some scores of millions of acres of new second-growth. At the proper stage we find these thickets full of deer. For this we naïvely thank the wisdom of our game laws.

In short, the reaction of land to occupancy determines the nature and duration of civilization. In arid climates the land may be destroyed. In all climates the plant succession determines what economic activities can be supported. Their nature and intensity in turn determine not only the domestic but also the wild plant and animal life, the scenery, and the whole face of nature. We inherit the earth, but within the limits of the soil and plant succession we also *rebuild* the earth—without plan, without knowledge of its properties, and without understanding of the increasingly coarse and powerful tools which science has placed at our disposal. We are remodelling the Alhambra with a steam-shovel.

Ecology and Economics

The conservation movement is, at the very least, an assertion that these interactions between man and land are too important to be left to chance, even that sacred variety of chance known as economic law.

We have three possible controls: legislation, self-interest, and ethics. Before we can know where and how they will work, we must first understand the reactions. Such understanding arises only from research. At the present moment research, inadequate as it is, has nevertheless piled up a large store of facts which our land using industries are unwilling, or (they claim) unable to apply. Why? A review of three sample fields will be attempted.

Soil science has so far relied on self-interest as the motive for conservation. The landholder is told that it pays to conserve his soil and its fertility. On good farms this economic formula has improved land-practice, but on poorer soils vast abuses still proceed unchecked. Public acquisition of submarginal soils is being urged as a remedy for their misuse. It has been applied to some extent, but it often comes too late to check erosion, and can hardly hope more than to ameliorate a phenomenon involving in some degree *every square foot* on the continent. Legislative compulsion might work on the best soils where it is least needed, but it seems hopeless on poor soils where the existing economic set-up hardly permits even uncontrolled private enterprise to make a profit. We must face the fact that, by and large, no defensible relationship between man and the soil of his nativity is as yet in sight.

Forestry exhibits another tragedy—or comedy—of *Homo sapiens*, astride the runaway Juggernaut of his own building, trying to be decent to his environment. A new profession was trained in the confident expectation that the shrinkage in virgin timber would, as a matter of self-interest, bring an expansion of timber-cropping. Foresters are cropping timber on certain parcels of poor land which happen to be public, but on the great bulk of private holdings they have accomplished little. Economics won't let them. Why? He would be bold indeed who claimed to know the whole answer, but these parts of it seem agreed upon: modern transport prevents profitable tree-cropping in cut-out regions until virgin stands in all others are first exhausted; substitutes for lumber have undermined confidence in the future need for it; carrying charges on stumpage reserves are so high as to force perennial liquidation, overproduction, depressed prices, and an appalling wastage of unmarketable grades which must be cut to get the higher grades; the mind of the forest owner lacks the point-of-view under-lying sustained yield; the low wage-standards on which European forestry rests do not obtain in America.

A few tentative gropings toward industrial forestry were visible before 1929 but these have been mostly swept away by the depression, with the net result that forty years of "campaigning" have left us only such actual tree-cropping as is under-written by public treasuries. Only a blind man could see in this the beginnings of an orderly and harmonious use of the forest resource.

There are those who would remedy this failure by legislative compulsion of private owners. Can a landholder be successfully compelled to raise any crop, let alone a complex long-time crop like a forest, on land the private possession of which is, for the moment at least, a liability? Compulsion would merely hasten that avalanche of tax-delinquent land-titles now being dumped into the public lap.

Another and larger group seeks a remedy in more public ownership. Doubtless we need it—we are getting it whether we need it or not—but how far can it go? We cannot dodge the fact that the forest problem, like the soil problem, *is coextensive with the map of the United States.* How far can we tax other lands and industries to maintain forest lands and industries artificially? How confidently can we set out to run a hundred-yard dash with a twenty foot rope tying our ankle to the starting point? Well, we are bravely "getting set," anyhow.

The trend in wild-life conservation is possibly more encouraging than in either soils or forests. It has suddenly become apparent that farmers, out of self-interest, can be induced to crop game. Game crops are in demand, staple crops are not. For farm-species, therefore, the immediate future is relatively bright. Forest game has profited to some extent by the accidental establishment of new habitat following the decline of forest industries. Migratory game, on the other hand, has lost heavily through drainage and over-shooting; its future is black because motives of self-interest do not apply to the private cropping of birds so mobile that they "belong" to everybody, and hence to nobody. Only governments have interests coextensive with their annual movements, and the divided counsels of conservationists give governments ample alibi for doing little. Governments could crop migratory birds because their marshy habitat is cheap and concentrated, but we get only an annual crop of new hearings on how to divide the fast-dwindling remnant.

These three fields of conservation, while but fractions of the whole, suffice to illustrate the welter of conflicting forces, facts, and opinions which so far comprise the result of the effort to harmonize our machine civilization with the land whence comes its sustenance. We have accomplished little, but we should have learned much. What?

I can see clearly only two things:

First, that the economic cards are stacked against some of the most important reforms in land-use.

Second, that the scheme to circumvent this obstacle by public ownership, while highly desirable and good as far as it goes, can never go far enough. Many will take issue on this, but the issue is between two conflicting conceptions of the end towards which we are working.

One regards conservation as a kind of sacrificial offering, made for us vicariously by bureaus, on lands nobody wants for other purposes, in propitiation for the atrocities which will prevail everywhere else. We have made a real start on this kind of conservation, and we can carry it as far as the tax-string on our leg will reach. Obviously, though, it conserves our self-respect better than our land. Many excellent people accept it, either because they despair of anything better, or because they fail to see the *universality of the reactions needing control.* That is to say their ecological education is not yet sufficient.

The other concept supports the public program, but regards it as merely extension, teaching, demonstration, an initial nucleus, a means to an end, but not the end itself. The real end is a *universal symbiosis with land,* economic and esthetic, public and private. To this school of thought public ownership is a patch but not a program.

Are we, then, limited to patchwork until such time as Mr. Babbitt has taken his Ph.D. in ecology and esthetics? Or do the new economic formulae offer a short-cut to harmony with our environment?

The Economic Isms

As nearly as I can see, all the new isms—Socialism, Communism, Fascism, and especially the late but not lamented Technocracy—outdo even Capitalism itself in their preoccupation with one thing: the distribution of more machine-made commodities to more people. They all proceed on the theory that if we can all keep warm and full, and all own a Ford and a radio, the good life will follow. Their programs differ only in ways to mobilize machines to this end. Though they despise each other, they are all, in respect of this objective, as identically alike as peas in a pod. They are competitive apostles of a single creed: *salvation by machinery.*

We are here concerned, not with their proposals for adjusting men and machinery to goods, but rather with their lack of any vital proposal for adjusting men and machines to land. To conservationists they offer only the old familiar palliatives: public ownership and private compulsion. If these are insufficient now, by what magic are they to become sufficient after we change our collective label?

Let us apply economic reasoning to a sample problem and see where it takes us. As already pointed out, there is a huge area which the economist calls submarginal, because it has a minus value for exploitation. In its once-virgin condition, however, it could be "skinned" at a profit. It has been, and as a result erosion is washing it away. What shall we do about it?

By all the accepted tenets of current economics and science we ought to say "let her wash." Why? Because staple land-crops are over-produced, our population curve is flattening out, science is still raising the yields from better lands, we are spending millions from the public treasury to retire unneeded acreage, and here is nature offering to do the same thing free of charge; why not let her do it? This, I say, is economic reasoning. *Yet no man has so spoken.* I cannot help reading a meaning into this fact. To me it means that the average citizen shares in some degree the intuitive and instantaneous contempt with which the conservationist would regard such an attitude. We can, it seems, stomach the burning or plowing-under of over-produced cotton, coffee, or corn, but the destruction of mother-earth, however "submarginal," touches something deeper, some sub-economic stratum of the human intelligence wherein lies that something—perhaps the essence of civilization—which Wilson called "the decent opinion of mankind."

The Conservation Movement

We are confronted, then, by a contradiction. To build a better motor we tap the uttermost powers of the human brain; to build a better country-side we throw dice. Political systems take no cognizance of this disparity, offer no sufficient remedy. There is however, a dormant but widespread consciousness that the destruction of land, and of the living things upon it, is wrong. A new minority have espoused an idea called conservation which tends to assert this as a positive principle. Does it contain seeds which are likely to grow?

Its own devotees, I confess, often give apparent grounds for skepticism. We have, as an extreme example, the cult of the barbless hook, which acquires self-esteem by a self-imposed limitation of armaments in catching fish. The limitation is commendable, but the illusion that it has something to do with salvation is as naïve as some of the primitive taboos and mortifications which still adhere to religious sects. Such excrescenses seem to indicate the whereabouts of a moral problem, however irrelevant they may be in either defining or solving it.

Then there is the conservation-booster, who of late has been rewriting the conservation ticket in terms of "tourist-bait." He exhorts us to "conserve outdoor Wisconsin" because if we don't the motorist-on-vacation will streak through to Michigan, leaving us only a cloud of dust. Is Mr. Babbitt trumping up hard-boiled reasons to serve as a screen for doing what he thinks is right? His tenacity suggests that he is after something more than tourists. Have he and other thousands of "conservation workers" labored through all these barren decades fired by a dream of augmenting the sale of sandwiches and gasoline? I think not. Some of these people have hitched their wagon to a star— and that is something.

Any wagon so hitched offers the discerning politician a quick ride to glory. His agility in hopping up and seizing the reins adds little dignity to the cause, but it does add the testimony of his political nose to an important question: is this conservation something people really want? The political objective, to be sure, is often some trivial tinkering with the laws, some useless appropriation, or some pasting of pretty labels on ugly realities. How often, though, does any political action portray the real depth of the idea behind it? For political consumption a new thought must always be reduced to a posture or a phrase. It has happened before that great ideas were heralded by growing-pains in the body politic, semi-comic to those onlookers not yet infected by them. The insignificance of what we conservationists, in our political capacity, say and do, does not detract from the significance of our persistent desire to do something. To turn this desire into productive channels is the task of time, and ecology.

The recent trend in wildlife conservation shows the direction in which ideas are evolving. At the inception of the movement fifty years ago, its underlying thesis was to save species from extermination. The means to this end were a series of restrictive enactments. The duty of the individual was to cherish and extend these enactments, and to see that his neighbor obeyed them. The whole structure was negative and prohibitory. It assumed land to be a constant in the ecological equation. Gun-powder and blood-lust were the variables needing control.

There is now being superimposed on this a positive and affirmatory ideology, the thesis of which is to prevent the deterioration of environment. The means to this end is research. The duty of the individual is to apply its findings to land, and to encourage his neighbor to do likewise. The soil and the plant succession are recognized as the basic variables which determine plant and animal life, both wild and domesticated, and likewise the quality and quantity of human satisfactions to be derived. Gun-powder is relegated to the status of a tool for harvesting one of these satisfactions. Blood-lust is a source of motive-power, like sex in social organization. Only one constant is assumed, and that is common to both equations: the love of nature.

This new idea is so far regarded as merely a new and promising means to better hunting and fishing, but its potential uses are much larger. To explain this, let us go back to the basic thesis—the preservation of fauna and flora.

Why do species become extinct? Because they first become rare. Why do they become rare? Because of shrinkage in the particular environments which their particular adaptations enable them to inhabit. Can such shrinkage be controlled? Yes, once the specifications are known. How known? Through ecological research. How controlled? By modifying the environment with those same tools and skills already used in agriculture and forestry.

Given, then, the knowledge and the desire, this idea of controlled wild culture or "management" can be applied not only to quail and trout, but to *any living thing* from bloodroots to Bell's vireos. Within the limits imposed by the plant succession, the soil, the size of the property, and the gamut of the seasons, the landholder can "raise" any wild plant, fish, bird, or mammal he wants to. A rare bird or flower need remain no rarer than the people willing to venture their skill in *building it a habitat.* Nor need we visualize this as a new diversion for the idle rich. The average dolled-up estate merely proves what we will some day learn to acknowledge: that bread and beauty grow best together. Their harmonious integration can make farming not only a business but an art; the land not only a food-factory but an instrument for self-expression, on which each can play music of his own choosing.

It is well to ponder the sweep of this thing. It offers us nothing less than a renaissance—a new creative stage—in the oldest, and potentially the most universal, of all the fine arts. "Landscaping," for ages dissociated from economic land-use, has suffered that dwarfing and distortion which always attends the relegation of esthetic or spiritual functions to parks and parlors. Hence it is hard for us to visualize a creative art of land-beauty which is the prerogative, not of esthetic priests but of dirt farmers, which deals not with plants but with biota, and which wields not only spade and pruning shears, but also draws rein on those invisible forces which determine the presence or absence of plants and animals. Yet such is this thing which lies to hand, if we want it. In it are the seeds of change, including, perhaps, a rebirth of that social dignity which ought to inhere in land-ownership, but which, for the moment, has passed to inferior professions, and which the current processes of land-skinning hardly deserve. In it, too, are perhaps the seeds of a new fellowship in land, a new solidarity in all men privileged to plow, a realization of Whitman's dream to *"plant companionship as thick as trees along all the rivers of America."* What bitter parody of such companionship, and trees, and rivers, is offered to this our generation!

I will not belabor the pipe-dream. It is no prediction, but merely an assertion that the idea of controlled environment contains colors and brushes wherewith society may some day paint a new and possibly a better picture of itself. Granted a community in which the combined beauty and utility of land determines the social status of its owner, and we will see a speedy dissolution of the economic obstacles which now beset conservation. Economic laws may be permanent, but their impact reflects what people want, which in turn reflects what they know and what they are. The economic set-up at any one moment is in some measure the result, as well as the cause, of the then prevailing standard of living. Such standards change. For example: some people discriminate against manufactured goods produced by child-labor or other anti-social processes. They have learned some of the abuses of machinery, and are willing to use their custom as a leverage for betterment. Social pressures have also been exerted to modify ecological processes which happened to be simple enough for people to understand;—witness the very effective boycott of bird-skins for millinery ornament. We need postulate only a little further advance in ecological education to visualize the application of like pressures to other conservation problems.

For example: the lumberman who is now unable to practice forestry because the public is turning to synthetic boards may then be able to sell man-grown lumber "to keep the mountains green." Again: certain wools are produced by gutting the public domains; couldn't their competitors, who lead their sheep in greener pastures, so label their products? Must we view forever the irony of educating our sons with paper, the offal of which pollutes the rivers which they need quite as badly as books? Would not many people pay an extra penny for a "clean" newspaper? Government may some day busy itself with the legitimacy of labels used by land-industries to distinguish con-servation products, rather than with the attempt to operate their land for them.

I neither predict nor advocate these particular pressures—their wisdom or unwisdom is beyond my knowledge. I do assert that these abuses are just as real, and their correction every whit as urgent, as was the killing of egrets for hats. *They differ only in the number of links composing the ecological chain of cause and effect.* In egrets there were one or two links, which the mass-mind saw, believed, and acted upon. In these others there are many links; people do not see them, nor believe us who do. The ultimate issue, in conservation as in other social problems, is whether the mass-mind wants to extend its powers of comprehending the world in which it lives, or granted the desire, *has the capacity to do so.* Ortega, in his "Revolt of the Masses," has pointed the first question with devastating lucidity. The geneticists are gradually, with trepidations, coming to grips with the second. I do not know the answer to either. I simply affirm that a sufficiently enlightened society, by changing its wants and tolerances, can change the economic factors bearing on land. It can be said of nations, as of individuals: "as a man thinketh, so is he."

It may seem idle to project such imaginary elaborations of culture at a time when millions lack even the means of physical existence. Some may feel for it the same honest horror as the Senator from Michigan who lately arraigned Congress for protecting migratory birds at a time when fellow-humans lacked bread. The trouble with such deadly parallels is we can never be sure which is cause and which is effect. It is not inconceivable that the wave phenomena which have lately upset everything from banks to crime-rates might be less troublesome if the human medium in which they run *readjusted its tensions*. The stampede is an attribute of animals interested solely in grass.

The story of the development of Aldo Leopold's philosophy of conservation is as fascinating as the philosophy itself. Susan Flader, author of a fine scientific biography of Leopold (*Thinking Like a Mountain*), offers a summary of her research in this essay. Flader shows that Leopold learned from his own experience and came to reject the firmly held beliefs of his youth. His example offers a valuable lesson for all of us in this time of changing demands and values in the forestry profession. Flader is professor of history at the University of Missouri.

CHAPTER TWENTY-SEVEN

Aldo Leopold and the Evolution of a Land Ethic

SUSAN FLADER

The last two decades have witnessed an explosion of interest in the ethical basis for people's relationship with their environment, as conservationists, scientists, political leaders, and philosophers have sought an undergirding for what some feared might otherwise be an ephemeral flurry of public concern about environmental quality. In the resulting literature on environmental philosophy, one name and one idea recur more frequently than any others: Aldo Leopold and his concept of a land ethic.

Leopold articulated his environmental philosophy most power-fully in "The Land Ethic," capstone of *A Sand County Almanac,* the slim volume of natural history essays for which he is best known today. But "The Land Ethic," and the shorter, lighter, more illustrative vignettes that help illuminate it draw their clarity, strength, and enduring value from a lifetime of observation, experience, and reflection.

At his death in 1948, Leopold was perhaps best known as a leader—*the* leader—of the profession of wildlife management in America. He was also a forester and is regarded as the father of the national forest wilderness system. In tracing the evolution of his land ethic philosophy, however, we must look beyond his interest in wild-life and wilderness to his lifelong effort to understand the func-tioning of land as a dynamic system, a community of which we are all members. Though his ethical philosophy was an outgrowth of his entire life experience, this quest for its origins focuses in particular on an aspect of his thought that has been less widely known, his concern about mountain watersheds and the problem of soil erosion.

This article reprinted by permission from *Aldo Leopold: The Man and His Legacy,* ed. T. Tanner (Ankeny, IA: Soil Conservation Society of America, 1987), pp. 3–24.

Foundations

Aldo Leopold was born in 1887 in Burlington, Iowa, in his grand-parents' home on a bluff overlooking the Mississippi River.[1] "Lugins-land"—look to the land—the family called the place, reflecting their German heritage and love of nature. The patriarch was Charles Starker, trained in Germany as a landscape architect and engineer, who instilled his keen esthetic sense in his daughter Clara, Aldo's mother. Clara had a pervasive influence on Aldo, eldest and admittedly the favorite of her four children, nurturing the esthetic sensitivity that would be so integral to his land ethic philosophy. Carl Leopold, Aldo's father, also had a formative influence on him. He was a pioneer in the ethics of sportsmanship at a time when the very notion of sportsman was taking shape in the United States.

Leopold frequently hunted with his father and brothers on weekends and arranged his classes so he would have weekday mornings free to cross the Mississippi for ducks or tramp the upland woods on the Iowa side. Summers he spent with the family at Les Cheneaux Islands in Lake Huron, fishing, hunting, and exploring. Despite the temptations of his avocation, he secured a remarkably good education in the Burlington schools, especially in English and history. Then, though he might have been expected to follow his father into management of the Rand & Leopold Desk Company, he went east to Lawrenceville Preparatory School in New Jersey, the Sheffield Scientific School at Yale, and eventually the Yale Forest School to prepare himself for a career in the new profession of forestry.

When he left for Lawrenceville in January 1904, it was with his mother's admonition to write frequently and "tell me everything." He wrote her from the train while passing through the mountains of Pennsylvania. He wrote again upon arriving in Lawrenceville, describing the lay of the land—"flat, but not so bad as I thought"—and yet again the next day, by which time he had taken an afternoon's tramp of some 15 miles and could pronounce himself "more than pleased with the country." That letter ran four pages with detailed accounts of timber species, land use practices, and birds, as well as of the dormitory, his class schedule, problems in algebra and prospects in German, everything. From then on for a decade, through his school years in the East and his early career in the Southwest until after his marriage, Leopold would write at least weekly to his mother, less frequently to his father, brothers, and sister. By rough count he penned something on the order of 10,000 pages of letters. He had learned grammar and sentence structure in the Burlington schools, but he learned to write by writing. His mother insisted on it.

At Yale, Leopold's letters grew longer with more elaborate narrative and keen observation, the result of continued encouragement from home and more frequent and extended tramps through the countryside. He frequented special places—Juniper Hill, Marvelwood, Diogenes Delight, the Queer Valley—and wrote in detail of his encounters with foxes, deer, all manner of birds, moods, and weather. But when he began technical studies in forestry—dendrology, mensuration, silviculture, forest economics—the forays became less frequent as his dedication to the scientific challenges of the new profession absorbed his attention.

Leopold as Forester

Yale had established the first graduate school of forestry in the United States in 1900 with an endowment from the family of the nation's leading forester, Gifford Pinchot. The school promoted Pinchot's doctrine of scientific resource management and what Samuel Hays has characterized as the Progressive Era's "gospel of efficiency" (2). When Leopold graduated with a master of forestry degree in 1909 and went to work for Pinchot in the U.S. Forest Service, he was one of an elite corps of scientifically trained professionals who would develop administrative policies and techniques for the fledgling agency charged since 1905 with responsibility for managing the national forests. He was assigned to the new Southwestern District embracing Arizona and New Mexico territories.

His private reflections, expressed in letters home and recalled in essays written some three decades later, indicate that he was thoroughly enamoured with the stark beauty of the country. But his first publications, which soon began appearing in local Forest Service periodicals, reveal the extent to which he bought into the utilitarian emphasis of the forestry enterprise. The earliest was a bit of doggerel titled "The Busy Season" (3), which he inserted anonymously into a newsletter he edited for the Carson National Forest in New Mexico:

> There's many a crooked, rocky trail,
> That we'd all like straight and free,
> There's many a mile of forest isle,
> Where a fire-sign ought to be.
>
> There's many a pine tree on the hills,
> In sooth, they are tall and straight,
> But what we want to know is this,
> What will they estimate?
>
> There's many a cow-brute on the range,
> And her life is wild and free.

> But can she look at you and say,
>   She's paid the grazing fee?
>
> All this and more, it's up to us,
>   And say, boys, can we do it?
> I have but just three words to say,
>   And they are these: "Take to it."

As he was coming to grips with the myriad management problems of the Carson, Leopold was also struggling with a personal dilemma—how to win the attention of a Spanish senorita he had just met, when he was "1,000,000 miles from Santa Fe" and his rival was right on the scene. He called on his best resource, his skill at expressing himself in letters.

> My dear Estella—
>   This night is so wonderful that it almost hurts. I wonder if you are seeing the myriads of little "Scharfchens volken" I told you about—do you remember the "little sheep-clouds"?—I have never seen them so perfect as they are tonight. I would like to be out in *our Canon*—I don't know how to spell it so you will have to let me call it that—and see the wild Clematis in the moonlight—wouldn't you? I wonder if we could find four more little bluebells for you to pin at your throat—they were beautiful that evening in the dim twilight as they changed with the darkness into a paler and more unearthly blue—and finally into that color which one does not see but *knows*—simply because they are Bluebells.

After six months of extraordinary letters, Estella decided on Aldo; a year later the two were married. For the rest of his life, Estella would inspire and respond to Leopold's special esthetic sensitivity, as his mother had earlier, keeping it alive even during the long years when he was otherwise preoccupied with more worldly, practical affairs. Six months after their marriage, Leopold was off settling a range dispute in a remote area when he got caught in a flood and blizzard. He became a victim of acute nephritis, a kidney ailment that nearly took his life. After 18 months of recuperation and long hours of reflection on how he might live whatever time remained to him, he was able to return to light office work at the district headquarters in Albuquerque.

It was at this junction that Leopold became involved in wildlife conservation, developing a new emphasis on cooperative game management that became a model for Forest Service activity around the nation. Though it was a line of work of his own choosing, an outgrowth of his early avocation, he approached it not in an esthetic mode but in the scientific, utilitarian spirit of the Forest Service. He would promote game management as a science, modeled on the principles and techniques of forestry. Game could bring nearly as much income to the region as timber or grazing uses of the forests, he calculated, if enough effort, intelligence, and money were committed to develop the resource. Extermination of wolves, mountain lions, and other predatory species was a key element in his early program, one he would live to regret (1).

By 1917 he had achieved national recognition for his successes in the Southwest and was beginning to publish his ideas on game management and forestry regularly in periodicals of nationwide circulation. From then until his death, he published frequently— never fewer than two articles a year, often more than a dozen. But for the next two decades his work was decidedly management-oriented. It was not until the last decade of his life that he began publishing the literary and philosophical essays for which he is best known today.

As one examines the contours of Leopold's life, it is ironic to find that the esthetic appreciation for wildlife that was so integral to his youth, so evident in his early letters, and so vital to his mature philosophic reflection was seemingly suppressed at mid-career, at least in his public persona, as he sought to develop a science and profession of wildlife management. The irony is compounded when one notes the extent to which he was pushing beyond traditional modes of thought in his understanding of the dynamics of southwestern watersheds by the early 1920s, developing an interpretation of the functional interrelatedness of virtually all elements of the system save wildlife. It was as if his effort to achieve parity for game animals within the Forest Service model of professional management limited his ken at the same time that he felt less constrained about challenging orthodoxy on larger issues. Thus, it is to his thinking about watersheds and soil erosion that we must turn if we would understand the evolution of his concept of a land ethic—his capacity to think about the system as a whole.

Southwestern Watersheds and Moral Obligation

Ever observant, Leopold had noted the condition of watersheds on national forests in the Southwest from his earliest days as a forest assistant. His first assignment had been to map and cruise timber along the route of a proposed wagon road that would have to clamber high over the Blue Range on the Apache Forest in eastern Arizona because erosion had foreclosed all possibility of a more logical route through the once-lush bottomlands of the Blue River. When he was promoted in 1919 to assistant district forester and chief of operations in charge of business organization, personnel, finance, roads and trails, and fire control on 20 million acres of national forests in the Southwest, Leopold had an opportunity to observe conditions anew as he crisscrossed the forests on inspection trips. By then erosion had washed out nearly 90 percent of the arable land along the Blue. Of 30 mountain valleys he tallied in southwestern forests, 27 were already damaged or ruined.

Leopold thought long and hard about soil erosion. He wrote about the problem in his inspection reports, trying to make foresters in the field more cognizant of changes occurring before their eyes. He spoke about it to the New Mexico Association for Science in a strongly worded warning, "Erosion as a menace to the Social and Economic Future of the Southwest," and reached for national attention to the problem through articles in the *Journal of Forestry.* He even grappled with the philosophical meanings of the phenomenon in a remarkable essay, "Some Fundamentals of Conservation in the Southwest," that was found in his desk after his death.

In "Some Fundamentals," written in 1923, Leopold probed the causes of the degradation that was reducing the carrying capacity of southwestern ranges and considered the implications for human ethical behavior. "The very first thing to know about causes," he wrote, "is whether we are dealing with an 'act of God,' or merely with the consequences of unwise use by man" (*10*). Through analysis of evidence from tree rings, archaeology, and history, he concluded that the deterioration of organic resources in the Southwest could not be attributed simply to climatic change (an act of God). But the nature of the climate, characterized by periodic drouth, had resulted in a delicately balanced equilibrium that was easily upset by man. Overgrazing, resulting from overstocking without regard to recurring drouth, was the outstanding factor in upsetting the equilibrium, in his view. This conclusion that human beings bore responsibility for unwise land use led him to a philosophical discussion under the subtitle "Conservation as a Moral Issue."

He began with an admonition from Ezekiel:

> Seemeth it a small thing unto you to have fed upon good pasture, but ye must tread down with your feet the residue of your pasture? And to have drunk of the clear waters, but ye must foul the residue with your feet?

Ezekiel seemed to scorn poor land use as something damaging "to the self-respect of the craft and society" of which one was a member. It was even possible, Leopold thought, that Ezekiel respected the soil "not only as a craftsman respects his material, but as a moral being respects a living thing." Leopold found support for his own intuitive feeling that there existed between man and earth a deeper relation than would follow from a mechanistic conception of nature in the organicism of the Russian philosopher, P. D. Ouspensky, who regarded the whole earth and all its parts as possessed of soul or consciousness. "Possibly, in our intuitive perceptions, which may be truer than our science and less impeded by words than our philosophies," Leopold suggested, "we realize the indivisibility of the earth—its soil, mountains, rivers, forests, climate, plants, and animals, and respect it collectively not only as a useful servant but as a living being."

Realizing that this premise of a living earth might be too intangible for many people to accept as a guide to moral conduct, Leopold launched into yet another philosophic issue: "Was the earth made for man's use or has man merely the privilege of temporarily possessing an earth made for other and inscrutable purposes?" Because he recognized that most people were heir to the mechanistic, anthropocentric scientific tradition or—like his wife, a devout Roman Catholic—professed one of the anthropocentric religions, he decided not to dispute the point. But he couldn't resist an observation: "It just occurs to me, however, in answer to the scientists, that God started his show a good many million years before he had any men for audience—a sad waste of both actors and music—and in answer to both, that it is just barely possible that God himself likes to hear birds sing and see flowers grow." Even granting that the earth is for man, there was still a question: "What man?" Four cultures had flourished in the Southwest without degrading it. What would be said about the present one?

> If there be, indeed, a special nobility inherent in the human race—a special cosmic value, distinctive from and superior to all other life—by what token shall it be manifest? By a society decently respectful of its own and all other life, capable of inhabiting the earth without defiling it? Or by a society like that of John Burroughs' potato bug, which exterminated the potato, and thereby exterminated itself?

We can only guess why Leopold decided not to publish "Some Fundamentals." One reason might have been a certain discomfort with the inconclusiveness of his philosophical arguments. Another might have had to do with criticism of his interpretation of the causes of soil erosion.

One colleague to whom he sent the draft for review, Morton M. Cheney of the lands division, took him to task for overemphasizing the destructiveness of erosion. Like most other foresters and scientists of the time, Cheney viewed erosion as a natural, ongoing geologic process, a "world-building factor" that would ultimately smooth the rough uplands of the Southwest and create an immense area of agricultural land. In retrospect, it is clear that Cheney was explaining to Leopold the classic geomorphological theory of landscape development of William Morris Davis, who described the stages through which mountains are uplifted and erode to peneplains. Davis's "cycles of erosion" had much in common with the prevailing model of forest ecology, the stages of plant succession to climax, which underlay forest policy in the Southwest. Both were developmental models, defining predictable stages leading to a stable endpoint or equilibrium that would persist indefinitely, unless assaulted by some force acting from outside the system. Thus Cheney articulated essentially the Forest Service prescription for management: protect the climax forests at the headwaters from fire, which will retard some erosion, and don't worry too much about erosion downstream, a natural process in any case.

Leopold didn't need a disquisition on erosion from Cheney. He had undoubtedly studied all that at Yale. But he knew it did not explain what he observed on the ground. He must also have been enormously frustrated by the unwillingness of his colleagues to recognize any responsibility on the part of the Forest Service to deal with the problem. One effect of Cheney's critique must have been to make Leopold realize that he was going to have to provide a much more persuasive analysis of the causes of erosion, starting from facts anyone could see. This he accomplished a year later in an extraordinary piece of observation and inferential reasoning dealing with the relationship between grazing, fire, plant succession, and erosion. "Grass, Brush, Timber and Fire in Southern Arizona" still stands as a landmark in ecological literature. In it he issued a direct challenge to Forest Service dogma: "Fifteen years of forest administration were based on an incorrect interpretation of ecological facts and were, therefore, in part misdirected" (5).

Leopold's theory, in brief, held that southwestern watersheds had maintained their integrity despite centuries of periodic wildfire set by lightning or Indians. Overgrazing by cattle beginning in the 1880s thinned out the grass needed to carry fire, and brush species, now free of both root competition and fire damage, began to "take the country," thus further reducing the carrying capacity for cattle. By the time brush and, higher up, yellow pine had grown dense enough to carry fire, the Forest Service had arrived on the scene to prevent it. Trampling by cattle along watercourses allowed erosion to start, and the grass was no longer sufficient to prevent devastation. Contrary to Forest Service doctrine, Leopold found evidence that fire, far from being an unmitigated evil, was natural and even beneficial, and that grass was a much better conserver of watersheds than were trees or brush. While the Forest Service was willing to acquiesce in some overgrazing and erosion in order to reduce the fire hazard, Leopold was willing to take an added risk of fire in order to maintain the integrity of the watersheds. The Forest Service was thinking of the commodity values of cattle and timber, Leopold was thinking of the whole system.

Although he still had not broken completely out of the mold in which he had been educated and there was much research yet to come, especially on such matters as fire ecology, climate change, and the role of wildlife, Leopold finally had a theory to explain the degradation of southwestern watersheds and was able to consider its implications for administrative policy and for human ethical behavior. Even as he was working out the new theory, he also prepared a comprehensive "Watershed Handbook" for the district, to train foresters to analyze and treat erosion problems, and he took his case for conservative land use to the public in an article, "Pioneers and Gullies," in *Sunset Magazine*. In addition to posing several short- and medium-range policy options—artificial works, better regulated grazing permits, a leasing system—he also addressed the long-range ethical issue, for the first time in print. But it was just a short prediction: "The day will come when the ownership of land will carry with it the obligation to so use and protect it with respect to erosion that it is not a menace to other landowners and to the public" (4). Frustrated as he was by the failure of the Forest Service to address the erosion problem, Leopold realized that the issue ultimately would have to be joined on private lands. Hence the need for a sense of ethical obligation—supported, he hoped, by a legal concept of contingent possession or some other public policy.

It would be another decade before Leopold would publish anything substantial on conservation as a moral issue. But when he did, it would relate integrally to his deepened ecological understanding of southwestern watersheds. Once he could explain in sufficient ecological detail the phenomenon of erosion, or other aspects of the dynamic functioning of ecosystems through time, he would no longer need to call on the authority and terminology of the philosophers. He could stand secure on his own ground. Was the earth living or not? Was the earth primarily for man's use or not? Definitive philosophical answers to such questions no longer seemed critical as the basis for an ethic. It was more important to grasp the interrelatedness of all elements of the system, physical and biological, natural and cultural, and to appreciate the extent to which human as well as biotic interests were dependent upon maintenance of the integrity of that system through time.

In 1924, Leopold left the Southwest for Madison, Wisconsin, to assume a new position as associate director of the Forest Products

Laboratory. The laboratory at that time was the agency's principal research center. Though he was offered the position in part because of his intense scientific curiosity and interest in research as demonstrated in his work on southwestern watersheds, the offer did not imply wholesale acceptance of his ideas within the Forest Service. Nor did the new job offer a fully satisfying scope for his interests. As its name implied, the laboratory dealt more with research concerning the uses of trees than with the living forest.

Four years later, Leopold left the Forest Service to venture full-time into the relatively new field of game management, which had been his hobby long before it became his profession. Drawing on his experience inspecting and analyzing the condition of watersheds in the southwestern forests, he began a series of game surveys in the north central states with funding from the Sporting Arms and Ammunition Manufacturers' Institute. The surveys augmented his understanding of interrelationships between wildlife and the land community and launched him on another major project, the writing of a comprehensive text for the new field. More or less unemployed during the darkest years of the depression, 1931 and 1932, while he was writing his now-classic *Game Management,* Leopold readily accepted a temporary position with Franklin Roosevelt's New Deal in the spring of 1933 and headed back to the Southwest. His assignment was to supervise erosion control work by the newly established Civilian Conservation Corps in dozens of camps in the national forests.

He found the problems as evident as when he had left the region a decade earlier. The causes and processes of erosion were still not well understood within the Forest Service, and there had as yet been no changes in grazing regulations. The worst problems were on private land, giving urgency to the need Leopold had identified in "Pioneers and Gullies" for institutional incentives to conservative land use. Small wonder then that he drew heavily on watershed problems in the Southwest when, for the first time in public or in print, he developed the case for a land ethic. The occasion was the annual John Wesley Powell Lecture to the Southwestern Division of the American Association for the Advancement of Science.

The Conservation Ethic

"The Conservation Ethic" [see previous essay] was by all odds the most important address of Leopold's career. . . . [It] is curiously transitional, much more economically based and management-oriented than his 1923 discussion of conservation as a moral issue and without appeals to Ouspensky or other philosophers. By the time the speech was published, Leopold had accepted a new chair of game management created for him in the Department of Agricultural Economics at the University of Wisconsin. There he would help devise new social and economic tools to deal with problems of land utilization and resource conservation. His emphasis in the address on ecologically based management by private owners on their own land was almost certainly in anticipation of his new position, where he would work not only with government agencies but also, and especially, with farmers. Even here, however, there is a mixture of older concepts with new: "It is no prediction but merely an assertion that the idea of controlled environment contains colors and brushes wherewith society may someday paint a new and possibly a better picture of itself." The "idea of controlled environment"—this confidence in the possibility of control, which permeates "The Conservation Ethic" and also *Game Management*, published the same year—is straight out of the Progressive Era conservation tradition of Gifford Pinchot. It assumes that scientific intelligence can learn enough about the system to exert complete control, an assumption that Leopold's invocation of an ecological attitude was even then beginning to challenge. The resolution would await yet another stage in his evolution of a land ethic.

In his new position at the university, he applied his concern about watershed integrity and soil erosion to southwestern Wisconsin. In a severely eroded watershed known as Coon Valley, Leopold and colleagues persuaded H. H. Bennett, chief of the federal Soil Erosion Service, to establish a pioneering demonstration on the integration of land uses—soil conservation, pasturage, crops, forestry, and wildlife—that would use trained technicians and hundreds of local farmers. Not long after the project began, Leopold expressed the difficulty and necessity of integrating land uses on private land in a wide-ranging critique of the single-track agencies and public-purchase panaceas of the New Deal. "Conservation Economics" is another classic Leopold essay, further substantiating the case for wise use of private as well as public land.

Philosophers speak of the need for a metaphysic, a theory of the scheme of things, as an undergirding for an ethic. Leopold no longer looked to philosophers for a metaphysic; he simply wanted to understand what was happening on the land. But he was still struggling to understand how the system functions and how human beings related to it. In 1935 he returned again to the erosion puzzle in the Southwest, trying his hand at working out a theory that would be mutually acceptable to foresters, ecologists, geologists, and engineers concerning the role of man, and other factors such as climate and topography, in causing erosion. In order to counter an explanation of climate-induced synchronous timing of erosion episodes proposed by the eminent Harvard geologist, Kirk Bryan, Leopold wrote "The Erosion Cycle in the Southwest," in which he came up with an elaborate theory of random timing that allowed a role for human agency. As with his 1923 effort, he sent the manuscript for review, including a copy to Bryan. And again, after receiving Bryan's critique, he decided not to publish; instead, he urged his son Luna to study at Harvard with Bryan to see if he could solve the problem.[2]

Throughout the second half of the decade, Leopold continued his intellectual struggle for a better understanding of the system. He sensed the inadequacy of prevailing models without being able to put his finger on the problem. Sometimes, as ideas grow and change, a certain dramatic experience can trigger a rearrangement of elements, resulting in a new theory of the scheme of things. For Leopold the trigger might have been the juxtaposition of several key field experiences in the mid-1930s: a trip to study game management and forestry in Germany, during which he was appalled by the highly artificial system of management that created a host of unanticipated problems; the acquisition of his own sand country farm, where he began to experience first-hand the imponderables of even the best-intentioned management; and, perhaps most vital, two hunting trips to the Rio Gavilan in the Sierra Madre of northern Chihuahua.

The Sierra Madre—just south of the border from the Southwest Leopold had struggled so long to understand, but protected from overgrazing by Apache Indians, bandits, depression, and unstable administration—still retained the virgin stability of its soils and integrity of its flora and fauna. The Gavilan River still ran clear between mossy, tree-lined banks. Fires burned periodically without any apparent damage, and deer thrived in the midst of their natural predators, wolves and mountain lions. "It was here," Leopold reflected years later, "that I first clearly realized that land is an organism, that all my life I had seen only sick land, whereas here was a biota still in perfect aboriginal health." The vital new idea for Leopold was the concept of biotic health. It was that idea that finally gave him a model, a way of conceptualizing the system, that could become the basis for his mature philosophy.

A Biotic View of Land

Leopold worked out his first comprehensive statement of the new scheme in a paper titled "A Biotic View of Land," which he delivered in 1939 to a joint session of the Ecological Society of America and the Society of American Foresters. The biotic idea represented a shift from the older conservation idea of economic biology, with its emphasis on sustained production of resources or commodities, to a recognition that true sustained yield requires preservation of the health of the entire system. Scarcely five years earlier, Leopold himself had asserted that "the production of a shootable surplus is the acid test of the sufficiency of a conservation system." Now he was distinguishing between the old economic biology, which conceived of the biota as a system of competitions in which managers sought to give a competitive advantage to useful species, and the new ecology, which "lifts the veil from a biota so complex, so conditioned by interwoven cooperations and competitions, that no man can say where utility begins or ends" (7). Though he had clearly been moving toward such a view years earlier in his work on southwestern watersheds, the 1939 statement marked the first significant publication in which his thinking about wildlife was fully integrated in the new conception. Thus, better than anything else he wrote, "A Biotic View of Land" signaled the maturity of Leopold's thinking.

It was here that he first presented the image of land as an energy circuit, a biotic pyramid: "a fountain of energy flowing through a circuit of soils, plants, and animals." It was here too that he drew ecological interrelationships into an evolutionary context. The trend of evolution, he suggested, was to elaborate and diversify the biota, to add layer upon layer to the pyramid, link after link to the food chains of which it was composed. Leopold posited a relationship between the complex structure of the biota and the normal circulation of energy through it—between the evolution of ecological diversity and the capacity of the land system for readjustment or renewal, what he would come to term land health.

Biotas seemed to differ in their capacity to sustain conversions to human occupancy. Drawing once again on his understanding of the Southwest, Leopold contrasted the resilient biota of Western Europe, which had maintained the fertility of its soils and its capacity to adapt to alterations despite centuries of strain, with the semiarid regions of America, where the soil could no longer support a complex pyramid and a "cumulative process of wastage" had set in. The organism would recover, he explained, "but at a low level of complexity and human habitability." Hence his general deduction: "the the less violent the man-made changes, the greater the probability of successful readjustment in the pyramid."

The biotic idea Leopold articulated, though deeply a product of his own thought and experience, was part of a larger conceptual reorientation in the biological sciences during the 1930s. These were the years, according to historians of science, when various strands of evolutionary and ecological theory, separated during the furor over Darwin's *Origin of Species* (1859), began to fuse into a broad unified theory. Among ecologists it has become known as the "ecosystem" concept, after a term suggested by the British ecologist A. G. Tansley in 1935.[3] The new conception postulated a single integrated system of material and energy. The system was not driven by any one factor, such as climate, and hence was not best understood as developing through predictable stages to a stable endpoint, as in the older models of forest succession and erosion that had troubled Leopold in the Southwest. Rather, it was in constant flux, subject to unpredictable perturbations that entailed a continual process of reciprocal action and adjustment. Agents of disturbance—whether fire, disease, grazing animals, predators, or man—which had been viewed by early forest ecologists as acting from outside the system to subvert the normal successional stages or the climax equilibrium, were now viewed as functional components within the system.

Not everyone who adopted the ecosystem model drew from it the same implications for land management or for ethical behavior. Human beings, an exceptionally powerful biotic factor, were now clearly located *within* the system; did that mean they could manipulate it to their own ends with moral impunity? Donald Worster, who has written about the history of ecological thought in *Nature's Economy*, has noted the emphasis in recent ecosystem analysis on concepts such as productivity, biomass, input, output, and efficiency, which he views as metaphors for the modern corporate, industrial system bent on economic optimization and control. He also views modern ecosystem analysis as reductionist, breaking down the living world into readily measurable components that retain no image of organism or community. He is, therefore, highly skeptical of the ecosystem concept as the basis for an environmental ethic (11).

A caveat is in order. Science alone is hardly an adequate basis for an ethic, and Leopold realized as much. For all his commitment to understanding how the system functioned, even Leopold did not insist that science provided all the answers. What he gained from the biotic view was a new humility about the possibility of ever understanding the system fully enough to exercise complete control. As he expressed it in a speech on "Means and Ends in Wildlife Management," some managers had admitted their "inability to replace natural equilibria with artificial ones," and their unwillingness to do so even if they could. The objective of management, as he now viewed it, was to preserve or restore the capacity of the system for sustained functioning and self-renewal. He would do this by encouraging the greatest possible diversity and structural complexity and minimizing the violence of man-made changes. The techniques of management might remain much the same, but the ends (of wildlife management, at least) were now fundamentally altered. The ends, he realized, were a product of the heart as much as of the mind.

Leopold came to a deeper personal understanding and appreciation of the new biotic idea and its implications for land management through his own participation in the land community at his "shack" in central Wisconsin's cutover, plowed up, worn out, and eroded sand country. All during the late 1930s and into the 1940s, while he was struggling to put the biotic idea and the land health concept on paper, he was also struggling to rebuild a diverse, healthy, esthetically satisfying biota on his farm. His journals of the shack experience record a daily routine of planting and transplanting—wildflowers, prairie grasses, shrubs and trees, virtually every species known to be native to the area. But the journals also record his tribulations. Take pines, for example, of which the family planted thousands every year. The first year, 1936, more than 95 percent were killed by drouth within three months. Another year, rabbits attracted to brush shelters he had built for the birds trimmed three-quarters of the white pines in the vicinity. Other times the culprits were deer or rust or weevils or birds alighting on the candles or vandals cutting off the leaders or flood or fire. Fire could be discouraging, especially if set by a trespassing hunter, but it also brought new life—sumac and wild plum, blackberry, bluestem, poison ivy and, most exciting to Leopold, natural reproduction of jackpines from cones undoubtedly opened by the heat.

The shack experience engendered in Leopold a profound humility in his use of the manager's tools, as he became acutely aware of the innumerable, ofttimes inscrutable factors involved in life and death, growth and decay. It also led him to ponder the basis for the individual decisions he found himself making every day—whether to plant something as useless as a tamarack (yes, because it was nearly extinct in the area and it would sour the soil for lady's-slippers), what to do about the sandblow on the hill (leave it as testimony to history and also as a habitat for certain species like little Linaria that would grow only there), whether to favor the birch or the pine where the two were crowding each other (he loved all trees, but he was "in love" with pines). He realized that ethical and esthetic values could be a guide for individual decisions, not a substitute for them. And he also gained a sense of belonging to something greater than himself, a continuity with all life through time. This he expressed in a series of vignettes that ultimately found their way into *A Sand County Almanac*.

Toward a Land Ethic

Toward the end of the 1940s, Leopold tried again to make the case for a conservation ethic. This new effort, for a 1947 address to the Garden Club of America on "The Ecological Conscience," was less ambitious than his 1933 "Conservation Ethic." He drew on four issues in which he had been involved in Wisconsin, including a wrenching debate about the state's "excess deer" problem that had preoccupied him for years. The public thought only about conserving deer because they were unable to see the land as a whole. As in each of his previous efforts to make the case for a sense of obligation to the community going beyond economic self-interest, he also addressed the problem of soil erosion, this time analyzing the failure of soil conservation districts in Wisconsin to achieve anything beyond those few remedial practices that were immediately profitable to the individual farmer. From his unpublished papers we know that he had been trying for several years to articulate the land health concept in relation to land use and erosion, but he made no effort to do so in "The Ecological Conscience" (8). The speech was not particularly significant—except for one pregnant sentence setting forth the criteria of an ethic, which would ultimately find a more appropriate context in his most famous essay, "The Land Ethic."

Later that year and in early 1948, Leopold substantially reshaped and revised the collection of essays for which he had been seeking a publisher all decade. It would begin with a selection of vignettes from the shack, arranged by month in the style of an almanac. That would be followed by sketches recounting various episodes in his career that taught him the meaning of conservation. Then he would conclude with several meatier essays, culminating in a comprehensive statement of his ethical philosophy. "The Land Ethic" incorporates segments from three previous essays, all thoroughly reworked and integrated with new material to reflect his current thinking. From "The Conservation Ethic" he drew the notion of the ecological and social evolution of ethics and the role of ecology in history; from "A Biotic View of Land" the image of the land pyramid, of land as an energy circuit; and from "The Ecological Conscience" the case for obligations to land going beyond economic self-interest. His efforts of a decade to articulate the concept of land health and the relationships between economics, esthetics, and ethics, filling numerous handwritten, heavily interlined pages, finally found compelling expression. And the whole came to a focus in the most widely quoted lines in the entire Leopold corpus: "A thing is right when it tends to preserve the integrity, stability, and beauty of the biotic community. It is wrong when it tends otherwise" (9).

Integrity, stability, beauty: the fundamental criteria of the land ethic. Integrity, referring to the wholeness or diversity of the community: the precept to retain or restore, insofar as possible, all species still extant that evolved together in a particular biota. Stability, embodying the concept of land health: the precept to maintain or restore an adequately complex structure in the biotic pyramid, so that the community has the capacity for sustained functioning and self-renewal. Beauty, the motive power of the ethic: the precept to manage for values going beyond the merely economic—and, probably also, an allowance for the subjective tastes of the individual. The three tenets were interrelated. Elsewhere in his writings Leopold had referred to an assumed relationship between the diversity and stability of the biotic community as "the tacit evidence of evolution" and "an act of faith." And as early as 1938 he had posited a relationship among all three tenets, and utility as well, in an unpublished fragment titled "Economics, Philosophy and Land": "We may postulate that the most complex biota is the most beautiful. I think there is much evidence that it is also the most useful. Certainly it is the most permanent, i.e., durable. Hence there is little or no distinction between esthetics and utility in respect of biotic objective."

The three cardinal tenets of the land ethic, first voiced in his address, "The Ecological Conscience," can also be discerned, in somewhat different terminology, in Leopold's earlier formulations of his ethical philosophy in 1923 and 1933. This supports a conclusion that his mature expression involved a deeper understanding of the functioning of the land system and a more cogent articulation rather than a change in fundamental values. All these formulations, and "A Biotic View of Land" as well, were premised on a conception of land as an interrelated, indivisible whole, a system that deserved respect as a whole as well as in its parts. In 1923 Leopold drew on his intuitive perceptions, buttressed by the concepts and terminology of Ouspensky and other philosophers and poets. Later he would base his conception of the land community on the findings of ecology, but his willingness to trust to intuition remained. In 1933, for example, he wrote, "Ethics are possibly a kind of advanced social instinct in-the-making," revising the phrase in 1948 to "community instinct." Each formulation also emphasized the notion of *obligation* to the whole, rather than focusing on the *rights* of individual constituents, whether human, animal, vegetable, or mineral. Leopold did not deny that nonhuman entities had rights, and occasionally he even referred to a species' "biotic right" to existence, but he was too much concerned with securing acceptance of his major premises to risk alienating people by entering the thicket of the rights debate.[4]

Leopold articulated his ethical philosophy out of a profound conviction of the need for moral obligation in dealing with the dissolution of watersheds in the Southwest. Hence his emphasis on the integrity of the system. Each successive reformulation of his philosophy was stimulated at least in part by his continuing concern for the erosion problem and advances in his understanding of the ecological processes involved. Remarkably, it was his compelling concern and curiosity about the phenomenon of erosion, which was never a major professional responsibility, rather than his lifelong interest in wildlife, which became his profession, that led him to his conviction of the need for a land ethic and his understanding of the biotic idea on which it was grounded. But once he had grasped the biotic concept, through which he finally integrated wildlife fully into his understanding of the functioning of the land community, it was his sensitivity to the esthetics of wildlife that would enable him to convey a sense of the land community and the land ethic to others. *A Sand County Almanac* is the case in point.

Leopold's fascination with the new biotic concept, especially the role of evolution, energy, and land health, led to an explosion of the classic literary essays for which he is best known today: "Marshland Elegy," with its haunting image of sandhill cranes, evolutionarily among the most ancient of species, standing in the peat bogs of central Wisconsin "on the sodden pages of their own history"; "Clandeboye," where the western grebe, also of ancient lineage, "wields the baton for the whole biota"; "Odyssey," the saga of two atoms cycling through healthy and abused systems; "Song of the Gavilan," where food is the continuum in the stream of life; "Guacamaja," a disquisition on the physics of beauty, recording the discovery of the numenon of the Sierra Madre, the thick-billed parrot; and "Thinking Like a Mountain," in which the wolf becomes metaphor for the functioning system.

These essays had a purpose with respect to Leopold's notion of the evolution of an ethic. Their purpose was to inspire respect and love for the land community, grounded in an understanding of its ecological functioning. Leopold would motivate that understanding of the whole by focusing the reader's attention on the subtle dramas inherent in the roles of wolf, crane, grebe, parrot, even atom, in the scheme of things. The essays were Leopold's attempt to develop a metaphysic, or an esthetic—to stimulate perception that might lead people to the transformation of values required for a land ethic. He would motivate not by inciting fear of ecological catastrophe or indignation about abused watersheds but rather by leading people from esthetic appreciation through ecological understanding to love and respect.

He had thus come full circle in his own development—from his youth, in which esthetic appreciation for wildlife provided his personal motivation to enter a career in conservation, through his professional experience in forestry and his concern about watersheds, which stimulated his consciousness of the need for an ethical obligation to land, to his maturity as an ecologist, when he successfully integrated all the strands of his previous experience. Reflecting on the process by which he himself had come to ecological and ethical consciousness, he would now inspire others along a similar route.

Notes

1. Biographical details and generalizations are based primarily on the Aldo Leopold Papers in the University of Wisconsin Division of Archives and interviews with family members and friends of Aldo Leopold. All quotations from letters and unpublished manuscripts are from items in the Aldo Leopold Papers.

2. World War II intervened and Aldo died before Luna was able to complete his Ph.D. dissertation on "The Erosion Problem of Southwestern United States" at Harvard in 1950.

3. Leopold's biotic idea, with its image of the land pyramid, probably owes more to another British ecologist, Charles Elton, and to an American scientist, Walter P. Taylor, both personal friends. Leopold did not use the term "ecosystem" in his writing, but he used "ecology" or "biotic idea" to express the same concept.

4. Because most philosophers are concerned with individuals rather than with communities and with rights rather than with obligations, Leopold's land ethic has often been distorted, disparaged, or dismissed in philosophical circles. Among those who have studied his work carefully enough to puzzle through some of the vagaries of language and seeming inconsistencies, Bryan Norton places Leopold in the tradition of American pragmatism and J. Baird Callicott identifies him as heir to the biosocial ethical tradition of David Hume and Charles Darwin. Both are reasonable in light of Leopold's education, reading, and experience.

References

1. Flader, Susan L. 1974. *Thinking like a Mountain: Aldo Leopold and the Evolution of an Ecological Attitude Toward Deer, Wolves, and Forests.* University of Missouri Press, Columbia, Missouri.
2. Hays, Samuel P. 1959. *Conservation and the Gospel of Efficiency: The Progressive Conservation Movement, 1890–1920.* Harvard University Press, Cambridge, Massachusetts.
3. Leopold, Aldo. 1911. *The busy season.* The Carson [National Forest] Pine Cone (July).
4. Leopold, Aldo. 1924. *Pioneers and gullies.* Sunset Magazine 52 (May): 15–16, 91–95.
5. Leopold, Aldo. 1924. *Grass, brush, timber, and fire in southern Arizona.* Journal of Forestry 22 (October): 1–10.
6. Leopold, Aldo. 1933. *The conservation ethic.* Journal of Forestry 31 (October): 634–643.
7. Leopold, Aldo. 1939. *A biotic view of land.* Journal of Forestry 37 (September): 727–730.
8. Leopold, Aldo. 1947. *The ecological conscience.* The Bulletin of the Garden Club of America (September): 45–53.

9. Leopold, Aldo. 1949. *A Sand County Almanac and Sketches Here and There*. Oxford University Press, New York, New York.

10. Leopold, Aldo. 1979. *Some fundamentals of conservation in the Southwest*. Environmental Ethics 1 (Summer): 131–141.

11. Worster, Donald. 1977. *Nature's Economy: The Roots of Ecology*. Sierra Club Books, San Francisco, California.

How do we apply the ideals of the Land Ethic in managing land or advising policymakers and landowners? In this essay, Albert C. Worrell, longtime professor at the Yale School of Forestry and Environmental Studies, grapples with this question. He interprets the Land Ethic in an economic vein in an attempt to give it more concreteness and clarity. Not all readers will support this approach, but Leopold's thinking is broad enough to accommodate many interpretations, because it recognizes that there are legitimate conflicts to be settled and that there are no easy answers. It is a particular pleasure to offer this essay here, as Worrell was one of my professors at Yale and a significant early mentor for me and many other young foresters.

CHAPTER TWENTY-EIGHT

Ethical and Legal Aspects of Forest Land Management

ALBERT C. WORRELL

The soil is clearly man's basic environmental resource. Despite the advances of technology, it is hard to imagine human beings living permanently in space capsules or marine modules.

The soil is also finite. Human ingenuity has increased output per hectare, but the fixed land area ultimately limits what people can have. Historically, they have met this problem by conquest and migration and developing technology. But the possible additional gains from such means are diminishing rapidly. The ultimate solution must be population control and perhaps reduction. In the meantime, land is scarce and becoming scarcer.

To complicate things, this limited land resource is controlled and managed by only part of the people. The rest have no direct control over what is done with it, though they are vitally affected. Those who do control the land are clearly able to benefit or harm many other people. So the land-use problem is basically a moral problem.

The Ethics of Land Use

Aldo Leopold's book of essays—*A Sand County Almanac*—has become a classic of the environmental movement. Much of its fame derives from the final essay—"The Land Ethic." This has always disappointed me. The insight is important and the concept is stimulating. But I never get any clear idea from it of what a land ethic is or how one might develop. I do believe, however, that such an ethic is cur-

Dr. Worrell is Edwin W. Davis Professor of Forest Policy at Yale University, New Haven, CT. This article reprinted by permission from Proceedings of the Fourth North American Forest Soils Conference, Laval University, Québec, Canada, August 1973 (Québec: Les Presses de l'Université Laval, 1974), pp. 661–667.

rently evolving and I would like to try to analyse this situation.

Moral philosophers have long discussed man's ethical duty to do what is right and avoid what is wrong. They have tried to make explicit an apparently general feeling that an individual has a moral obligation to the others in his society. What has bothered them is how does one decide what is good and what is bad?

The problem shows up in property tax assessment. Most laws specify that land be valued at its "highest and best use" as indicated by market price. But many states have modified their laws so land can be assessed on its value for agriculture or forestry and not pushed into urban development. Apparently the highest-and-best use is not always good land use.

Economists blame this on a weakness in the market system. They feel that competitive market prices are good indicators of comparative social value. But they recognize that some costs do not fall directly on the producers. Their suggestion is to force producers to pay such external costs. A "pollution fee," for example, would make the producer bear costs he imposes on other people by polluting air or water. If this made the total cost of his product higher than what the consumers would pay, he would not produce it. Or he would find some cheaper means of handling the effluent.

But suppose the market price is high enough to cover all production costs including the pollution fee? Does this mean it is good to continue polluting? This might be reasonable if the fee were used to immediately clean up the pollution—as might be done with litter in a public park.

But suppose the external effects cannot be completely offset by any corrective measures—as would seem to be true of strip mining? No matter how high a "land restoration fee" is charged, some environmental effects will be permanent. Economists would suggest that the strip miner might compensate people for these effects. He might, for example, provide a free pool to replace the old swimming hole that is fouled by acid from the mine. If the value of coal, as indicated by the market price, were high enough to cover the costs of mining, restoring the land, and providing the pool, would not strip mining be desirable?

Environmentalists would reply that there are some external damages for which no compensation is possible. Take the case of nuclear tests which would destroy life on a Pacific island. I am sure we all agree there is no conceivable way in which any human inhabitants could be compensated for being annihilated. But suppose they and their possessions were moved to a similar island where their original conditions were reproduced?

Some environmentalists would still insist that destruction of the first island is morally wrong. Some apparently feel that any degradation of natural conditions is wrong. But how can this be true? Evolution and ecological change go on continuously. Lightning fires, floods, and insect outbreaks are natural phenomena. Where does environmental disruption by man differ?

The difference must lie in the fact that people can perceive and understand the consequences of their actions. They make conscious choices and can therefore consider the moral implications of what they do.

The moral question appears easy to resolve in the case of wanton or unnecessary environmental disruption. Purposeless destruction for destruction's sake is clearly immoral. The same can surely be said of unnecessary disruption, regardless of the purpose.

When disruption is necessary to accomplish some purpose, the problem becomes more complicated. But all purposes are not the same. It cannot be right to disrupt the environment for unworthy or immoral purposes.

But what about worthy purposes? Merely to survive, people must disrupt the environment to some extent. I see no way of arguing that faced with a choice between starvation and environmental destruction, mankind should choose to starve. This only becomes a moral problem if some people by refraining from destroying the environment can make it possible for others to survive. But the alternatives will seldom be this drastic.

The usual question will be: does the end justify the means? The only way I see to answer this question is in terms of the effects on people. There are those who feel that most of our current environmental problems result from being too man-centered. I suggest that they stem more from a what-is-to-my-personal-advantage-in-the-short-run attitude. A comparison of land-use benefits and costs should be valid if it includes all tangible and intangible effects that can be identified. We obviously need more knowledge before we can make precise determinations of these effects. But I believe people know enough today to express the judgment that it is immoral to disrupt the environment unless the total long-run costs will be justified by the benefits.

The moral aspects of land use are not limited to environmental costs, however. Those who control the land decide what it is to produce and thus affect the well-being of other people. They therefore have some obligation to manage the land in the best way.

This is not a simple obligation. It would seem immoral to withhold land from production if people want and need the things it might produce. But does this mean it is the duty of Canada and the United States to try to meet needs everywhere in the world? An economist might say: Needs do not indicate an effective demand unless people have the purchasing power to buy what they want. But does my moral obligation to my fellow men extend only to those things they are able to pay me for? Philanthropy and welfare programs do not support this.

Land usually can produce a variety of benefits. Most of us would consider it unethical to grow flowers on land which could produce wheat for hungry people. But suppose there are adequate sources of other grains and people just happen to prefer wheat? Would it then be unethical to grow flowers? An economist would say: grow the one that will contribute most to people's utility. If they will be happier eating rye bread with flowers on the table than white bread from bare tables, the land should grow flowers, even though the people prefer white bread to rye.

Our solution to such complex choices in consumption and production has been to depend on the competitive market system. We use market prices to indicate comparative values and net revenue (or profit) as the criterion of efficiency. And, in general, it works reasonably well. But the market system is far from perfect, especially in the environmental area. Most criticisms of "economics" result from too rigid adherence to the concepts of a free market system. The land manager's moral obligation is to enhance the welfare of society, not to conform to an economic system. He must think about the effects of his actions on people in broader terms than net revenue.

Social welfare depends not only on what land produces, but also on who receives it. We probably agree it would be unethical for the owners to consume all of the land's output and make other landless people do without. It seems obvious the total welfare would be greater if the benefits were more widely distributed. But distribution is many sided. There are choices between living consumers and future generations, between those who desire different kinds of products, and between those of different wealth and power.

The moral issue seems to center on the right of all people to share in the benefits produced by the resources of their country (and perhaps the world). It has proved difficult to define "equal rights" because individual people differ so much. The current approach is to avoid discriminating against individuals or groups because of their particular characteristics, attitudes, desires, or present condition. On this basis, a land manager's moral obligation might be not to discriminate against any particular segments of society.

I am convinced that a definite moral attitude toward land use is growing and spreading among our people. In the next section I will illustrate how this is being reflected in the legal area. But first let us summarize the moral issues.

The current land ethic, as I perceive it, considers it morally wrong:

1. To wantonly destroy or degrade land resources without any purpose.

2. To impose land-use costs on anyone for unworthy or immoral purposes.

3. To destroy or degrade land resources when it is not necessary, regardless of the purpose.

4. To impose land-use costs on anyone unless the resulting benefits to someone will be greater than the total costs.

5. To manage land resources in ways which deprive people of things they could enjoy were the resources managed for the benefit of society as a whole.

6. To manage land resources in a pattern which benefits primarily the present generation or certain segments of society and discriminates against future generations or other segments, no matter how small these may be.

Legal Aspects

The legal system in North America has allowed a high degree of individual freedom in the management of property. The *laissez faire* philosophy assumed that society would benefit most if each individual did what was best for him personally. A landowner's moral obligation, therefore, was to manage his land for his own maximum benefit, so long as this did not create a nuisance to the neighboring property owners. The laws were mainly concerned with assuring his freedom to do so.

This did not work too well with forest land, and eventually laws had to impose an obligation not to destroy this resource. First came restrictions on burning. Then encouragement to protect the forest. And finally, in some states, compulsory regeneration after harvest. Of greater importance was the decision to stop transferring forest land into private ownership and to return some of it to public ownership.

A private person never owns all the property rights in land. Some always remain with the public, which can therefore control his actions to some extent through the police power, taxation, or eminent domain. The public as an owner does have full property rights. These are exercised for it by the relevant unit of government. The actual management is delegated to particular government agencies, and it is the administrators and personnel of these agencies who decide how the public lands are used.

The moral responsibility for the use of public forests is therefore somewhat complicated. It would appear it must ultimately rest on the people who actually manage the lands. But if they comply with legislative and higher administrative directives, they are fulfilling their duty. And this also seems true of the government itself, if it conforms to the mandates of the constitution.

The administration of our permanent public forests fell largely into the hands of highly motivated people who were devoted to the public interest. They emphasized protection of the land and timber and managed the forests for use and a sustained yield for the future. It appeared the government was fulfilling its moral obligations by giving professional managers freedom to decide how the land should be used.

As population continued to increase and the effects of environmental degradation became more apparent, legal actions also expanded. Effective water and air pollution laws were passed in 1948 and 1955. Restrictions were gradually imposed on the use of chemical pesticides. And Congress had second thoughts about the freedom of public-forest managers. In 1960, they gave specific instructions that the national forests be managed for multiple uses. In 1964, they took away power over the use of certain lands by incorporating them in a National Wilderness Preservation System. And, in 1970, The National Environmental Policy Act laid a basis for future control over public and private actions which affect the environment.

We have now reached a point where misuse and degrading effects on the environment are so obvious and threatening that the public is ready to legally impose moral obligations on those responsible for them. It is already happening in the area of land use. I would like to demonstrate this briefly with four areas of current legal action which are clearly related to the moral obligations discussed earlier and which will have a major impact on forest managers in the future.

The National Environmental Policy Act requires all federal agencies to prepare a detailed statement describing the environmental impact of any major proposed action which might significantly affect the quality of the environment. This statement must be submitted to other relevant agencies for comment. And then copies of the statement and the comments must be made available to the Council on Environmental Quality and the public.

Preparing an Environmental Impact Statement forces an agency to think through the justification of its proposed action and to submit to outside criticism. This is bound to exert pressure on the agency to recognize its moral obligations. But it also alerts the public and other agencies and gives them a chance to try to modify the proposal or stop it altogether.

A number of states have passed laws extending this requirement to state agencies and this seems likely to spread. Environmental impact statements are not yet required of private land managers. But laws are appearing which require such studies in power plant siting and public utility developments. And the idea is being modestly incorporated in subdivision regulations. My guess is that, within a decade, this will be one of the responsibilities of almost all forest managers.

The administration of government programs has been subjected to a wave of objections in recent years. In the past, actions of government agencies were shielded from aggrieved citizens by sovereign immunity. Only when the government itself recognized it was at fault, would it accept legal action to determine damages. A citizen could seek compensation from the legislature for a specific injury. But it was difficult to get a government agency to change its management practices. This situation has been changing rapidly.

One of the bases for this change is the Public Trust Doctrine. This asserts that a government which holds and manages important natural resources has a duty of care and responsibility to the general public similar to that of a trustee to a beneficiary. The point is that a trustee's duty is enforceable at law. The public trust doctrine is well established in the case of submerged coastal lands and the courts are cautiously feeling their way into its application to other public lands.

Standing to sue has been recognized for conservation organizations and a number of states have passed laws giving individual citizens the right to bring suit over environmental damages. Appeals procedures are being improved to handle complaints through administrative reviews within government agencies. And the courts are becoming more willing to engage in judicial review. Individual complaints cover the whole range of moral obligations and it seems clear that public land managers will face increasing review of how they are fulfilling these obligations.

Straws in the wind suggest that the trustee concept will eventually be extended to private landowners. The right of a Connecticut water company to sell or develop lands which were originally obtained through eminent domain is currently being tested in court. Society may well take the future attitude that the exclusive private property rights it permits and protects carry with them a trustee responsibility to society for the land.

Regulations restricting the use of fire and pesticides are already familiar to forest managers. Restrictions on the siltation and blocking of streams and smoke emission are becoming more common and noise limits on chain saws and other machines loom on the horizon. More important for the future perhaps are moves toward rural zoning.

Zoning has been practiced largely in urbanized areas, but today the line between rural and urban is often difficult to define. Flood-plain zoning is becoming common as a means of keeping people from building there. Tidal wetlands laws in a number of states establish a zone in which landowners' actions are severely limited. The recent inland wetland laws are less restrictive, but they require permission for any but compatible uses. These appear to be first steps toward inevitable statewide zoning.

Finally, integrated land-use planning is gradually being required by law. Many local governments have official planning commissions and some are hiring full-time planners. They are joining in regional planning agencies to coordinate their efforts. And state governments are feeling their way into statewide planning. This is not a phenom-enon peculiar to megalopolis, either. Congress is currently con-sidering a national land-use policy act which will encourage and assist the states in developing land-use plans and in coordinating these on a national basis.

Most suburban areas now require landowners and developers to submit detailed plans for approval before they can undertake land development. Many local planning agencies are requiring the main-tenance of open space and aesthetic values as essential parts of these development plans. The day of freedom to develop land resources as one pleased is rapidly coming to an end.

In summary, I think the public is telling forest managers that they are trustees of a vital natural resource and have a moral obligation to manage it in the best long-run interest of society as a whole. In order to be sure that they meet this obligation, the public is going to require explicit consideration of environmental impacts, submit their actions to judicial and other review, regulate their management by zoning and other means, and require that their management conform to inte-grated land-use plans. The forest manager who truly subscribes to a land ethic should not find this onerous.

Whatever your own religious orientation, the importance of the Bible's teachings in Western culture makes A. J. Fritsch's excellent summary of the Scriptural basis for an environmental ethic a valuable addition to this book. In Fritsch's analysis of the stewardship concept you will find a strong parallel to Leopold's and Nash's widening circle of rights. The notions of sparing the earth, repairing the earth, and sharing the earth are key themes. They connect directly to the Progressive Era conservation tradition that created the National Forests, National Parks and Monuments, Wildlife Refuges, early forestry schools, and a host of state conservation agencies. Mr. Fritsch is a Catholic social activist.

CHAPTER TWENTY-NINE

Theological Foundations for an Environmental Ethics

A. J. FRITSCH

> The wolf lives with the lamb,
> the panther lies down with the kid,
> calf and lion cub feed together
> with a little boy to lead them.
> The cow and the bear make friends,
> their young lie down together.
> The lion eats straw like the ox.
>
> ISAIAH 11:6–7

The Judeo-Christian tradition is the cornerstone of Western civilization and has furnished the foundations of the ethical conduct of that culture. The scientific and technological revolutions of the past few centuries were stepchildren of the tradition, thus implicating it as a cause of the current environmental crisis. A much quoted essay by Lynn White, Jr., concludes that "we shall continue to have a worsening ecologic crisis until we reject the Christian axiom that nature has no reason for existence save to serve man."[1] Many others have joined in the debate. Lewis Moncrief holds that the Judeo-Christian tradition is only one of many cultural factors contributing to the environmental crisis.[2] Others contend that it is a perversion of that tradition which has precipitated the present crisis.[3]

This [article] is not the place to continue the debate; an environmental ethics should attempt to search out in this tradition both the meaning of our actions and ways of correcting the faults committed. Several points stand out when looking into the sacred Scriptures and the living experience of people steeped in the Judeo-Christian tradi-

This article reprinted by permission from his book, *Environmental Ethics* (New York: Doubleday, 1980), pp. 233–254. Copyright © 1980 by the Science Action Coalition.

tion: a prophetic witness to the need to reform; exemplary lifestyles demonstrating the need for harmony with the earth; and a stewardship that stresses the major elements of environmental conservation. Perhaps there are many more such points that theologians could fruitfully develop, some of which were only germinally operative in biblical times.

Prophetic Witness

> The spirit of the Lord has been given to me,
> for he has anointed me.
> He has sent me to bring the good news to the poor,
> to proclaim liberty to captives
> and to the blind new sight,
> to set the downtrodden free,
> to proclaim the Lord's year of favour.
>
> LUKE 4:18–19

The prophetic voice constantly sounds throughout the Judeo-Christian tradition, awakening people to their wrongdoings, pointing out the dire consequences of continuing in their ways, and calling them to reform and give justice to the oppressed. When heard, this voice often sparks resistance from the established order; when listened to, it produces a *metanoia,* or change of heart. When it is not heeded, the predictions come true. The prophetic voice is public, to be heard by all; it often runs counter to a prevailing culture and lifestyle; it offers an option for change. It is direct and concrete:

> You are the ones who destroy the vineyard
> and conceal what you have stolen from the poor.
>
> ISAIAH 3:14

It attacks not just violence but also idolatry, or the act of giving undue attention or value to the material things around us (Isaiah 40:18–20 or Jeremiah 10:1–9). The consequence of continuing such practices are shown:

> A workman made the thing,
> this cannot be God!
> Yes, the calf of Samaria shall go up in flames.
> They sow the wind, they will reap the whirlwind.
>
> HOSEA 8:6–7

However, the invitation is also present:

> Sow integrity for yourselves,
> reap a harvest of kindness.
>
> HOSEA 10:12

Positive acts of violence and wrongdoing are not the only prophetic issues:

> Woe to those who add house to house
> and join field to field
> until everywhere belongs to them
> and they are the sole inhabitants of the land.
>
> ISAIAH 5:8

The modern prophetic voice continues as part of that tradition, speaking for human rights and against greed and avarice. It is direct and balanced; it is often dire in that it paints the world as it is—if change is not made, something worse will happen; it is filled with hope by pointing out the possibility of that change. That voice extends to consumer and environmental matters, exposing the glorification of modern idols such as automobiles and aerosol sprays and the waste of natural resources.

A prophetic voice does not speak in the abstract or generalities;[4] it is concrete and relevant and addresses issues of major moment to people, such as nuclear power and human rights. For example:

> As a result of these inherent properties of the plutonium and other materials associated with the nuclear fuel cycle, use of this source gives rise to several fundamental problems:
> —The potentially catastrophic releases of plutonium and nuclear wastes . . .
> —The diversion of nuclear fuel . . . by criminal elements . . .
> —The necessity for perpetual, reliable containment of the nuclear wastes. . . .[5]

Another example is:

> The right to have a share of earthly goods sufficient for oneself and one's family belongs to everyone. . . . If a person is in extreme necessity, he has the right to take from the riches of others what he himself needs.[6]

While the environmental crisis began in the West, it is the religious tradition of these lands that influenced such international meetings as the Stockholm UN Conference on the Environment in 1972 and the Vancouver UN Habitat Conference in 1976. At these meetings key people included Barbara Ward, René Dubos, and Margaret Mead, also steeped in the Judeo-Christian tradition.

Perhaps the unique contribution of the prophetic voice today is in the continuity between ecojustice (justice for the whole order of created things and beings) and social justice. It cautions that a national energy plan must be one where none make an unfair sacrifice and none reap an unfair benefit;[7] but it adds that this is impossible while inequities exist among these who control the supply and information. Vested interests make any fairness principle a sham.

Prophetic voices say that all have a right to the basic necessities and that the needy have a right to take from the rich. One of these necessities is fuel for cooking and heat, which is often obtained from nonrenewable resources. But matters do not rest with the utterance of rights. It follows that in some way the resources must belong to all, and thus some form of control different from what is commonly accepted must be recognized. Thus the message calls for a funda-mental rethinking of social and economic structures, so that worse consequences will not follow. The prophetic voices give to environ-mental ethics urgency, vision, and promise of rational solutions.

Exemplary Community

The tradition that fostered the prophetic word realizes that it must be coupled with deed. But the deed does not necessarily have to be confrontation. It may be the lived experience of peace, harmony, promise, and joy that is held out as a rational alternative to the present state of affairs.

> The faithful all lived together and owned everything in common; they sold their goods and possessions and shared out the proceeds amoung themselves according to what each one needed.
>
> ACTS 2:44-45

The short-lived early Christian communistic community included faithfulness to the word of God; common life, which enhances self-discipline; ownership in *common,* which generally takes the form of simpler lifestyle; getting rid of excesses to be freer for ser-vice; and sharing proceeds with the needy. The religious community life has been expressed in a multitude of forms through the centuries such as the Essene community, desert hermitage, monastery, con-vent, rural commune, or settlement house. Some forms have explicit religious traditions, while a great number in our day—though guided by religious principles—are actually secular in character. Others, like the Taizé community in France, try to blend the rich expressions of the tradition into ecumenical life.

The variety of communities complements the ongoing prophetic witness; when authentic the community shows love in action—see how the members love one another! The community symbolizes what the entire human family is seeking; it attracts those searching for a ground for belief or a haven; it presents concrete examples of a balanced spiritual "ecology," where all parts work together.

This harmony of all the parts—which include plants, animals, and all of creation together with human beings—is expressed in many ways:

as brotherhood:	How good, how delightful it is for all to live together like brothers. <div align="right">PSALM 133:1</div>
as shalom:	Great is Yahweh who likes to see his servant at peace! <div align="right">PSALM 35:37</div>
as messianic age:	The wolf lives with the lamb, the panther lies down with the kid, calf and lion cub feed together with a little boy to lead them. <div align="right">ISAIAH 11:6</div>
as joy:	O soil, do not be afraid; be glad, rejoice, for Yahweh has done great things. Beast of the field, do not be afraid; the pastures on the heath are green again, the trees bear fruit, vine and fig tree yield abundantly. <div align="right">JOEL 2:21–22</div>

The harmony or peace does not consist in mere prosperity and well-being, but essentially in righteousness.[8] It is a completeness, perfection, a condition in which nothing is lacking. The oneness of all creation can be thus reflected in a community of persons who practice righteousness together. It is prophetic word through lived experience. It presents a model of what our earth must become so that a balanced environment might prevail.

The exemplary community exudes power, but not power in physical or ordinary political terms. While not denying a power expressed in such ways as "sons of God," "Christ-in-power," "re-creating the earth," "transforming bread and wine," the community still views that power and dominion (Genesis 1:28) in terms of loving service to those within and outside of its boundaries.

You call me Master and Lord, and rightly; so I am. If I, then, the Lord and Master, have washed your feet, you should wash each

other's feet. I have given you an example so that you may copy
what I have done to you.

JOHN 13:13–15

While the power of human beings working in community is very
real, it still demands constant self-examination and self-discipline.
And this is precisely the importance of a community-in-faith during
the heady times of a technological revolution. Examination shows us
where we have drifted from our beginning commitment; discipline
helps us continue in a spirit of humility and freedom. The community becomes the instrument for maintaining commitment and
humble and loving service as keystones in the exercise of a spiritual
power of the children of God. Without commitment, the vital symbol
of God-always-faithful-to-his-people is lost; without humility, power
degrades into pride, haughtiness, insensitivity, and disrespect for
God, fellow human beings, and all of creation. Individuals who break
with their Judeo-Christian tradition and leave a self-correcting community soon cast about for meaning and value; their system of checks
and balances is often lacking, and they replace loving service with
cynicism and pointless action.

Toward Understanding Stewardship

Always consider the other person to be better than yourself, so
that nobody thinks of his own interests first but everybody thinks
of other people's interests instead.

PHILIPPIANS 2:3–4

Bruce Birch says the key to a biblical understanding of humanity
and nature is relationship.[9] God, humanity, and nature are to be in
harmonious relationship, with nature itself having intrinsic worth
before God (Psalm 24:1) and being witness to God's work. Humanity
is at one with nature in our standing before God in a harmony characterized in the Old Testament as *wholeness*, or shalom. For this perspective of harmony and interdependence, Birch says a greater stress is
placed on the realization of human limitation, of humanity's rooted-
ness in creation, on sin as relational, and on the redemptive task as
including both human and nonhuman creation.

As human beings, we recognize our interdependence with other
creatures on this earth. We are from dust and are what we are because
of our earthly origins. Our relationship both with the Creator and
with other creatures is weakened and even broken through sin, which
is an act of disharmony. However, our limitations and weaknesses can
be overcome and we can be filled by the power to take on our human

responsibilities to other creatures and the environment. But this power is a delegated power, for we are stewards and not owners of the thing of creation (Deuteronomy 8:7–19). If on the one hand we are part of creation, on the other we are empowered to give Christlike service to other creatures. Periodically, the prophetic word recalls us to the responsibility of being creatures-with-power; the faithful communities offer ongoing checks on our practices; the Bible gives us clues to how to exercise stewardship properly.

Sparing the Earth

Many of the basic notions of how to treat nonhuman creatures come from the Book of Genesis. The Genesis account shows:

—God is the Creator of all things of the earth. (1:1–25)
—Earth is good and enters into the whole of human history. (1:26–31)
—Men and women are created in the image of God. (1:27)
—Everything God makes is good. (1:31)
—Humanity has dominion over the things of earth. (1:28)
—Humanity is placed on earth to cultivate and take care of it. (2:15)
—Humanity has power over creatures in the act of naming the animals. (2:20)
—Humanity is obliged to refrain from excessive use (tree of good and evil). (2:16–17)

The Genesis passage shows our relationship with both God and the rest of creation. Creation is in no way evil; it is to be lived with, cultivated, developed, and spared. None of these actions gives us as much trouble as that of "dominion" over the creatures. In this passage may come a philosophy of unwarranted environmental damage. But our human dominance must involve exercising a power of ministry to protect, guard, and enhance creation. Even our way of "dominating" is subject to growth and development.

A false or political notion of dominance can be harmful to those exercising the power. For if humanity is part of this earth and needing to be in harmony with it, then harm to the earth is harm to human beings. Autocratic dominance can lead to damage or destruction to those parts of creation which we might think we have a right to use, consume, or damage. Whether we perceive of reverence and respect for creation as flowing from reverence for God and/or humanity, or even from something within nature itself, it contains a homocentric or self-preservational element.

The cardinal precept not to eat the seed grain, no matter how hard the times, also holds for our relationship to natural resources in general and to all other living species. These are the seeds of life to future generations. Humanity's venture into the future is conditioned by a healthy and varied environment, which in turn is dependent upon many of our present actions. Destruction of parts of the environment is as suicidal as imposing grave risks on future generations; it is a desecration and an act of irreverence to both God and creation. From the biblical account we sense our power to build and to destroy, for contained in our power are both promise and peril.

Recognizing the ambiguous nature of this power is paramount. Here the need for self-discipline is quite important, for power used properly can build, but misused can destroy. Disciplining oneself can often be best done by refraining from use of certain good things, whether these be food, drink, or natural resources. The operative principle *Use resources sparingly* is a way to control our human power and a beginning point of spiritual growth and development.

> Hence, one is to make use of them [created things] in as far as they help him in the attainment of his end, and he must rid himself of them in as far as they prove a hindrance to him.[10]

The practice of sparing resources shows our own commitment to the future. We believe in a tomorrow when another generation will have authentic needs. We are opposed to the philosophy "Let us eat and drink today; tomorrow we shall be dead." (I Corinthians 15:33) An all-absorbing activity of sequestering goods and accumulating wealth is counter to the sparing principle—for after the barns are filled "this very night the demand will be made for your soul." (Luke 12:20)

With the rise of modern technology, people acquired the ability to dominate in ways never before dreamed of, such as through drugs, computer information retrieval, television commercials, and the obvious array of modern weaponry. Thus the need to become self-disciplined and to spare resources has become more evident. A profound ethical issue arises: exactly how much is needed for attaining human goals? Needs may be relative, for the luxuries of one generation may become the necessities of another. Constant reexamination is required lest the list of needs continue to grow.

> Do not weary yourself with getting rich,
> and have nothing to do with dishonest gain.
> PROVERBS 23:4

A quantitative limit to what should be gathered for one's living is shown by a generous, providing God:

> Everyone must gather enough of it [the manna in the desert] for his needs, one omer a head, according to the number of persons in your families.
>
> EXODUS 16:16

But enough is sufficient, and keeps the gatherers from becoming greedy.

The concept of sparing the earth's resources is further seen in the following:

> Land must not be sold in perpetuity, for the land belongs to me [God], and to me you are only strangers and guests.
>
> LEVITICUS 25:23

God had absolute ownership of the Holy Land, and the Chosen People were guests on the land. Thus the resources of the land were not theirs to squander. They were periodically to remember their situation, and even give the soil a rest. "But in the seventh year the land is to have its rest." (Leviticus 25:4) Not only did this sabbatical for the land reflect the need for self-discipline and respect for the bounty God had given, it was also good agricultural practice.

In the jubilee year (every fifty years) a liberation of slaves was enforced, along with a general enfranchisement of people and goods—rejoining clans and recovery of ancestral property (Leviticus 25:24–31). It was a way of redistributing wealth and controlling individual greed. The fact of the matter is that such operations seldom occurred, but the basic principle of sparing resources was set forth.

Repairing the Earth

The constant call for restraint and self-discipline in the Bible has a very sound reason: The human inclination to sin and exceed bounds. The relationship of harmony with God and creatures is broken by human misconduct; the sinner is driven from the primitive state of innocence and obliged to toil for a living (Genesis 3:23). An environmental catastrophe results in the form of a flood (Genesis 6–9), but God saves his faithful remnant and makes a convenant with all of us:

> Here is the sign of the Covenant I make between myself and you and every living creature with you for all generations.
>
> GENESIS 9:12; SEE ALSO 9:16

When human beings sin, the earth is wounded; when God forgives, the earth is healed—for humanity and earth form a single unity.

From the beginning till now the entire creation, as we know, has been groaning in one great act of giving birth.

ROMANS 8:22

But part of human fidelity to a forgiving God involves repairing the damage done through sinfulness. Sin creates a disharmony, but repentance reestablishes wholeness, or environmental balance.

Thus we might develop a second operative principle: *Repair environmental damage.* We each have both a social and an individual responsibility to make restitution for damages done. If we disturb the earth, we are responsible both to human beings and to all of creation.

Fathers may not be put to death for their sons, nor sons for fathers. Each is to be put to death for his own sin.

DEUTERONOMY 34:16

The concept of personal responsibility was slow to develop in human history. And in environmental matters further development is necessary, not only for the many who litter and pollute, but for the corporate "persons" who potentially are able to do far greater damage than individual consumers. Polluters commit a "social" sin, a disharmony that damages the entire social fabric. Often, the perpetrators—groups of individuals or institutions—do not realize their wrongdoing. In such cases, once the wrongdoing is recognized, the culprit—not a governmental agency or the citizenry as a whole—should be made to repair the damage. Corporate executives should bear the responsibility for ecological crimes as much as should individual litterers, and that might even include jail sentences.

Scripture tells us that with repair come forgiveness and reestablishment of harmony with the earth and the Creator. A fragile ecology demands rapid repair, so that the earth also might be quick to forgive. The repentant sinner returns to a land flowing with milk and honey, and therein lies a promise that the earth is resilient and can be repaired, though not returned to a state of ecological purity. The wrongdoer can learn through repairing damage to live closer to nature.

Caring for the Earth

Involved in social sin, we must recognize the obligation to preserve the earth for future generations (Yahweh is the God of Abraham, Isaac, and Jacob—an enduring God), the fragile nature of this finite earth, and our capacity to damage and destroy this earth-gift. Infidelity is forgetfulness of the Creator, of human responsibility, and of the earth itself. Caring for the earth includes a social responsibility for this and future generations and for the earth's resources. To care is to show concern, especially when our neighbor fails to make necessary repairs. For each environmental abuse is a social sin and hurts the entire ecological structure. But social responsibility is never equally shared by all people. Thus we must imitate Christ and take on ourselves the social sin and make restitution through care and concern.

> ...and in my own body to do what I can to make up all that has still to be undergone by Christ for the sake of his body, the Church.
>
> COLOSSIANS 1:24

Caring is part of the duty of the entire human family. The caring person must be observant of where and when damages arise, prompt at responding to the damage, and willing to go out of the way to make restitution. The third operative principle is: *Caring for the earth is the concern of all.* This concern is not just for the culprit but is a common responsibility. Damage is always greater than what can be allocated to individual perpetrators. Thus concern must be greater. An energy waster or an extravagant person is wasting our resources, and a realization of the common heritage of those resources makes us act firmly and promptly to expose and punish the culprit. By our neglect and silence we become accomplices in the social crime.

Fidelity to the Judeo-Christian tradition includes a number of actions such as: reporting polluters of air, water, or land; sponsoring ordinances that restrict noise pollution; demanding proper labeling of indoor pollutants; teaching the family not to litter. Silence in the face of wrongdoing is an expresson of infidelity; it denies the prophetic call; it weakens the community.

Sharing the Earth

The earth does not belong to us absolutely. We receive a trust for a specified time, and then we pass on like the withering grass, but the earth remains. Tillers of the soil, who are in communion with it, understand the transitory nature of crops and even eroding soil—it takes a century to make an inch of soil and a moment to lose it.

Some of us have greater trusts than others, through either birth, acquisition, gift, or accident. Those with less deserve more in justice, not charity. Thus we have an imperative to share both in time (with future generations) and in space (with less fortunate human beings). Furthermore, we must share our earth with the other living cohabitants. A fourth principle emerges: *Share the fruits of the earth with the less fortunate.*

Scripture says that *all* human beings are created in the image of

God, not just those of one's family, social group, religion, or nation. The Good Samaritan parable extends neighborhood from backyard to the ends of the earth. Modern communications make this awareness even more acute, but even in St. Paul's day there was a sense of sharing with distant people; he asked Christians in Greece to support the church of Jerusalem.

The Scriptures contain many examples of the need to share. The parable of Dives and Lazarus in Luke 16:19–31 speaks of punishment for failing to give to the poor. Exhortations to share with others abound:

> When reaping the harvest in your field, if you have overlooked a sheaf in that field, do not go back for it. Leave it for the stranger, the orphan and the widow, so that Yahweh your God may bless you in all your undertakings.
>
> DEUTERONOMY 24:19; SEE ALSO 24:20–21

> If one of the brothers or one of the sisters is in need of clothes and has not enough food to live on, and one of you says to them, "I wish you well; keep yourself warm and eat plenty," without giving them these bare necessities of life, then what good is that? Faith is like that: if good works do not go with it, it is quite dead.
>
> JAMES 2:15–16

> Tell them that they are to do good, and be rich in good works, to be generous and willing to share.
>
> I TIMOTHY 6:18

> If your brother who is living with you falls on evil days and is unable to support himself with you, you must support him as you would a stranger or a guest, and he must continue to live with you.
>
> LEVITICUS 25:35

Sharing is an integral part of the biblical message, and many passages show God's favor to those who imitate the God who shares with his people. Sharing concretizes the unity of faith and justice, wherein believers show fidelity to a generous God and manifest it by establishing justice for their neighbors throughout the world. Sharing is closely associated with sparing the earth, for to share is to require less for more people; sharing taps the human resources that are wasted by those scratching for the bare essentials of life, and gives them an opportunity to develop talents and further the glorification of God's creation. Sharing is a form of communication—the best form within the Judeo-Christian tradition—between those who have something to give and those who are in need.

Responding to the question that Christ will be asked: "When did we see you hungry and feed you?" (Matthew 25:37) means giving in charity to the needy. But there is perhaps more—a hunger for justice that must be fulfilled. *Ecojustice*, or the need to deal justly with the environment, is not removed from the problem of justice and human rights that is felt throughout the world. For humanity and earth are united, and what is a disharmony for one is so for the other. So to share with the oppressed in such a way that they are empowered is really to give and share something with human beings and the earth itself. Thus the ultimate sharing that we can do is to help the oppressed gain control over their own community, and, in our so doing, both the social [and natural] environment will surely benefit.

Developing the Concept of Stewardship

Stewardship in its rudimentary meaning refers to management, a management which uses no more of the available resources than needed (a sparing), which does not allow damage to go unattended (repairing), which includes a proper dominion (caring), and which looks out for others' needs (sharing). Wealth, power, or resources are held in trust and include serious social responsibilities.

> What is expected of stewards is that each one should be found worthy of his trust.
>
> I CORINTHIANS 4:2

Stewardship need not be considered only in economic terms but embraces the entire complexus of human living in our environment. As William Byron says, stewardship means wise use, and this can also apply to the ideas needed to develop renewable energy alternatives.[11] Max Stackhouse extends stewardship to include three areas: care for persons, care for spirituality, and care for social institutions.[12]

The oldest Christian traditions understood the use of property to be in common, with private ownership being a concession to human weakness.[13] Holding things in common was an ideal of the early Christian community (Acts 2 and 4), and the tradition was followed in subsequent ages in religious communities and even down to this day. To use things in common has a direct bearing on our environment, for air and water and energy resources are held by all and are not for the exclusive use of a few. So those who damage the water quality or pollute the air or waste energy are really taking from what belongs to a common heritage, and in some way are stealing from their fellow human beings.

Christians have struggled with the concept of economic activity as either a reluctantly tolerated necessity or a potential positive activity to be integrated into everyday living. Stackhouse says the

Historical Development of Stewardship Concept

Century	Historical Situation	Ideal Christian Ethical Response
1st–4th	Hostility of pagan Empire	Learn to live in persecuted circumstances and to conserve spiritual resources.
5th–7th	Barbarian invasions of West	Adopt flexible educational method (monasteries) in order to civilize and retain culture.
8th–12th	Arab and Turkish conflicts	Integrate ancient philosophy and literature into current Christian thought
13th–15th	Trading with the East	Expand vision of the West to entirely new and respected cultures
16th	Circumnavigation of globe	Accept the concept of a finite globe on which a limited number of people live
17th and 18th	Scientific revolution	Develop cosmic consciousness, with the earth as one part of the universe
18th–20th	Industrial revolution	Conserve human labor and make the machine work for human betterment
19th	Evolutionary theory	Be aware of the vast expanse of time to develop the species
20th and on	Age of scarcity	Become aware of the need to share resources and power with less fortunate members of the human family.

early Church recognized the tension between Christianity and private property by requiring the vow of poverty from the clergy while allowing the ordinary believer to own property. The later Middle Ages showed great suspicion of economic activity, considering money both unnatural and unspiritual.[14] Stackhouse believes that "this two-level morality and anti-material spirituality was broken by the Reformation, which generated new economic attitudes and institutions, but no new clear and sustained economic ethic."

The understanding of stewardship has evolved during the Christian era, with much effort centered on the use of money and, to a lesser extent, real property. Often internal church discipline was the reason, but by no means the only one. In fact various historical conditions and events are responsible for expanding the Christian's self-understanding of being a steward and the Church's consciousness in developing and fostering proper stewardship. While the Church's awareness of its role was not perfect, it does advance beyond biblical times, as indicated on the preceding chart.

With a growing understanding of stewardship, one may expect improved personal attitudes about the environment: less waste, more reclamation, better plant and animal protection, more sharing of resources. But with this change of heart must come the fullness of the message of the Tradition, and that is the vision of a "new heaven and a new earth" (Isaiah 65:17; Revelation 21:1). The old is not good enough, even though creation was good. The creative act is an ongoing process, and the believers partake in that act. It is part of nature that improvement occurs, and it is part of human nature to help bring plants and animals as well as other people to a better quality of life.

References

1. L. White, Jr., "The Historical Roots of Our Ecologic Crisis," *Science* 155 (1967), p. 1207.

2. L. W. Moncrief, "The Cultural Basis for Our Environmental Crisis," *Science* 170 (1970), p. 508.

3. A. Fritsch, *Theology of the Earth* (Washington, D.C.: CLB Publishers, 1972).

4. "This Land Is Home to Me: A Pastoral Letter on Powerlessness in Appalachia by the Catholic Bishops of the Region" (Prestonsburg, Ky., 1974), p. 5.

5. "The Plutonium Economy, Study Material for the Proposed Policy Statement," National Council of the Churches of Christ in the USA (New York, 1976), p. 2.

6. *Documents of Vatican II:* No. 69, "The Church Today," W. M. Abbott, ed. (New York: Guild Press, 1966), p. 278.

7. President Jimmy Carter, "Preface to The National Energy Plan" (April 29, 1977), Washington, D.C., pp. x–xi.

8. J. L. McKenzie, *Dictionary of the Bible* (Milwaukee: Bruce Publishing Co., 1965), p. 651.

9. B. C. Birch, "Nature, Humanity and God," Subcommittee on Energy Ethics, National Council of Churches Taskforce (May 21, 1977), p. 14.

10. L. J. Puhl, *The Spiritual Exercises of St. Ignatius* (Westminster, Md.: Newman Press, 1954), p. 12.

11. W. Byron, *Toward Stewardship: An Interim Ethic of Poverty, Pollution and Power* (New York: Paulist Press, 1975), p. 14.

12. M. L. Stackhouse, "Toward a Stewardship Ethics," *Theology Digest* 22 (Autumn 1974), p. 231.

13. W. Byron, op. cit., p. 16.

14. M. L. Stackhouse, op. cit. p. 229.

15. R. Dubos, *The God Within* (New York: Charles Scribner's Sons, 1972).

16. R. Dubos, "Think Globally, Act Locally," *Environmental News* (Boston, Mass.: U.S. Environmental Protection Agency, New England Regional Office, May 1978), p. 11.

Robert Selle's short essay from *American Forests* supplies a light comment on the more serious approach taken by Fritsch. You can identify the various themes of Fritsch's essay in Selle's message from the Almighty. Mr. Selle served as a Lutheran missionary in the Philippines.

CHAPTER THIRTY

A "Divine Land Ethic":
On Subduing the Earth

ROBERT SELLE

Dear AFA:

I am the owner of a substantial plot of real estate called Earth, and I read with interest an invitation to your readers/members to share their own personal land ethic.

Since the bulk of your readers are tenants of my forests and fields, and since there's been a terrible misunderstanding of my environmental stance, please permit me to set the record straight. Unfortunately, my own words have been quoted and abused, particularly those recorded in Genesis 1:28: "Be fruitful and multiply, and fill the earth and subdue it; and have dominion over the fish of the sea and over the birds of the air and over every living thing that moves upon the earth."

One example of this misuse of my words appeared recently in the pages of your respected publication:

> Our forebears thought it was their theological duty to subdue the American wilderness. Wild places were worse than useless; they were an embodiment of some evil. Trees were weeds, wild animals were unclean and expendable. Cleared land was not only a physical need but a moral need. They were simply carrying out the scripture of Genesis, "fill the earth and subdue it." It was this view, nourished on these words, that has fed continously on the American forest, the wild prairies, and today the township woodlots, which contain more than half the nation's timber base.

If words such as these must be printed as a historical reflection of human error, please take care to issue a resounding rebuttal on behalf of God! Scriptural theology has too long absorbed the blame for a

This article reprinted from *American Forests*, December 1983, pp. 183–184.

polluted and abused environment. If some misguided souls have adopted a philosophy such as described above, they are feeding on their own ignorance, and not on my clear teachings!

Accusing fingers point to disappearing species of flora and fauna, to eroded and overgrazed pastures, to an industrial dump bubbling with noxious chemicals, to a clearcut mountainside cluttered with slash, and then indict my divine command as the cause of all this. What injustice!

I intended my creation to be used with love and respect. Yes, I want it to be *used*, for the good of humanity and for the glory of the Heavenly Father, to be *used* for shelter, food, clothing, medicines, refreshment, and inspiration. Remember the psalmist who rhapsodized the mountains around Jerusalem as an analogy of the divine and powerful presence "round about his people from this time forth and forevermore"?

In addition, I carefully designed my creation as a sustained-yield entity, something that could be *reused* again and again if given the chance, if treated gently. That's why plants yield seed, each according to its kind. That's why living creatures multiply according to kind. That is precisely how I renew, in a cycle of centuries.

Surely I did *not* wish my creation abused and destroyed by human greed, nor did I plan for it to be paved over, poisoned, polluted, or frittered away. The fault for a sick world should be placed squarely where it belongs—on humans who use land and trees, water and air for their own selfish purposes. I'm sick and tired of getting the blame for the rape of my earth by ignorant pioneers, greedy industrialists, and quick-profit farmers. And no conservationist who confesses the Almighty God should ever be shamed with an inference that his God is anti-environment.

After all, I *made* the earth, didn't I? And when I looked at what I had accomplished, I saw that it was good. I intended harmony. In fact, I *made* harmony, a harmony that your ancestors corrupted and destroyed. Put the blame where it belongs—on yourselves!

Don't use the feeble excuse that you misunderstood the word "subdue," since I gave clear and definitive guidelines for land-use in Leviticus 25: "You do not own [your land]; it belongs to God, and you are like foreigners who are allowed to make use of it."

I tried hard to help Job recognize the awe that all of you should feel as you consider proper stewardship of my creation:

Were you there when I made the world: If you know so much, tell me about it. Have you ever in your life commanded a day to dawn? Have you walked on the floor of the ocean? Who makes rain fall where no one lives? Who waters the dry and thirsty land so that grass springs up? Do you find food for lions to eat? Who is it that feeds the ravens when they wander about hungry, when their young cry to me for food? Do you know when mountain goats are born? Who gave the wild donkeys their freedom? I gave them the desert to be their home. Does a hawk learn from you how to fly when it spreads its wings toward the south? Does an eagle wait for your command to build its nest high in the mountains? Stand up now like a man and answer my questions. Are you trying to prove that I am unjust—to put me in the wrong and yourself in the right? . . .

(EXCERPTS FROM JOB 39 AND 40)

Please note that I didn't inspire Scripture primarily as an environmental textbook. The Bible is basically a work that speaks of man's relationship to God and to his fellowmen. Yet you can still find numerous clues as to how I regard my Earth.

Look at divine care for the birds, and how beautifully I clothe the wildflowers (Matthew 6). Examine my concern for agricultural land, and how I commanded my Jewish children to give the earth a complete rest every seventh year, to let the soil lie fallow (Leviticus 25). Consider how I make the springs to flow so the wild animals can drink and the nearby trees be nourished (Psalm 104). Yes, I appointed man to be ruler over everything I have made. I placed him over all creation (Psalm 8), but I continually have hoped that he would use this earth to honor my greatness, not as a monument to his own greed and pettiness.

Most members of AFA are stewards of the woodlands. That is good. Who would impute that my sole purpose in the creation of a tree is for its destruction? I made trees to be used, certainly, in a creative and appreciative way. Olive trees give oil, fig trees give fruit, grapevines produce wine, and thornbushes offer shade. This honors me and benefits you (Judges 9:7–13). I delight that some trees are utilized for medicine, cosmetics, graceful ships, balms, paper, food, warmth, hedges that fence a field, and even the walls of a church. I am happy that other trees are preserved as a den for my raccoon, a nest for the pileated. I am pleased that you sit in the shade of another for thought and meditation, and that you save a twisted and ostensibly worthless grandfather of the forest to remind you of life's lessons.

My divine land ethic can be summarized in two words: *loving use.* I have always meant for land to be used wisely for the good of humanity and for the glory of God. If the use of a plot of ground fulfills these two conditions, that's great!

I recognize that sometimes, the world being what it is, the well-intentioned human may be caught in a no win situation as he attempts to use land in a healthy fashion. Centuries of injustice and greed have made solutions complicated at best, and occasionally impossible.

I think, for instance, of a certain tropical forest where the hornbills sing and orchids bloom for my eyes only. There is a forester there, hired by a logging concession, who plans to cut on a sustained-yield basis. There is a village on the edge of the forest whose hungry residents follow the loggers and make their slash-and-burn farms to raise sweet potatoes and upland rice for their children. There is an environmentalist, committed to preserving the rare monkey-eating eagle that nests high in the forest canopy. Three different motivations intertwine: lumber production for a wood-hungry world, hunger of malnutritioned children who cluster around village tables, and the admirable desire to save one of my unique creations from extinction through the obliteration of its habitat. Any of these three, pushing plans to completion, will cause an evil. The lumber will rot unused, or a child will starve, or a species will die. Perhaps there is room for all three, but usually our society does not allow it, and God gets the blame.

The means to a discerning land ethic is the same means toward a happy life: justice, wisdom, sharing, mercy, love. There is no law against these things, and if they are consistently applied, the questions of proper land use would not even arise, for there would then be no hunger or hate, no waste or wastelands, no pollution or tears. (But now I am straying to a different topic: Heaven.)

If you would permit a piece of fatherly advice, and if you are one of those concerned citizens who looks around and despairs over the foolhardy contamination and waste of air and water, marsh and woodland, soil and sea, don't first try to change the whole world. First, change yourself. And if you need help, remember that I'm here.

Sincerely,
GOD

Notes

1. See the December 1982 and April and May 1983 issues [of *American Forests*]

2. *American Forests*, December 1982, "In Defense of Woodlots" by Justin Isherwood, page 33.

The question in Robert Heilbroner's title is an old joke, but his purpose is serious. He states that, from a purely rational perspective, posterity holds only a tenuous claim upon us. But he rejects that position, arguing instead that humanity has a moral duty to provide for the survival and sustenance of future generations. Compare Heilbroner's views with Klemperer's views on sustained yield for a better understanding of the contrast between the economic and a moral perspective on these questions.

Heilbroner is a noted author and essayist on social problems who has spent many years on the faculty of the New School for Social Research in New York.

CHAPTER THIRTY-ONE

What Has Posterity Ever Done for Me?

ROBERT L. HEILBRONER

Will mankind survive? Who knows? The question I want to put is more searching: Who cares? It is clear that most of us today do not care—or at least do not care enough. How many of us would be willing to give up some minor convenience—say, the use of aerosols—in the hope that this might extend the life of man on earth by a hundred years? Suppose we also knew with a high degree of certainty that humankind could not survive a thousand years unless we gave up our wasteful diet of meat, abandoned all pleasure driving, cut back on every use of energy that was not essential to the maintenance of a bare minimum. Would we care enough for posterity to pay the price of its survival?

I doubt it. A thousand years is unimaginably distant. Even a century far exceeds our powers of emphathetic imagination. By the year 2075, I shall probably have been dead for three quarters of a century. My children will also likely be dead, and my grandchildren, if I have any, will be in their dotage. What does it matter to me, then, what life will be like in 2075, much less 3075? Why should I lift a finger to affect events that will have no more meaning for me seventy-five years after my death than those that happened seventy-five years before I was born?

There is no rational answer to that terrible question. No argument based on reason will lead me to care for posterity or to lift a finger in its behalf. Indeed, by every rational consideration, precisely the opposite answer is thrust upon us with irresistible force. As a Distinguished Professor of political economy at the University of London [wrote] in the winter [1974] issue of *Business and Society Review:*

Suppose that, as a result of using up all the world's resources, human life did come to an end. So what? What is so desirable about an indefinite continuation of the human species, religious convictions apart? It may well be that nearly everybody who is already here on earth would be reluctant to die, and that everybody has an instinctive fear of death. But one must not confuse this with the notion that, in any meaningful sense, generations who are yet unborn can be said to be better off if they are born than if they are not.

Thus speaks the voice of rationality. It is echoed in the book *The Economic Growth Controversy* by a Distinguished Younger Economist from the Massachusetts Institute of Technology:

> . . . Geological time [has been] made comprehensible to our finite human minds by the statement that the 4.5 billion years of the earth's history [are] equivalent to once around the world in an SST. . . . Man got on eight miles before the end, and industrial man got on six feet before the end. . . . Today we are having a debate about the extent to which man ought to maximize the length of time that he is on the airplane.
>
> According to what the scientists now think, the sun is gradually expanding and 12 billion years from now the earth will be swallowed up by the sun. This means that our airplane has time to go round three more times. Do we want man to be on it for all three times around the world? Are we interested in man being on for another eight miles? Are we interested in man being on for another six feet? Or are we only interested in man for a fraction of a millimeter—our lifetimes?
>
> That led me to think: Do I care what happens a thousand years from now? . . . Do I care when man gets off the airplane? I think I basically [have come] to the conclusion that I don't care whether man is on the airplane for another eight feet, or if man is on the airplane another three times around the world.

Is it an outrageous position? I must confess it outrages me. But this is not because the economists' arguments are "wrong"—indeed, within their rational framework they are indisputably right. It is because their position reveals the limitations—worse, the suicidal dangers—of what we call "rational argument" when we confront questions that can only be decided by an appeal to an entirely different faculty from that of cool reason. More than that, I suspect that if there is cause to fear for man's survival it is because the calculus of logic and reason will be applied to problems where they have as little validity, even as little bearing, as the calculus of feeling or sentiment applied to the solution of a problem in Euclidean geometry.

If reason cannot give us a compelling argument to care for posterity—and to care desperately and totally—what can? For an answer, I turn to another distinguished economist whose fame originated in his profound examination of moral conduct. In 1759, Adam Smith published "The Theory of Moral Sentiments," in which he posed a question very much like ours, but to which he gave an answer very different from that of his latter-day descendants.

Suppose, asked Smith, that "a man of humanity" in Europe were to learn of a fearful earthquake in China—an earthquake that swallowed up its millions of inhabitants. How would that man react? He would, Smith mused, "make many melancholy reflections upon the precariousness of human life, and the vanity of all the labors of man, which could thus be annihilated in a moment. He would, too, perhaps, if he was a man of speculation, enter into many reasonings concerning the effects which this disaster might produce upon the commerce of Europe, and the trade and business of the world in general." Yet, when this fine philosophizing was over, would our "man of humanity" care much about the catastrophe in distant China? He would not. As Smith tells us, he would "pursue his business or his pleasure, take his repose or his diversion, with the same ease and tranquillity as if nothing had happened."

But now suppose, Smiths says, that our man were told he was to lose his little finger on the morrow. A very different reaction would attend the contemplation of this "frivolous disaster." Our man of humanity would be reduced to a tormented state, tossing all night with fear and dread—whereas "provided he never saw them, he will snore with the most profound security over the ruin of a hundred millions of his brethren."

Next, Smith puts the critical question: Since the hurt to his finger bulks so large and the catastrophe in China so small, does this mean that a man of humanity, given the choice, would prefer the extinction of a hundred million Chinese in order to save his little finger? Smith is unequivocal in his answer. "Human nature startles at the thought," he cries, "and the world in its greatest depravity and corruption never produced such a villain as would be capable of entertaining it."

But what stays our hand? Since we are all such creatures of self-interest (and is not Smith the very patron saint of the motive of self-interest?), what moves us to give precedence to the rights of humanity over those of our own immediate well-being? The answer, says Smith, is the presence within us all of a "man within the beast," an inner creature of conscience whose insistent voice brooks no disobedience: "It is the love of what is honorable and noble, of the grandeur and dignity, and superiority of our own characters."

It does not matter whether Smith's eighteenth-century view of human nature in general or morality in particular appeals to the modern temper. What matters is that he has put the question that tests us to the quick. For it is one thing to appraise matters of life and death by the principles of rational self-interest and quite another *to take responsibility for our choice.* I cannot imagine the Distinguished Professor from the University of London personally consigning humanity to oblivion with the same equanimity with which he writes off its demise. I am certain that if the Distinguished Younger Economist from M.I.T. were made responsible for determining the precise length of stay of humanity on the SST, he would agonize over the problem and end up by exacting every last possible inch for mankind's journey.

Of course, there are moral dilemmas to be faced even if one takes one's stand on the "survivalist" principle. Mankind cannot expect to continue on earth indefinitely if we do not curb population growth, thereby consigning billions or tens of billions to the oblivion of non-birth. Yet, in this case, we sacrifice some portion of life-to-come in order that life itself may be preserved. This essential commitment to life's continuance gives us the moral authority to take measures, perhaps very harsh measures, whose justification cannot be found in the precepts of rationality, but must be sought in the unbearable anguish we feel if we imagine ourselves as the executioners of mankind.

This anguish may well be those "religious convictions," to use the phrase our London economist so casually tosses away. Perhaps to our secular cast of mind, the anguish can be more easily accepted as the furious power of the biogenetic force we see expressed in every living organism. Whatever its source, when we ask if mankind "should" survive, it is only here that we can find a rationale that gives us the affirmation we seek.

This is not to say we will discover a religious affirmation naturally welling up within us as we career toward Armageddon. We know very little about how to convince men by recourse to reason and nothing about how to convert them to religion. A hundred faiths contend for belivers today, a few perhaps capable of generating that sense of caring for human salvation on earth. But, in truth, we do not know if "religion" will win out. An appreciation of the magnitude of the sacrifices required to perpetuate life may well tempt us to opt for "rationality"—to enjoy life while it is still to be enjoyed on relatively easy terms, to write mankind a shorter ticket on the SST so that some of us may enjoy the next millimeter of the trip in first-class seats.

Yet I am hopeful that in the end a survivalist ethic will come to the fore—not from the reading of a few books or the passing twinge of a pious lecture, but from an experience that will bring home to us, as Adam Smith brought home to his "man of humanity," the personal responsibility that defies all the homicidal promptings of reasonable calculation. Moreover, I believe that the coming generations, in their encounters with famine, war, and the threatened life-carrying capacity of the globe, may be given just such an experience. It is a glimpse into the void of a universe without man. I must rest my ultimate faith on the discovery by these future generations, as the ax of the executioner passes into their hands, of the transcendent importance of posterity for them.

The contrast between economic and ethical mandates for resource conservation has long been with us. While Worrell offered an economic twist on the Land Ethic, Mark Sagoff, in this selection, finds economic reasoning inadequate to the task. Instead, he focuses on land use regulation as a policy problem, offering a stimulating critique of the economic assumptions that are applied to natural resources, but in the end he does not fully reject economics.

Aldo Leopold was accustomed to rising at dawn, boiling a pot of coffee on a fire in front of his Sand County cabin, taking in the sunrise, and enjoying a walk in the cool summer morning. Suppose he could invite Worrell and Sagoff for a visit. Imagine a three-way discussion between these authors on the Land Ethic. Think of the areas of agreement and difference they would find.

Sagoff is a philosopher at the University of Maryland who has written extensively on environmental policy and ethics.

CHAPTER THIRTY-TWO

Do We Need a Land Use Ethic?

MARK SAGOFF

Economists often argue that environmental problems are economic problems and require economic solutions.[1] They say that when externalities are internalized, the commons divided, and "fragile" values priced, the environment will be adequately protected.[2] Most environmentalists believe that this is only half true. Many of our environmental problems do have economic causes. Environmentalists tend to believe, however, that the solutions had better be ethical ones. An economic cure may be worse than the disease.

A policy is economically "efficient" insofar as it satisfies preferences which consumers reveal or would reveal in markets. These preferences call for environments that can be supplied and maintained at a profit. These include trailer parks, fast-food restaurants, golf courses, condominiums, and pinball arcades. Demand for these environments is assured because it is created along with the environments themselves.[3] Brian Harry, chief naturalist at Yosemite National Park when the MCA Corporation took over its management, put the matter succinctly. "People used to come here for the beauty and the serenity," he said. "Those who come now don't mind the crowds; in fact, they like them. They come for the action."[4]

In this essay, I argue that economic approaches to environmental policy succeed, in principle, to the extent that they satisfy aggregate consumer demand. Consumers are likely to demand what Harry calls "action." By *action* may be meant anything that an attractive man and woman can be photographed enjoying together in a state of undress or semi-undress. It may include virtually any activity of consumption, especially conspicuous consumption, that is shown in TV or written up in glossy magazines. Economic solutions to environmental problems could, then, provide the "action" that consumers

Reprinted by permission from *Environmental Ethics*, Vol. 3 (Winter 1981), pp. 293–308.

demand. This might be done on the grand scale, for example, by converting wilderness areas into resort complexes and amusement parks. It happens on a smaller scale each time a woods or a marsh is replaced by a health spa, a gas station, or a fast-food stand.

This conversion of the natural into the efficient conflicts with many principles and convictions that ought to influence environmental policy, whether consumers act on them or not. These principles, which have nothing conceptually to do with what consumers are willing to pay for, may require us to protect even "useless" endangered species and their habitats. These principles and ethical convictions suggest that we owe more to a million year old wilderness than to regard it as a future site for strip mines and singles bars. A market or an economic approach to environmental policy, if it remains consistent and true to itself, is likely to convert nature into a place "where the action is." A policy based on moral conviction and aesthetic principle, however, would tend to preserve some environments for their own sake.

The land use ethic we have is primarily economic. It assumes that the *chooser* is always the *consumer*. It stands on the premise that consumers reveal their preferences in what they buy or would buy if the price were right. The goal is to satisfy the greatest number of the most intense consumer preferences at the least cost. Another goal may be to redistribute wealth or opportunities so that even the least wealthy people can buy what they most want. In a world in which resources are plentiful and demand is scarce—in Eden, for example—everyone can have a mansion by the sea. But in our world, it has to be a motel room. In our world you have to franchise enough pizza huts and build gas stations to keep up with demand. You subdivide the old estates and unload the duplexes and the split foyers.

This land ethic takes it as a premise that consumer preferences reveal the values of our society. People want what they want—not what some public official thinks they ought to want. And they should have what they are willing to pay for: Winnebagos, powerboats, bowling alleys, movies. What argument can be given for failing to satisfy the wants of the vulgar mass . . . and pleasing the tastes of an affluent elite instead?

Once the question has been asked in this way, I believe, there is no way to answer it, but answers are attempted. Economists remind us that exploitation of the natural environment is often inefficient. They describe externalities, spillover effects, and pollution costs. They talk about the "shadow" price of "intangible" or "fragile" values. They speak in terms of the "problem of the commons"—as if that shows, somehow, the folly of translating natural beauty into urban blight.

I believe that this sort of thinking supports an economic approach to land use and land management. To worry about fragile values or the problem of the commons is not necessarily to criticize the economic ethic; it may be to defuse criticism. It is to make environmental law in the image of economic efficiency. It is to rest public policy on a platform of analytical sophistication in the service of consumer demand.

In this essay, I argue that an economic approach to land use leads to, or at least may justify, the destruction of our remaining waterfront and wilderness areas, even if fragile values and the problem of the commons are taken into account. Consumers do not share merely a natural commons; they also share roads, trailer parks, and fast-food stands. And they may find a lot of amenity in a cheap motel if it has its own swimming pool. The satisfaction of consumer preferences may involve the protection of highways as much as the protection of habitats—for we travel in off-road vehicles, snowmobiles, trailers, pickups, caravans, cabin cruisers, motorcycles, trail bikes, busses, houseboats, campers, broncos, vans, and hobicats. The effect is obvious. The sprawl of the city will spread to the sea. If you do not believe me, look around.

Let me begin with the view that land has value primarily not *as land* but *as property.* We owe this thought directly to John Locke, particularly, to chapter 5 of the *Second Treatise of Government.* There, Locke writes:

> God, who hath given the World to Men in Common, hath also given them reason to make use of it to the best advantage of Life, and convenience. The Earth, and all that is therein, is given to Men for the Support and Comfort of their being. And though all the Fruits it naturally produces ... belong to mankind in common, ... yet being given to the use of Men, there must of necessity be a means *to appropriate* them some way or other before they can be of any use, or at all beneficial to any particular Man.[5]

Locke rests his argument on the observation that land has little or no value except when labor changes its character and thus, as it were, forces its favors. Locke says:

> Land which is wholly left to Nature, that hath no improvement of Pasturage, Tillage, or Planting, is called, as indeed it is, *wast;* and we shall find the benefit of it amount to little more than nothing.[6]

And, again:

. . . labor makes the far greatest part of the value of things, we enjoy in this World: And the ground which produces the materials, is scarce to be reckoned in, as any, or at most, but a very small part of it. . . .[7]

And, again:

Tis *labour . . . which puts the greatest part of Value upon Land.*[8]

Although Locke adheres to a labor theory of value, he does not, therefore, deny that there are moral limitations on land ownership. To be sure, Locke contends that land, water, air, and minerals are virtually worthless in their natural state. Yet he did not infer from this that a person may rightfully possess *any* unowned resource into which he "mixes" his labor. Locke restricts ownership, at least at first, to that which an individual can use without waste or spoilage.[9] Second, Locke allows a rightful original claim to land only when there exists "enough and as good" for others.[10] If all, or almost all, of a resource is already owned, then an individual has arrived late: he must buy from others. And this he can do if he has *money.*

If the labor theory causes Locke to place these two moral limitations on the acquisition of property, the theory of money allows him to overcome these restrictions. As soon as people can trade for money, scarcity and spoilage no longer limit the amount of property one person may accumulate. As for the spoilage limitations, Locke says:

. . . a man may fairly possess more land than he himself can use the product of, by receiving in exchange for the overplus, Gold and Silver, which may be hoarded up without injury to any one, these materials not spoiling or decaying in the hands of the possessor.[11]

The use of money, so Locke argues, also permits a person to acquire resources rightfully even after they have become scarce. A person has only to buy them from someone who has a rightful title. And that a person will do, at least in theory, only if he can make a more profitable, more efficient, and therefore more beneficial use of the resource. Accordingly, Locke reasons that a person can "heap up" as much land and wealth as he can use or cause to be used economically—"the *exceeding of the bounds of his* just *property* not lying in the largeness of his possession, but the perishing of anything uselessly in it."[12]

A labor theory of value and a money theory of value—these Locke uses to transcend the moral limits he himself has placed on property ownership. Land, like any other resource, is worth only what you can get for it. It is worth what you can do with it or perhaps to it; its value is what you can sell it or its products for. Many texts in resource economics repeat this message. "In principle, the ultimate measure of environmental quality," says one standard introduction today, "is the value people place on these . . . services or their *willingness to pay*."[13]The concept of economic efficiency—the idea that resources should be allocated to those who derive the greatest benefit from them—is not mentioned by Locke explicitly. Locke seems to believe, however, that something like this concept allows him to transform a natural right to the property one needs into an unlimited right of acquisition.

I wish Locke could see Ocean City today. I wonder what he would say about the commercial holocaust that sweeps from Baltimore, Maryland to Norfolk, Virginia. I would tell him that it is all private property—all owned and properly franchised. I would show him that we have mixed our labor with the land—and our exhaust, sludge, cans, bottles, shopping carts, newspapers, tires, boxes, cigarettes, fenders, furniture, toxic chemical wastes, and everything else we can bury or throw away. Locke plainly had farms in mind as his paradigm of private property. What would he say about tract developments, chemical dumps, and commercial strips?

Locke, perhaps, might deplore what seem to be the consequences of his theory of property. Many of us would agree with him. Many have said that Locke's theory of natural property rights—and his view that government exists to protect those rights—rests on a mistake. But what mistake? Where did he, or we, go wrong?

Many writers have proposed answers to this question. They have suggested that Locke's theory is inadequate because it:

- leads to inequity and injustice;
- assumes that man has domination over nature or that nature exists simply for his sake;
- works only when resources are plentiful;
- ignores the ecological or biological function of the land;
- fails to consider the problem of the commons;
- neglects "fragile," "intangible," or "aesthetic" values.[14]

I believe that something is to be said for each of these criticisms. I argue, however, that none of them gets to the essential shortcoming of the Lockean ethic for the use of land.

That *property* is the origin of *inequality* is a thesis commonly argued—and brilliantly argued, for example, by Rousseau. His *Discourse on the Origin and Foundations of Inequality* makes the following observation against Locke:

> ... from the moment one man needed the help of another, as soon as they observed that it was useful for a simple person to have the provisions of two, equality disappeared, property was introduced, labor became necessary; and vast forests were changed into smiling fields which have to be watered with the sweat of men, and in which slavery and misery were soon seen to germinate and grow with the crops.[15]

If the only criticism to be made of Locke's theory of property is that it leads to inequality and injustice, then, it seems, we would know how to correct it, to make it perfect. We would outlaw slavery. We would redistribute wealth. We could restructure taxation, inheritance, education, health insurance, and other institutions to mitigate the effects of the concentration of wealth, or alleviate them entirely. The fact that we fail to do this is not a consequence of a Lockean approach to land as property. It is to be blamed, rather, on our unwillingness to distribute justly the wealth which a Lockean approach to land use creates.

If we took Locke to Ocean City, moreover, what would he see? Would he see a lot of people starving while a privileged few eat salmon and drink Chateaux Margaux? No. Locke would see just about everyone lined up for steamed crabs, ice cream, and beer. Not bad—for $3.95. To show Locke Ocean City is not to present to him the horrors of social inequality. It is to show him the horrors of social equality. Schlock on every block. K-Mart lowers the price. Who can complain when $250 on the used-car market buys an eight-cylinder, four-on-the-floor '73 Impala with mags and stripes? It goes from 0 to 80 in ten seconds flat. Not even the rich in the Honda CVCCs can do that. So what if it needs a muffler? Locke might point out that it is no accident that the last bastions of beauty on the Eastern Shore are the estates of the rich. There are a lot of compelling reasons to redistribute wealth in the United States—but environmental quality is not one of them. We already have plenty of people this side of the poverty line to crowd every Ho Jo, Go Go, Disco, and peep show that can be built between Rehoboth and Virginia Beach.

If social injustice is not the problem, what is? Lynn White, a historian of science and technology, in a well-known essay, has said that our environmental woes derive, in large part, from "Judeo-

Christian attitudes toward man's relation to nature," attitudes which permit us to regard ourselves as "superior to nature, contemptuous of it, willing to use it for our slightest whim."[16] I am not sure to what extent the Judeo-Christian tradition represents nature as an object for man's domination, and to what extent it represents nature as an object for respect and veneration, "for nature," as Sir Thomas Browne has written, "is the art of God."[17] Locke, as we have seen, does introduce his theory of property with the remark that God gave the Earth to men "for the Support and Comfort of their being."[18] May we infer from this that Locke's theory fails because it assumes that man dominates and is the steward of nature?

I do not think that this is the problem. Locke refers to Genesis in a rhetorical way; this does not show, however, that his theory depends upon a traditional or religious conception of man's relation to nature. Indeed, I do not believe that any conception of man's "place" in the universe is required by Locke's theory or entailed by it. Locke seems to start from the unavoidable needs and the natural rights which belong, or so he argues, to every individual.

Locke's theory depends, as I think, on the unobjectionable view that individuals have a right to secure their freedom and to pursue their happiness however they will as long as they respect the same right in others. They are to be constrained by government only to the extent necessary to protect the rights of others. This, anyway, seems to be part of Locke's conception of freedom: freedom, in general, is the ability of a person to do as he wishes and to get what he wants, except insofar as he is constrained by rules of competition and cooperation common to all.[19] And what is happiness? Happiness, we may think, is bound up with the satisfaction of desire. This is the reason that an economically efficient land use policy—one that maximizes the satisfaction of consumer wants over the long run—appears to many of us to be an ethical policy as well. It gives people what they desire. Locke's theory of property has this ethical appeal. It does not rest on a conception of man's place in nature. It asks us only to make such use of nature—for example, offshore drilling and highway construction—which, over the long run, gives more consumers more of whatever they demand.

Writers sometimes argue that Locke's theory of property may work in a world in which resources are plentiful and await development, but it breaks down when resources become scarce. William Ophuls, for example, has written:

> His [Locke's] argument on property by appropriation is shot through with references to the wilderness of the New World, which only needed to be occupied and cultivated to be turned into property for any man who desired it. Locke's justification of

original property and the natural right of men to appropriate it from nature thus rests on cornucopian assumptions: there is always more left; society can therefore be libertarian.[20]

Much as I admire Ophuls' book, *Ecology and the Politics of Scarcity,* I believe that his criticism of Locke misses the point. Locke understands that it is not the availability but the scarcity of resources that creates property. Government protection of property is needed when people want things that are *not* free for the taking. "The great and *chief end* therefore, of Mens uniting into Commonwealths, and putting themselves under government, *is the preservation of their property.*"[21] The power of government to protect property rights hardly rests on "cornucopian assumptions"; it rests on the fact that scarcity requires the division between mine and thine.

Some people have argued that Lockean notions of property, if enforced, are the best response we have to the problem of scarcity. William Baxter, for example, writes that "good conservation practices are implicit in carefully defined and enforced situations of ownership."[22] The reason is this. Those who own a resource in an enforceable sense, and can exclude others from its use, will exploit it gradually and carefully, in order to maximize future return. Thus, if you own an oil well, you would probably hold on to it, assuming that the price of oil will continue to rise, so that conservation is more profitable than exploitation. If you need quick cash you may mortgage the resource or you may sell it to someone who will hold it for long-term capital gain.

Any conceptual connection that may exist between conservation and well-defined and enforced situations of ownership, however, tends to break down in practice. Two problems seem formidable. The first is the problem of *agency.* Those who make decisions at the relevant levels in large corporations have their own personal success or careers in mind, rather than the long-term profitability of the organization. Thus, a short-term "good showing" is more important to them than is the eventual capital growth of the firm. Accordingly, executives may squander resources to make themselves look good on the balance-sheet, no matter what happens to the corporation or to the environment in the long run.

Second, while holding a resource may be more profitable than present consumption, an alternative investment may be even better. If the government offers risk-free securities at a twenty percent return annually, one is likely to sell everything one can to today's consumers in order to buy bonds; one might also liquidate resources to invest in fast-food retailing or some other business. To think that property rights can be a basis for conservation is to assume that conservation of resources is more profitable than *any* other investment that might be made with gains from immediate consumption. As long as "quick buck" investments exist, however, a lot of opportunities will be more attractive.

So far, I have discussed criticisms of Locke's theory from four points of view: first, that it leads to inequality and injustice; second, that it assumes man's "domination" of nature; third, that it works only when resources are plentiful; and fourth, that it fails to conserve for future generations. All of these criticisms are plausible but none is entirely convincing. We have yet to put our finger on a sure reason to believe that Locke's theory of property is inconsistent with a principled and dignified policy for the natural environment.

Every well-socialized individual understands that one does not have a right to use one's property in a way that causes harm or injury to someone else. My property right to my hammer does not permit me to hit your head with it. Can we make Locke's theory of property consistent with environmental goals by hedging it with a doctrine of harm? Is the police power a sufficient ground on which government can limit the use of private property to favor public ends?

At first we may think that the answer to this question is yes. A doctrine of harm—especially ecological harm—is called for as a supplement to Locke's labor theory of value. Locke, after all, believed that land, in the state nature leaves it in, is practically worthless. Readers almost three hundred years ago could believe Locke when he wrote:

> We see in *Commons*, which remain so by compact, that 'tis the taking any part of what is common and removing it out of the state nature leaves it in, which *begins the Property;* without which the Commons is no use.[23]

Now we know that land, air, and water, in the "state nature leaves them in" are enormously valuable—and that the labor which changes the state of these resources may destroy this value. Does this knowledge provide a way to apply Locke's theory to land use today?

I do not believe so. The reason is that our knowledge, while helpful in a general way, is often inadequate to make out specific causal connections. If I fill in my wetland or build a shopping mall over a salt marsh, what harm, exactly, have I done? Suppose your house floods ten years later and a hundred miles away. Am I at fault? Who is? Any paved area, any farm, any housing project may have also contributed. How much harm did each property owner cause? To ask this question is to see that no one can possibly answer it.

The Chesapeake Bay is the largest and richest estuary in the

United States. It undergoes enormous changes even in a year. The crabs become scarce and public hearings are held. Who is at fault? The crabs become plentiful and the hearings end. The oysters are few and we are sure that pollution or overharvesting or *something* is the cause. But they may, as inscrutably, come back again. Islands disappear; shoreline washes away; the bay itself fills in. Is this our fault or is it "natural"? How much does the builder of a new bowling alley contribute to this?

Within the past few years, most farms in the Chesapeake region have followed an instruction to practice "no-till" agriculture in order to combat the erosion of the soil.[24] To do this they depend upon the use of herbicides, notably atrazine. Recently, several species of underwater vegetation in the bay have begun to disappear. Traces of atrazine have been found in some of these plants. Is atrazine in runoff destroying underwater vegetation? Is Hurricane Agnes the culprit? Could a predator be doing the job? Is some combination of these—and any number of other conditions—responsible? All of these are possibilities.[25] How important is the vegetation to the ecology of the bay? Is it better to have the atrazine than the erosion? What is the amount of harm done by a farmer in Chestertown who tills half his land and applies simazine to the other half? (The question is grossly oversimplified, since, for example, different levels of nitrogen fertilizers, also pollutants, are used in no-till and traditional agriculture.) The research on these questions is likely to continue for a long time. Meanwhile, the underwater vegetation, for reasons perhaps known only to it, may return to the bay.

I submit that the problem of "pricing" or "internalizing" externalities in an ecological context, in many or most instances, simply defies solution. It can't be done. Linear chains of causality are not often found; events are the results of any number of interacting causes. Allied Chemical Corporation paid a $13 million fine for dumping Kepone in the James River in 1976. News reports at the time quoted statements to the effect that the organochloride pesticide could severely damage life in the Chesapeake Bay.[26] What damage has occurred or is likely, in fact, to occur? We simply do not know. It depends on too many conditions—e.g., on the market for oysters (now depressed); the occurrence of hurricanes; and the rate at which the chemical sinks under the sediment. Item: preliminary studies indicate that there is a potential for microbial degradation of Kepone.[27]

The doctrine of harm will not make Locke's conception of property rights workable in an ecological context. Harms exist: the use we make of our property may cause or contribute to these harms. But *which* use of *whose* property accounts for *how much* of what *specific harm?* You will sooner discover the philosopher's stone than find a way to answer that question.

Earlier, I suggested that Locke's theory of property might be criticized for presupposing an "anthropocentric" view of the natural world. A better and more common criticism directs itself to the moral psychology which underlies the Lockean tradition. This represents us as possessive individuals each intent on maximizing his or her self-interest. Once this criticism has been made, it is easily developed in terms of the problem or "tragedy" of the commons. The words ring in our ears:

> Ruin is the destination toward which all men rush, each pursuing his own best interest in a society that believes in the freedom of the commons. Freedom in a commons brings ruin to all.[28]

Readers of this essay are likely to be familiar with the way Garrett Hardin relates the "logic" of the commons to the problem of pollution. Hardin writes:

> The rational man finds that his share of the costs of the wastes he discharges into the commons is less than the cost of purifying his wastes before releasing them. Since this is true for everyone, we are locked into a system of "fouling our own nest," so long as we behave only as independent rational, free-enterprisers.[29]

The solution which Hardin suggests—mutual coercion mutually agreed upon—intends to prevent the would be "maximizer" from realizing the gains of pollution while distributing the costs. Is this what we need? Is a solution to the problem of the commons the key by which we can overcome the shortcomings of a Lockean land use ethic? Does it bring us closer to an alternative ethic?

I do not believe that Hardin goes much beyond Locke in providing a land use ethic or a guide for environmental policy. This may be seen when we ask what Hardin would have us to do, say, about the fellow who is tearing around in that '73 Impala without the muffler. What does Joe Macho have to gain by agreeing to a law which makes us all drive quiet CVCCs? The bomber is what he can afford. He may *like* the noise and not mind the pollution. *You* do. How can you stop him? You could *pay* him to stop driving—or to drive something else. But what gives you the right to fine him or to force him to do what suits *your* tastes, *your* interests?

You might argue that his noise and pollution cause harm—but

that is what is sometimes called "subjective." To be annoyed at the noise may mean no more than you do not like it. The epidemiological evidence, moreover, will convince no one that there is more cancer, say, in Ocean City than on Nantucket, where there is less pollution. Hardin, incidentally, suggests we sell the nation's parks or charge for their use, to keep crowds away.[30] But the crowds enjoy what they find—otherwise they would not come. They can smell the lavatories—but this may be *efficient*, not inefficient, park use. An inefficient policy would charge a high admittance price so that only a few could enjoy the luxury of "unspoiled" nature. (One can do that, incidentally, on a much vaster scale, simply by contemplating the stars.) Hardin favors and would protect his own interests. One person's trailer home is another's fouled nest.

Hardin's conclusion is this: we should organize competition so as not to get in each other's way. This advice lies well within the tradition of Hobbes, Locke, and Adam Smith. Hobbes called for coercion, in the form of the Leviathan, to provide one public good, security; Hardin adds environmental quality to this list. Locke did not believe that people would voluntarily respect property rights; he thought governments had to exist to enforce them. And Adam Smith did not argue that the Invisible Hand operates *in nature* but *in markets*. And markets are created and maintained by the enforcement of property and contract. This *is* mutual coercion mutually agreed upon. It is hard to find anything in Hardin—which is not also in Hobbes and Locke.

What is crucial in the tradition of Hobbes and Hardin is the appeal to rational self-interest. Rational self-interest is supposed to make us agree to rules which govern competition and enhance cooperation in ways that benefit all. Sounds good. But rational self-interest may be satisfied when a magnificent wilderness is converted into a tacky amusement park. Rational self-interest is served by technologies that make us independent of natural processes and therefore more able to destroy them. I know that without fusion power or something like that we must rely on photosynthesis and the nitrogen cycle. If we accept an economic land use ethic, however, the only limits that stand between us and the worst excesses of exploitation are technological. When a safe, clean, abundant, and inexpensive source of energy is found, then, apparently, these limits will vanish. Nothing then prevents the self-interest or preference of consumers from creating an environment completely dominated by highways, shopping centers, housing projects, motels, and endless commercial strips.

Rational self-interest, at any rate, is an untenable basis for public policy because it is unenforceable. Suppose we agree to laws regulating pollution. Who would enforce them? The policeman. The inspector. But if they were self-interested, they would take bribes. We could bring them before a judge. But she would take gifts, stock, free passes, sexual favors, or something, if she acted with a single eye to her self-interest. We might appeal to the President—but the regress is obvious. At some point, someone has to do something not because it is in his or her self-interest but because it is right. Why not start this at the beginning? Why not introduce principles, not merely preferences, into a public policy? Why not protect the environment not because that will satisfy insatiable and often contemptible consumer demand, but because it is what we can believe in and take pride in for reasons of an utterly different kind?

We come now to the last refuge of the liberal mind. It involves the shadow pricing of "intangible" or "soft" variables.[31] This is the attempt to price not only our *interests* but our *principles* and *beliefs* as market externalities. It may be understood, in the context of the Lockean tradition, as a last effort to interpret political issues in economic terms. It is a way of representing *contradiction* as *competition.* It attempts to represent differences of *opinion,* which should be settled through debate, ending, if necessary, in a vote, as if they were conflicts of *interest,* requiring a cost-benefit analysis, ending in a bottom line. Economists rescue the principle of efficiency from its likely ugly consequences for the environment not by introducing other principles or views, but by the incredible ruse of giving approaches to environmental policy that oppose theirs a surrogate market or shadow price. They may do this, for example, by asking environmentalists how much they are willing to pay for the mere knowledge that a species is protected or that a wilderness is not despoiled. Economists who know how to frame and whom to ask these questions can "show" that any policy that is morally, aesthetically, or culturally important is "efficient" as well.

This approach by the economist to environmental policy has a fascinating relation with classical utilitarian theory. The utilitarians were concerned about the fact that many intense pleasures come from tawdry or tasteless sources. (Prostitution is an example.) Few were as hardy as Bentham: few would agree that poetry is less valuable than Pushkin if it produces less pleasure. More might agree with Mill that Socrates dissatisfied is better than a pig satisfied. But if this is true, why do we make "satisfaction"—e.g., consumer satisfaction—the basis of public policy? Why should markets—even if we assume that they maximize "satisfaction"—determine what we do to the environment?

The answer economists sometimes give is this. Even if people enjoy pinball and do not read Pushkin they may still have a certain respect for Pushkin. This deserves a market price. The mere availability of poetry, in other words, is a benefit even for those who never read it. People enjoy knowing that some members of their society are artists, or whatever; therefore, there is an interest, although not one expressed in markets, in the arts. Transaction costs, free rider problems, or something of that sort, prevent people from "buying" these "aesthetic" or "moral" benefits. Economists have a way to "correct" this market failure. It is perfectly bizzare. Since they recognize that Socrates dissatisfied is better than a pig satisfied, they give Socrates' dissatisfaction a shadow price.

I do not have to criticize this tactic here.[32] Suffice it to say that it is the emotive theory of value run wild. Suffice it to say that it confuses what people believe in and care about with what they desire and will spend money on. Suffice it to say that economists who price the opinions of others need to listen, therefore, only to their own. Suffice it to say that market analysis, when carried on in these terms, is a subversion of public debate. Economic analysis, carried on in these terms, can do nothing to reveal or clarify values other than those of economists themselves. Cost-benefit analysis does not open up the "back room" of policy making to the light of day. It only explains away the loud knocking at the door.

I have now exhausted the remedies and stratagems by which economists and others have tried to defend market-based and property-based solutions to environmental problems. I have argued that in spite of these remedies the cure remains at least as bad as the disease. It *is* the disease. A worn out and no longer useful form of thought approaches societal problems as if efficiency and equality were the only values involved. These values are important to our society, but they are not our only ones, and in the environmental area, they often function as red herrings drawn across the path we should follow. This is the path of public discussion and debate in which we assess the ethical principles and aesthetic ideals that make us, in a sense, "Nature's Nation."[33]

Those who are developing the field of environmental ethics contribute to this discussion and debate. The best contribution they may make, I believe, is to define and argue for principles other than those of efficiency and equality, which, as I have proposed, may have as much to do with causing as with resolving environmental problems. Those who argue that other ethical and aesthetic convictions should guide environmental policy, however, invite the charge that they are elitists, trying to protect the privileges of the affluent few by criticizing the demands of the less affluent many.[34] The only way to answer this attack, I believe, is to find arguments to show that the tastes of "elitists" (if that is what they are) can be argued for and are not merely the sort of preferences that markets create and satisfy. They must be shown to have a basis all of us can recognize and respect.

The way to do this may be to emphasize the distinction between objects that are valuable *as individual things* and objects that have value because of some *purpose* that they serve. The distinction between intrinsic and instrumental value is one of the oldest in philosophy. Those who love and admire the environment value it for what it is and "for its own sake"; they do not value it simply because of the "satisfaction" or utility it provides. This is also the way we appreciate friendship, freedom, truth, indeed, anything we believe has an instrinsic worth. This is the essence of appreciation. To appreciate is not to value the objects one enjoys; it is to enjoy the objects one values.

What is the value of a magnificent environment such as wilderness? What is a species worth, e.g., the Colorado squawfish or the snail darter? As well ask what is the value of a work of art, a friendship, or a significant moment in the past. These things may have no use; indeed, one who uses friends soon no longer has them. The value of art, the value of history, the value of friendship, depend not on the uses they have (though these may be important), but on the meanings they have. They express what we are, what we believe in, and what we care about. We respect these things, and their value consists, perhaps, in what this respect tells us about ourselves. It assures us that we are not mere bundles of preferences, but human beings capable of more than desire; we are not merely self-interested consumers bent on achieving the lowest common denominator of satisfaction. To develop a land use ethic is to elaborate categories for public policy beyond those of efficiency and even equality. It is to bring into environmental law a concept of dignity to balance the concept of price.

Notes

1. For an excellent annotated bibliography of literature arguing this point, see A. Fisher and F. Peterson, "The Environment in Economics: A Survey," *Journal of Economic Literature* 14 (1976): 1–31.

2. For a popular book representing many others, see William Baxter, *People or Penguins? The Case for Optimal Pollution* (New York: Columbia University Press, 1974).

3. For discussion, see Martin Kreiger, "What's Wrong With Plastic Trees?," *Science* 179 (1973): 446–55.

4. Steven V. Roberts, "Visitors are Swamping the National Parks," *New York Times*, 1 September 1969, p. 15. Quoted in Joseph Sax, *Mountains Without Handrails* (Ann Arbor: University of Michigan Press, 1980), p. 74.

5. *Second Treatise of Government,* chap. 5, sec. 26. (In *Locke's Two Treatises of Government,* Peter Laslett, ed. [Cambridge: Cambridge University Press, 1963]).

6. Ibid., sec. 42.

7. Ibid.

8. Ibid., sec. 32.

9. Ibid., sec. 31.

10. Ibid., sec. 33.

11. Ibid., sec. 51.

12. Ibid., sec. 46.

13. A. M. Freeman, Robert Haveman, and Allen Kneese, *The Economics of Environmental Policy* (New York: Wiley & Sons, 1973), p. 23.

14. A large literature relating property and inequality includes the Rousseau essay mentioned below and Thorstein Veblens' *The Theory of the Leisure Class* (Boston: Houghton Mifflin, 1973). For Locke and man's dominion over nature, see, for example, Kathleen Squadrito, "Locke's View of Dominion," *Environmental Ethics* 1 (1979): 255–62. Robert Nozick discusses Locke and the problem of scarcity in *Anarchy, State, and Utopia* (New York: Basic Books, 1974), pp. 175–82. R. Bryant discusses the problem of relating property regimes to ecology in *Land: Private Property, Public Control* (Montreal: Harvest House, 1972). The problem of property and the commons is widely discussed, e.g., in Garrett Hardin and John Baden, eds., *Managing the Commons* (San Francisco: W. H. Freeman, 1977). For a discussion and survey of attempts to relate fragile values or benefits to property data, see A. Myrick Freeman III, "Property Values and Benefit Estimation," in A. M. Freeman, *The Benefits of Environmental Improvement: Theory and Practice* (Baltimore: Resources for the Future, 1979), pp. 108–64.

15. J. J. Rousseau, *The First and Second Discourses,* Roger D. Masters, ed. (New York: St. Martin's Press, 1964), pp. 151–52.

16. Lynn White, "The Historical Roots of the Ecological Crisis," *Science* 155 (1967): 1204.

17. *Religio Medici* I. 16.

18. Locke, *Second Treatise,* sec. 26. Locke suggests that "natural *Reason,* which tells us that Men, being once born, have a right to their Preservation," is as good a basis as revelation to show that the Earth is "given" to man for his use (sec. 25).

19. I paraphrase Locke, *Second Treatise,* sec. 22.

20. William Ophuls, *Ecology and the Politics of Scarcity* (San Francisco: W. H. Freeman, 1977), p. 155.

21. Locke, *Second Treatise*, sec. 124.

22. Baxter, *People or Penguins*, p. 34.

23. Locke, *Second Treatise*, sec. 28.

24. See R. E. Phillips et al., "No-Till Agriculture," *Science* 208 (1980): 1108–14.

25. See *Proceedings of the Bi-State Conference on the Chesapeake Bay.* 27–29 April 1977, Commonwealth of Virginia Publication CRC #61, pp. 46–47, 121f.

26. Marvin Zim, "Allied Chemical's $20 Million Ordeal with Kepone," *Fortune.* 11 September 1978, p. 91.

27. *Proceedings,* p. 136.

28. Garrett Hardin, "The Tragedy of the Commons," *Science* 162 (1978): 1244.

29. Ibid., p. 1245.

30. Ibid., p. 1245.

31. I comment extensively on the relevant literature in "Economic Theory and Environmental Law," *Michigan Law Review* 9 (1981): 1393–1415.

32. For criticism, see the article listed in previous note.

33. Perry Miller, *Nature's Nation* (Cambridge: Harvard University Press, 1967).

34. For discussion and bibliography, see Richard Andrews, "Class Politics or Democratic Reform: Environmentalism and American Political Institutions," *Natural Resources Journal* 20 (1980): 221–41, esp. 222.

One leader in the revival of interest in Leopold's Land Ethic has been Professor James Coufal of the State University of New York's College of Environmental Science and Forestry at Syracuse. In this article, Coufal makes the case for including a Land Ethic Canon in the Society of American Foresters Code of Ethics. He relates the need for such a canon to the emerging "New Social Paradigm," which places a high value on nature for its own sake. A Land Ethic Canon was adopted by the Society's membership in Fall 1992.

CHAPTER THIRTY-THREE

Wilderness Management, Environmental Ethics, and the SAF Code of Ethics

JAMES E. COUFAL

The Society of American Foresters' (SAF) Code of Ethics [reprinted in this volume] should contain an explicit statement on environmental or land ethics. This paper will support this claim by looking at the reasons for codes in general, and for an explicit statement on environmental or land ethics in the SAF Code in particular, making a suggestion for such a statement. It will also provide reasons why the Wilderness Management Working Group of the SAF should take a leadership role in seeking a land ethics statement.

To provide context, an ethic is part of any value system that is used to judge the rightness or wrongness, and the desirability or wisdom of our objectives and actions (Strong and Rosenfield 1981). Aldo Leopold's "land ethic" (Leopold 1966[see "The Conservation Ethic" in this volume]) is the prototype of an environmental ethic, and the terms "environmental" and "land" ethic are used synonymously. A land ethic should, thus, help us to judge the rightness or wrongness and the desirability or wisdom of our objectives and actions related to the land, and in Leopold's view, the "land" includes rocks, soil, water, air, and all the plants and animals of the ecosystem in question. His land ethic was most succinctly put when he said that, "a thing is right when it tends to preserve the integrity, stability, and beauty of the biotic community. It is wrong when it tends otherwise" (Leopold 1966).

In his call for individuals to extend their ethical considerations beyond individual-to-individual and individual-to-societal relationships to the land, Leopold appears to have described "conservation," taken here in its classical sense of "wise use," in at least three ways. Conservation in Leopold's (1966) view is a *relationship* between man

345

and the land—harmony; an *intellectual process* (understanding and education); and an *action* (preservation of the land's capacity for self-renewal). His view is based primarily on the science of ecology, and not so much on religion and sentimentality, which do enter the views of others. These three themes will be returned to in looking at wilderness in relation to the need for a land ethic.

Reasons for Professional Codes

Describing attempts by civil engineering to put a statement on environmental ethics into their professional code, Vesilind (1987) listed three reasons for professional codes in general:

1. To enhance the profession's public image; or to promote public relations.

2. To establish rules of conduct and a system of enforcement of these rules.

3. To promote the public welfare, especially by placing public good ahead of personal gain.

Earlier, Flanagan (1981) had presented an additional reason:

4. To promote the pride of practitioners, especially professionals, in their occupations.

Codes, Public Image, and Professional Pride

Public image and professional pride are important in relation to environmental ethics, so we can begin with what might at first glance appear to be a tangential approach and later return to the main question. When Leopold wrote, nearly 60 years ago, in the *Journal of Forestry*, it is unlikely that he thought that he would become a prophet of the environmental movement; but so he has (Leopold 1933).

Leopold believed in government incentives, government regulations, and government example on public lands that would provide a role model of proper land management for private landowners, all to the end that the land would be given ethical consideration. But most of all, and very strongly, he believed that a proper human relationship with the land, what he called harmony, would only come about when *individuals*, both landowners and land users, had made a land ethic part of their value system (Leopold 1966). He believed, in other words, in the stewardship of individuals as being the only real foundation of persons living in harmony with the land. Others have come to believe much the same; Rolston (1986), for example says, "An environmental ethic ought to be incarnated as a way of life." Leopold (1966) thought that government action might be necessary, but not sufficient to this end. And, despite the "environmentalist's" emphasis on his belief that some land needs to be preserved, Leopold also seemed to feel that preservation was necessary but not sufficient to the practice of conservation, and that intensive management could be compatible with the land ethic.

Significantly, what Leopold thought was necessary and what he prophesied so long ago, the belief in an environmental ethic by individuals, might very well be happening at the present. Despite our obvious failures [e.g., one-passenger vehicles, the NIMBY approach (not in by backyard), the RIGM syndrome (Regulate, I've got mine), our reaction to the "end" of the "oil shortage," etc.], there are many signs that environmentalism is becoming an accepted and highly regarded value, and that forests are receiving increasing attention. Shands (1988) believes, for example, that public concern with below-cost sales on National Forests is, "the recent manifestation of a broad, deep, and enduring change in public attitude toward the forests." One study will be noted in some detail to support the contention of changing public attitudes regarding nature.

Milbrath (1984) did a survey study involving the U.S., England, and West Germany, repeated after a three-year interval, looking at the environmental attitudes of groups he defined as significant. It is of interest that he did not hold resource managers as a separate and significant group! He described every society as having a Dominant Social Paradigm (DSP), a belief structure that organizes the way people perceive and interpret the functionings of the world around them. To greatly simplify his explanation, Milbrath says that the DSP of our modern world is one of dominion, a belief in technological development combined with fierce competition for unceasing, unlimited progress, especially economic progress.

In both years of the study, this DSP was adhered to most closely by business and political leaders, with the general public holding the middle ground, and the environmentalists holding views most disparate from the DSP. There were country-by-country differences, but the pattern was the same in all three countries. The major finding of the second study was that both the general public and business and political leaders had moved toward the environmentalists in the three-year interval between studies, confirming Spitler's experience that "the modern industrialist more and more accepts the need for environmental controls and demands only a fair and reasonable approach" (Spitler 1988).

Milbrath called the beliefs at the environmentalist end of the scales the New Social Paradigm (NSP), as listed below:

1. A high valuation of nature.

2. A sense of empathy which generates compassion for other species, other peoples, and future generations.

3. A desire to carefully plan and act so as to avoid risk to humans and nature.

4. A recognition of limits to growth and the need to adapt our beliefs and actions to them.

5. A belief that we need a new society that incorporates new ways to conduct our economic and political affairs.

These beliefs, especially the first four, do not seem radical, but the degree to which they are held translates into the fifth, a willingness to reshape our society and its institutions. It is also important to note how much of this NSP is related to social welfare or social ethics through such questions as who benefits and who pays, how do we handle the world's uneven distribution of resources, and how can forestry involve the people living on the land, and learn from them (at least in third world countries, forestry recently has responded through the activities of "social forestry"). The NSP also reflects a greater valuation of forests as spiritual and philosophical resources than as economic ones, an idea forestry may acknowledge but to which it has not necessarily responded well.

Milbrath believes that social changes begin fundamentally and are most widely expressed in the beliefs and values of persons, and that the NSP has a strong and real chance of becoming the DSP, even to the extent of titling his book "Environmentalists: Vanguard For A New Society." To extrapolate from a more recent work by Milbrath, each of us, and for purposes of this paper, the profession of forestry, is being forced to choose, by conscious action or through default, some position on the NSP-DSP continuum and to deal with the implications and issues this creates (Milbrath 1989). This kind of growing belief system is, in part, responsible for Flood's call for the 1990s to be "The Decade of Human Forestry" (Flood 1990).

What has this shift to do with environmental ethics as a way to promote the profession's public image? First, one of the social changes taking place is a lack of trust in "decisions by experts"; in our case, read "decisions by foresters." William Shields, Chairman of The American Forest Resource Alliance, recently said, "they (the public) begin by examining resource issues with the presumption that the resource must be protected from us" (Shields 1989). It is this public belief that has led to what Fortmann (1986) has called the "last legal form of indoor blood sport" that many of us now participate in more frequently than we might wish—public hearings. But, to put a statement on environmental ethics in the SAF Code to enhance public relations seems to be doing the right thing for the wrong reason, and is reminiscent of Magill's charge that foresters seem more interested in changing the public's image of forestry rather than in responding to the public's goals, needs, desires, and values (Magill 1988).

It does seem important for foresters to know whether they have underlying differences in value systems as compared to the public and other resource professions (Spitler 1988), or if they share a broad set of underlying values but differ in interpretation of facts and in the means to reach common goals (Davos 1988). Discussion of an environmental ethic for forestry will serve to reveal such commonalities and differences.

Proclaiming a land ethic that voluntarily sets higher standards for our profession than those expected for others should also provide a level of self-esteem and a sense of special relationship to the forest values we protect and manage (Flanagan 1981). Thus, while a statement of environmental ethics in our SAF Code of Ethics might be necessary as a symbolic action, like Martin Luther's 95 theses proclaiming, "Here we stand," it is not sufficient, and the enhancement of our public image must and will come about only through our on-the-ground actions. Such actions, what SAF Vice-President Ross Whaley (1990) calls "demonstrated exemplary stewardship of the resources," should be framed, nonetheless, in a shared philosophy or wisdom, and an environmental ethic in the SAF Code can certainly serve as part of this shared wisdom.

Codes to Establish Rules of Conduct

The second reason for having professional codes is to establish enforceable rules of conduct (Vesilind 1987). Enforcement cannot be an end in itself, but has to be the instrumental means to the more basic reason for codes, that of enhancing the public welfare. But, establishing enforceable rules of conduct is of such great concern that it leads to several objections to a statement on environmental ethics, objections that apply to ethical statements and professional codes in general.

The first objection says that ethics are strictly a personal responsibility: a matter of honor (Vesilind 1987). If this were accepted, there wouldn't be any codes of ethics. But many professions, including forestry, do have such codes, and it seems, therefore, that the majority of professionals feel a need to codify relationship and rules of conduct, which to a large extent still remain personal and matters of honor, the code being only (but importantly) a means to internalize accepted standards. There are often practices that work only if they are widely agreed on via standards, rules of thumb, or orienting principles, and some sense of consensus is required for any ethical code that interrelates groups within society. Further, the values managed on forest lands—soil, water, air, wildlife, etc.—are more often than not public goods, even where forests are privately owned; and a personal ethic is inadequate for corporate goods that must be managed by persons acting in concert.

A second objection, and one noted in regard to a statement on environmental ethics in the SAF Code, is that, in our litigious society, such a statement would open the door to many contentious, costly, and time-consuming lawsuits. Unfortunately, this may be true, but it is like saying let's not establish any rules of conduct because we might have to enforce them! The potential for lawsuits emphasizes the need to be careful and clear with the wording of any statement or canon on environmental ethics.

Since lawsuits are also possible with the current canons, and since there is no great outcry about this possibility, the objection noted seems implicitly to recognize the current public awareness of land management issues and the public concern for the proper ethical treatment of the land as being separate and distinct from the individual and societal relationships covered in the current SAF Code. Such public recognition should at least give the profession cause to debate the need for a statement on environmental ethics.

In regard to possible litigation, it is also of interest that, unlike the starship *Enterprise*, we cannot boldly go where no man has gone before, because other professional resource management groups have statements regarding environmental ethics in their codes. For example, the Code of Ethics of The Wildlife Society sets out four objectives in a preamble to the canons. The first two are:

1. To develop and promote sound stewardship of wildlife resources and of the environment upon which wildlife and humans depend;

2. To undertake an active role in preventing human-induced environmental degradation.

The American Institute of Certified Planners, a group growing in signficance, has a canon that says, "A planner must strive to protect the integrity of the natural environment." These two approaches suggest opportunities for the SAF, and should cause us to ask, "What makes forestry so different from wildlife or planning that we shouldn't or couldn't have a statement on environmental ethics in our Code?" Reversing this question and putting it in a more positive mode, we might ask, "What makes forestry unique so that it should have such a statement, especially when we consider Aldo Leopold's role in the modern environmental movement?" This will be examined in another section.

Finally, a third objection related to enforceable rules of conduct is that environmental ethics, or ethics of any sort, are subjective, while science is purportedly objective; or a related but not identical issue is that science is rational, while environmental ethics is emotional and, therefore, irrational.

But the opposite of emotional is not necessarily irrational; rather it is indifferent, stoic, insensitive. One can be emotional (passionate, excited, demonstrative) and rational. When foresters and other resource professionals equate the emotionalism of environmentalists with irrationality, I believe they fall into the trap of stereotyping. Environmental ethics is emotional, and, therefore, it is something environmentalist "do-gooders" have, while foresters have rational (unemotional?) science. With such a belief, a statement on environmental ethics in itself might seem irrational. Yet, an unguided applied science, such as forestry, is irrational.

It might be argued that a pure science is objective and rational, or value free, but an applied science, such as forestry, is value laden because the decisions about what to apply it to, what goals and benefits to obtain by applying it, who benefits and who pays, and other similar questions, are value laden and involve subjective emotions, such as whose preferences to satisfy.

At the extreme, but not unusual to be heard in resource professionals' conversations, the environmentalists are depicted as biocentric egalitarians, naturalist no-growthers, sentimental tree-huggers, and dickie bird lovers, who imply, if they don't say it outright, that humans are always the aliens and nature is always right. But, in turn, foresters and other resource professionals are often seen by the environmentalists as technological heroists, nature-conquerors, and land-rapers, scientists who see technology and not ethics, uses and not values, means and not ends, as the basic answers to problems, and who say that the ability to do something is reason enough to do it. If foresters find this description unflattering and wrong, there are two important questions that need to be asked. First, why are we perceived in such a manner? And second, if the environmentalists' stereotype of us is wrong, can it be possible that our view of them is wrong as well?

Regarding the subjectivity of ethics and the objectivity of science, there are some in modern science, especially quantum physics, who have come to the conclusion that the structures and phenomena we observe in nature are nothing but the creations of our measuring and organizing mind (Capra 1984). While this is not universally agreed upon, there is a consensus that what we choose to measure, how we choose to measure it, and the very act of measuring it creates changes and produces biases in the results we obtain. Davos (1988) discusses how our Baconian tradition causes us to subordinate values to facts, and goes on to show how the state of our art to measure and model natural systems, the subjectivity involved in choosing integration models and the weights given to factors in the models, and the uncertainty associated with natural systems make subjectivity inherent in "factual analysis." The old saw about science being a search for closer approximations to the truth is more real than ever as scientists recognize that the world cannot be analyzed or managed as independently existing parts, but only in greater or lesser interconnectedness, a complicated web of relationships of which humans and their ethics are a part. Science, in other words, just isn't as objective as we try to make it out to be, and foresters can't hide behind it in some pseudo-religious act of faith. Experience also shows us that the basic ethical maxims have withstood the test of time for thousands of years, whereas the "truths" of science change yearly, if not more often (Hargrove 1987).

Codes and the Public Welfare

The first canon [now the second] of the SAF Code of Ethics says, "A member's knowledge and skills will be utilized for the benefit of society"; and the most recent edition of the SAF's "Ethics Guide" says, "This is the canon which underscores the members' ultimate responsibility to serve the long-term interest of society as a whole" (SAF 1989). Synonyms for "ultimate" include "maximum" and "supreme," and if foresters' maximum and supreme responsibility is to serve the long-term interest of society as a whole, can SAF's Code of Ethics be complete without an expression of philosophy that includes behavior toward the land? We claim, after all, to be *land managers,* and as Wolf pointed out, the current SAF Code could easily be applied to plumbing (Wolf 1989); in other words, what is in the current Code that reflects our special relationship to the land? I believe there is nothing.

Most foresters learned early in their education that trees are not only the product, but also the factory, not only the interest, but also the principal, and that one cannot injure or destroy the factory or deplete the capital without having long-term adverse effects. We have long espoused and practiced, sustained yield and multiple use, albeit with exceptions, concepts contained in the modern themes of sustainable development and social forestry and in the modern ideas of environmentalism. It is interesting to speculate whether forestry has not created some of the public dissatisfaction with its practices by advertising such concepts to a greater level of success than the actual on-the-ground practice has attained. And, without doubt, we take pleasure in saying that foresters are the first environmentalists and in using slogans like "For A Forester, Every Day Is Earth Day" (SAF 1990), supporting and using movements that are attuned to the idea of environmental ethics.

The current Code deals with human relationships; foresters and society, colleagues, clients, and even bosses. It has become fashionable to stress the human dimensions of forestry; forestry students are told that we can't escape people by hiding in the trees because we practice forestry *for* people, *with* people, *through* people. Perhaps this has clouded our view so that we pay less heed than we should to the fact that the land, in the Leopoldian definition, is our ultimate (maximum, supreme) colleague, client and boss. In this sense, the need for a statement on environmental ethics comes about, I think, because we know the relationship it expresses is right. Whether for anthropocentric reasons, like believing that the environment is the foundation for the practice of forestry and must be protected to guard human interests, or for ecocentric reasons, such as believing in the inherent or intrinsic worth of the soil, water, air, the plants, and the wildlife, and the network of complex interactions that exists among them (and humans), that must therefore be protected for themselves, we not only know but we feel and are convinced it is right.

The Ethics Committee of the SAF has not pursued the idea of a statement on environmental ethics in the SAF Code because they believe that the "SAF Forest Policies & Positions" (SAF 1990) deals sufficiently with the issue. The substance of these policies and positions is not at question, but just as the positions flow logically from the policies, so it seems the policies should flow from a more fundamental statement of mission or a philosophy of the role of forestry and foresters. The Code of Ethics provides the opportunity to make such a fundamental statement in a concise, easily disseminated form. The following suggestions for additions and changes to the SAF Code are based on and offered in the spirit of the above.

Including these changes, or something similar, in the SAF Code of Ethics provides a publicly stated professional commitment to living in harmony with the land, a task made ever more difficult with increasing populations, improving technology, and inequitable distribution of resources. Leopold believed in the need for individual commitment to land ethics, and foresters should exemplify such individual commitment. The time is appropriate, if not overdue, for forestry to synthesize that individual commitment into a powerful profession-wide commitment of service to the land as the basis for service to the people. But how is this related to wilderness and the SAF Wilderness Management Working Group?

Suggested Additions and Changes to the SAF Code of Ethics

The canons of the current codes could continue to be interpreted in their current human relations mode, as well as extending to a more environmentally sensitive mode if the Preamble were expanded as follows:

Preamble:
SAF Code of Ethics

Current	Proposed
The purpose of these canons is to govern the professional conduct of members of the Society of American Foresters in their relations with the public, their employers, including clients, and each other as provided in Article VIII of the Society's Constitution. Compliance with these canons helps to assure just and honorable professional and human relationships, mutual confidence and respect, and competent service to society.	This Code of Ethics is based on the belief that foresters must strive to accomplish three major tasks simultaneously: 1. provide & implement alternatives to help landowners reach their objectives; 2. insure an appropriate flow of goods, services, & values to meet society's needs; and 3. maintain & enhance the integrity of the ecosystems they work with, including both the land & human elements of these ecosystems, all by practicing good stewardship of the land.
	The purpose of these canons is to govern the professional conduct of members of the Society of American Foresters in their relations with the public; their employers, including clients; each other; and with the land entrusted to their care as provided in Article VIII of the Society's Constitution. Compliance with these canons helps to assure just and honorable professional, human, and environmental relationships, mutual confidence and respect, and competent service to society.

Wilderness, Environmental Ethics and the
SAF Wilderness Management Working Group

In a sentimental sense, wilderness managers and the SAF Wilderness Management Working Group should be interested in environmental ethics because Aldo Leopold was a forester, the father of modern wildlife management, a founder of The Wilderness Society, and a prophet of the modern environmental era. More fundamentally, since its earliest conception, wilderness has been a discussion of values, and it could be argued that the concept of wilderness was a foundation stone in the evolution of the New Social Paradigm (NSP) discussed earlier. Wilderness people seem to have tapped a vein, or better, they share one with a growing number of the public, of beliefs and values fundamental to the NSP. One way to examine this vein is to look at parallels between Leopold's three descriptions of conservation and compare them to wilderness ideas.

Leopold (1966) described conservation in at least three ways. First, he set it in the context of a *relationship* when he said, "Conservation is a state of harmony between men and land" (p. 243). Second, he talked of conservation as an *intellectual process*, for example, when he said, "conservation is our effort to understand and preserve" the capacity of land for self-renewal (p. 258), and in noting, "one of the requisites for an ecological comprehension of land is an understanding of ecology" (p. 262). Finally, he saw conservation as an *action*, as in preserving the capacity of land for self-renewal (p. 258), or in the sense of husbandry that is "realized only when some art of management is applied to land by some person of perception" (p. 293). Wilderness, as much or more than other forest values, is a relationship, an intellectual process, and an action.

Wilderness as Relationship

Wilderness is a relationship in that it provides identity, is an artifact, and enhances harmony. Wilderness, and its companion wildness, are the crucible in which we, as a species, were forged. As we have come to value the cultural and ethnic diversity of our society and desire to preserve it, so too have we recognized the need to preserve the roots of the heritage of our species. As individuals wilderness also provides us identity by offering a challenge against which to test ourselves, even if that challenge lies only in knowing that the wildness exists, lurking near or far, just as the wildness within each of us lurks near or far and tests our ability to endure and grow. But wilderness is not only a test; we seek it also because we enjoy it, and that which we enjoy is one of the strongest marks of our identity. Wilderness has come to be seen as a home with intrinsic values; one which we can't truly leave and which we need to develop our evolving values.

Wilderness and wildness stand in another relationship to humans; they are identified only in comparison to culture or civilization, and they are artifacts in that they "can only survive by human understanding and forbearance that we now must make. The only thing we have to preserve nature with is culture. The only thing we have to preserve wildness with is domesticity" (Berry 1987).

When all was wild, there was nothing to measure wildness against, nor likely was there any concern to make such measure. Now wildness is measured against civilization, and like the art treasures of ancient civilization, its rarity provides value that emerged with civilization.

Finally, wilderness enhances a relationship of harmony. The harmony spoken of is not just the tranquility or internal calm that wilderness brings to many, for others find anxiety, agitation, or excitement in the presence of wild surroundings. The harmony of wilderness is that of an arrangement of parts in pleasing and functional relationships to each other, parts that can only be fully understood in relation to the whole system. The action and importance of this wholeness in wilderness is implicit in Rolston's statement that, "It is not form (species) as mere morphology, but the formative (speciating) process that humans ought to preserve" (Rolston 1986). A danger of test-tube speciation is that it is speciation out of context.

Wilderness as an Intellectual Process

Wilderness is an intellectual process in at least three ways; as a scientific baseline, as a source of re-creation, and as a philosophical stimulus.

Change of any sort must be measured against something; change to something is only accomplished with change from something. Change is fraught with danger if one doesn't know the starting point, and even more so if one doesn't know how the starting point was arrived at. Management of forest ecosystems deals with change; purposeful change to obtain a value or purposeful reduction of change to maintain a value. Change in forested ecosystems must be measured against those systems that are the wildest, the least influenced by man. These wilderness ecosystems also serve as reservoirs of genes, species, and of the systems themselves.

Wilderness is first creation, in the base sense of the above: the starting point of evolution. It is re-creation in the human sense because we seek it for pleasure and challenge, and, as Leopold noted, recreation in this case "is not the outdoors, but our reaction to it" (Leopold 1966). The pleasure and challenge may be on-the-ground, but it may also be vicarious, and it goes beyond substantive and economic needs. Rolston (1986) describes two kinds of positive recreational values: those that involve activities which allow us to demonstrate skills, and those that provide the opportunity to contemplate nature's shows. The abundance of National Geographic specials, Nova, and other similar television programs indicates the great interest in wildness and wilderness in our society.

Wilderness is a philosophical stimulus as philosophers discuss and debate such seemingly esoteric issues as the objectivity and/or subjectivity of values in nature (Callicott 1987); whether ecosystems can have moral considerability (Salthe and Salthe 1989); deep ecology; eco-feminism; the morality of hunting; animal rights; and so on. While such issues may seem esoteric, we each have a dominant paradigm through which we filter and frame our world-view, and whether this is well thought out or not, it contains a philosophical stance on all of these and other issues that permeates our perspectives of those disciplines that we must understand and deal with to practice resource management: science, law, economics, sociology and psychology, education, and theology. The movement to preserve wild areas was one of the first, and continues to be one of the most common, debates in the area of environmental philosophy and ethics. If, as Berry (1989) suggests in the very first issue of Earth Ethics, "both religion and economics need to establish the ecosystems of planet earth as normative for their own proper functioning," wilderness advocates must be at the forefront of the debate.

Wilderness as Action

Wilderness is action in that in itself it is dynamic, it requires a long-term commitment, and it requires management. "Preservation" is a misnomer in its implication that it maintains the status quo. Succession, including the effects of natural disasters, means that wilderness can be more honestly said to preserve a process rather than a form (ecosystem), and it is this process we are asked to consider ethically.

But the form may go on apparently unchanged for many decades, and to preserve either the form or process requires a long-term commitment even for the shortest of early succession forests. It was the call to beware of short-term expediency, especially economic expediency, that Leopold stressed.

If, in Leopold's terms, health is the capacity of the land for self-renewal, and conservation is our effort to understand and *preserve* this capacity, then conservation and preservation are not polar extremes, but rather preservation is conservation in being wise use. The battles between conservationists and preservationists have tended to solidify positions instead of finding common ground, creating confusion among the various publics, and among resource professions, as well. Recently, Wood (1990) declared:

> As representatives of the land we are a house divided. The cleavage between those who see management as total control and those who see conservation as total restraint is clear. Less clear is the separation among those who see their first duty as short-term service to people and those who give their first service to land that it may serve people.

A land ethic could serve as common ground, and the common ground that is becoming more apparent is encompassed by the concept of stewardship. Conservation, including its form "preservation," is stewardship. We are entrusted with the land not only for ourselves, but for current and future others, and to leave the land in as good or better shape than we received it, we must manage it. The public's desire for good stewardship, and the need to manage wilderness if stewardship is to be achieved, was never more evident than in the Yellowstone fires of recent years. Further, some in the profession feel that forestry is in the midst of a paradigm shift that will put more emphasis on land health (Behan 1990), although others feel there is no need for such a shift, nor is one occurring (O'Keefe 1990).

As an ethical concept, stewardship will be better understood by all if resource professionals have defined and stated principles to work up to, and wilderness managers once again have the historical bases for providing leadership to see that this comes about by working toward the inclusion of a statement of environmental ethics in the SAF Code of Ethics.

Conclusions

If "the challenge of a 'revolution' in wilderness management" on existing wilderness areas must be the focus of the next 25 years (Fege and Corrigall 1990), the incipient revolution in individual environmental ethics that seems to be taking hold in our country also offers wilderness managers the opportunity to continue, and to enhance, their role in bringing credibility to resource professions by stressing values rather than uses, commitment rather than expediency. The common scientific and intellectual grounds already exist; it is the resolution of differences in attitudes and values that must occur. Agreement on a land ethic for the SAF would be one small step toward this resolution. It is fitting to close with words from Aldo Leopold (1966), who said, "That land is a community is the basic concept of ecology, but that land is to be loved and respected is an extension of ethics."

References

Behan, R. W. 1990. Multiresource forest management: a paradigmatic challenge to professional forestry, Journal of Forestry 88(4): 14–18.

Berry, T. 1989. Planetary progress. Earth Ethics. 1(1):6–8.

Berry, W. 1987. Amplifications: preserving wildness. Wilderness 50(176):39–40, 50–54.

Callicott, J. B. 1987. Just the facts, Ma'am. The Environmental Professional 9(4):279–288.

Capra, F. 1984. The Tao of Physics. New York: Bantam Books.

Davos, C. A. 1988. Harmonizing environmental facts and values: a call for co-determination. The Environmental Professional 10(1):46–53.

Fege, A. S.; Corigall, K. H. 1990. The enduring resource of wilderness. Journal of Forestry 88(3):27–29, 39.

Flanagan, D. T. 1981. Legal considerations of professional ethics. Paper at Society of American Foresters, New England Section. Portland, ME.

Flood, P. 1990. Securing a future resource: what are the pressures? Paper at New York Society of American Foresters meeting. Oriskany, NY.

Fortmann, L. 1986. At issue: people and process in forest protest. American Forests 93(3&4):12–13, 56–57.

Hargrove, E. C. 1987. The value of environmental ethics. The Environmental Professional 9(4):289–294.

Leopold, A. 1933. The conservation ethic. Journal of Forestry 31(6):634–643.

Leopold, A. 1966. A Sand County Almanac: with essays on conservation from round river. New York: Ballantine Books.

Magill, A. W. 1988. Natural resource professionals: the reluctant public servants. The Environmental Professional 19(4):295–303.

Milbrath, L. W. 1984. Environmentalists: Vanguard for a New Society, SUNY Series in Environmental Public Policy. Albany, NY: State University of New York Press.

Milbrath, L. W. 1989. Envisioning A Sustainable Society: Learning Our Way Out. SUNY Series in Environmental Public Policy. Albany, NY: State University of New York Press.

O'Keefe, T. 1990. Holistic (new) forestry: significant differences or just another gimmick? Journal of Forestry 88(4):23–24.

Rolston III., H. 1986. Philosophy Gone Wild. Buffalo, NY: Prometheus Books.

SAF. 1990. Forest Policies and Positions. Bethesda, MD: Society of American Foresters.

SAF. 1989. Ethics Guide. Bethesda, MD: Society of American Foresters.

Salthe, S. N.; Salthe, B. M. 1989. Ecosystem moral considerability: a reply to Cahen. Environmental Ethics 11(4):355–361.

Shands, W. E. 1988. Beyond multiple use: managing national forests for distinctive values. American Forests 94(3&4):14–15, 56–57.

Shields, W. 1989. Outlook: the public and natural resources. In Focus. National Forest Products Assoc.

Spitler, G. 1988. Seeking common ground for environmental ethics. The Environmental Professional 10(1):1–7.

Strong, D. H.; Rosenfield, E. S. 1981. Ethics or expediency: an environmental question. In: K. S. Schrader-Frechette, ed. Environmental Ethics. Pacific Grove, CA: The Boxwood Press.

Whaley, R. S. 1990. Securing a future resource: what is the future and how do we reach it? Paper at New York Society of American Foresters meeting. Oriskany, NY.

Wolf, T. M. 1989. Land ethic responses. In: Letters. Journal of Forestry 87(10):4.

Wood, G. W. 1990. The art & science of wildlife (land) management. Journal of Forestry 88(3):8–12.

Vesilind, P. A. 1987. Environmental ethics and civil engineering. The Environmental Professional 9(4):336–342.

V. GOVERNMENT SERVICE
AND PUBLIC POLICY

Those of us of a certain age will never forget the spectacle of former President Richard M. Nixon waving to the crowd on the White House Lawn as the presidential helicopter waited for the last time following his resignation as President. The Watergate scandal revealed behavior by high public officials that had been at the least shabby and deceitful and in several cases was proven to be in violation of laws. Several prominent White House staffers and one Cabinet member consequently served prison terms. Many of the Watergate offenders were attorneys, whose loyalty to their client had clearly gotten the best of their ethical instincts.

Corruption in government in the US predates the 1972 election campaigns, however. Citizens had become jaded by revelations of bribes and payoffs in corrupt city governments. High officials had gone to prison before, but a president's resignation under suspicion of obstruction of justice brought home to Americans that mere hope and good wishes were insufficient to ensure the ethical conduct of government, and that ethics in the public service needed a good deal more emphasis.

What does this event mean to those of us trying to understand forestry ethics? To me it means that we need to reflect carefully on the specific ethical requirements of public service. There is no clear difference between the ethical problems arising for land managers in government compared to private service. Public and private foresters both sell timber. Public and private foresters can face similar questions of scientific integrity, conflict of interest, or abuse of position. Government agencies and private corporations are large bureaucracies; both provide tests for an individual's ethical sensitivities on a regular basis. Some people seem offended by the suggestion that a different standard ought to apply to public employees, as if this suggests a higher standard of holiness for some, and a denigration of standards in the private sector. Yet citizens expect higher standards of government officials in some respects, particularly if they are in appointive or policy-influencing positions.

Sunshine laws, financial disclosure requirements, and rules concerning employment after government service all provide disincentives to improper behavior. The practices of making decisions according to detailed procedures, of documenting decisions, and according strict equality of treatment to those dealing with government are further examples of strict standards in government. These practices are described in Willbern's essay as service orientation and procedural fairness. Such insistence on procedure, openness, and equal treatment is not considered necessary in private work. These requirements emerge from efforts to ensure that the routine machinery of government is not completely annexed to the needs of the political party currently in power.

In government, appearances of propriety are important, since appearances and perceptions are at the root of legitimacy in democratic government. For this reason, prudent government officials always act on conservative interpretations of ethical standards. Nonetheless, double standards remain. As the Commission on Government Ethics commented, "What the Framers rightly considered the most powerful branch bids fair to be the least accountable branch." In Washington, ethical leadership by example is often in the wrong direction. Senior members of Congress ask voters to believe that by accepting $10,000 from an interest group for delivering a 30-minute ghost-written speech, they are placing themselves under no obligation. Few of these people have so much to say, or are so eloquent, that they could earn $10,000 speaking fees if they did not serve in the Congress. (Note: after these words were written, Congress limited its ability to accept these honoraria in exchange for a pay raise. I leave the words standing, since Congress acted because honoraria were unpopular, not because they are unethical.) The "Keating Five" were prominent Congressmen and Senators who solicited and received massive campaign contributions from Savings and Loan looter Charles Keating. They pressured regulators on his behalf, calling this "normal constituent service." Keating's actions later earned him a term in federal prison. At an ethics proceeding hearing to determine whether the Five had violated ethics rules, one of them opined that "everybody does it," so it must be acceptable.

Young people serving in federal and state conservation agencies, when faced with this kind of ethical blindness, may find it difficult to tell what the standards really are, and all too easy to rationalize shabby behavior. I believe the ethical standards in public service in the conservation field are generally high, but special alertness is needed when the standards espoused by some leading members of Congress are so abysmally low.

The difficulty of finding ethical guidance in government employment, and the potentially serious consequences of errors, make finding a mentor all the more important. If you work in government, your agency has attorneys assigned to its legal tasks, who are usually glad to advise on ethical questions. They can keep your inquiry confidential. Develop the habit of asking questions. Frequently, an informal discussion of a situation will help you clarify your own view.

It is all too common for younger and less experienced people to be placed in complex situations without adequate ethical guidance. Shortly after I became director of the Maine Bureau of Public Lands, I saw a need to train our staff on ethical matters. I sought reading materials to facilitate this. I found that none existed. Hence, all too many years later, this book.

Maxims for Foresters

Gifford Pinchot, first president of the Society of American Foresters, offered the following advice to guide the behavior of foresters in public office. Although they originated during one of his lectures on forest policy at the Yale School of Forestry in the early 1900s, many of these maxims may be even more applicable today (*Journal of Forestry* 92(2) February 1994).

1. A public official is there to serve the public and not to run them.

2. Public support of acts affecting public rights is absolutely required.

3. It is more trouble to consult the public than to ignore them, but that is what you are hired for.

4. Find out in advance what the public will stand for. If it is right and they won't stand for it, postpone action and educate them.

5. Use the news media first, last, and all the time if you want to reach the public.

6. Get rid of an attitude of personal arrogance, of pride of attainment or superior knowledge.

7. Don't try any sly or foxy politics. A forester is not a politician.

8. Learn tact simply by being absolutely honest and sincere, and by learning to recognize the point of view of others. Meet them with arguments they will understand.

9. Don't be afraid to give credit to someone else. Encourage others to do things. You may accomplish many things through others that you can't get done on your single initiative.

10. Don't be a knocker. Use persuasion rather than force, when

possible. Plenty of knockers to be had. Your job is to promote unity.

11. Don't make enemies unnecessarily and for trivial reasons. If you are any good, you will make plenty of them on matters of straight honesty and public policy and will need all the support you can get.

Additional Reading

Cooper, Terry L. *The Responsible Administrator: An Approach to Ethics for the Administrative Role.* Port Washington, NY: National University Publications/Kennikat Press, 1982.

Johnson, J. T. *Can Modern War be Just?* New Haven: Yale University Press, 1984. Provides a valuable introduction.

Kennan, George. Morality and foreign policy. *Foreign Affairs* 64(1985):205–218. An eloquent essay by one of America's leading diplomats and foreign policy thinkers.

Pastoral Letter on War and Peace, The Challenge of Peace: God's Promise and Our Response. Origins—NC Documentary Service. May 19, 1983. The famous *Pastoral Letter* by the US Catholic Bishops.

Thompson, D. F. The possibility of administrative ethics. *Public Administration Review* 45(1985):555–561.

Wachs, Martin. Ethical dilemmas in forecasting for public policy. *Public Administration Review* 42(1982):562–567.

_____ , ed. *Ethics in Planning.* New Brunswick, NJ: Rutgers—The State University of New Jersey, 1985.

U.S. General Accounting Office. *Employee Conduct Standards: Some Outside Activities Present Conflict-of-interest Issues.* Washington, D.C.: U.S. General Accounting Office. GAO/GGD-92-34, 1992.

Quite a number of books are available on ethics and morality in public policy and foreign policy. The 40th anniversary of Hiroshima undoubtedly contributed in part to this increase in applying moral thinking to foreign policy. The 1990–1991 Gulf War revived debate over moral issues of war, and rekindled debate nationally over the theory of a "just war."

York Willbern's essay provides an excellent introduction to ethics in government. Note his list of six levels of public morality, which concern primarily the process of how decisions are reached, in contrast to the Land Ethic, which concerns the substance of decisions. Willbern returns to a theme raised earlier by Boulding when he contrasts public and private action, noting that the pursuit of self-interest is not the operative standard within government. Particularly relevant in these times of intense controversy over management priorities are Willbern's ethic of democratic responsibility and his ethic of compromise and social integration. These highlight the age-old conflict between the scientific-bureaucratic concept of decisionmaking in the public interest versus the resolution of conflict through political compromise. In the scientific-bureaucratic model of government decisions, actions are taken on the basis of a concept of the public interest, as derived through objective calculations made by technically qualified experts. According to the ethic of compromise and social integration, decisions are made in ways that seek to harmonize competing interests through the political process.

Willbern was Emeritus Professor of Public and Environmental Affairs and Political Science at Indiana University, Bloomington, at the time this article was published.

CHAPTER THIRTY-FOUR

Types and Levels of Public Morality

YORK WILLBERN

Students of government and public administration, from Plato to Wilson and from Weber to the proponents of the "new public administration," have nearly always known that what public officials and employees do has a central and inescapable normative component, involving values, morality, and ethics,[1] although they may have differed as to the degree to which this component could be separated, either analytically or in practice, from aspects of administration involving facts, science, or technique. Discussions about moral considerations involving public officials, however, frequently deal with significantly different types of forces and phenomena.

The most obvious distinction is that between consideration of the ethical behavior (honesty, rectitude) of the official and consideration of the moral content of the public policy or action the official promulgates or carries out.[2] Most public criticism of public ethics focuses on the former; the concerns of adherents of the "new public administration" were on the latter.

Serious attention to the ethical and moral components of public officialdom suggests that there are other important distinctions also. This essay is primarily one of taxonomy; it attempts to identify and characterize particular components or facets of official ethics and morality in an effort to lay out a rough map of the terrain.[3] Classification is difficult in this area not only because of the overlapping of the concepts and the activities, but because of the ambiguities of the words used to describe them. Nevertheless, such an effort may be helpful in joining discussion and in making it more likely that people talk about comparable rather than different things.

While for some purposes it would be valuable to distinguish the

This article reprinted by permission from *Public Administration Review,* March/April 1984, pp. 102–108. Copyright © 1984 by the American Society for Public Administration (ASPA), 1120 G Street NW, Suite 500, Washington, DC, 20005. All rights reserved.

ethical problems involving elected or politically appointed officials from those involving civil servants, or among those in particular aspects of public service (i.e., public works, social work, university teaching, or others), no concentrated attention will be given here to such differences.

It is suggested that six types, or levels, of morality for public officials can be discerned, with, perhaps, increasing degrees of complexity and subtlety. There are, of course, substantial interrelationships among these levels, but they are different enough to be analytically interesting. They are (1) basic honesty and conformity to law; (2) conflicts of interest; (3) service orientation and procedural fairness; (4) the ethic of democratic responsibility; (5) the ethic of public policy determination; and (6) the ethic of compromise and social integration.

In general, the first two or lower levels (and in some degree the third) concern aspects of personal morality and, hence, the ethical conduct of the individual public servant; the other, or higher, levels deal more with the morality of the governmental decisions or actions taken by the official or employee. Public scandals and outrage at unethical behavior focus mainly on the lower levels. Most public officials and employees are confronted with choices, as individuals, which involve ethical concerns at these levels. At the higher levels, most actions and decisions and policies in modern complex governments and bureaucracies are more collective, corporate, and institutional in nature, in which the individual moral responsibility is shared with others in complex ways. This is true even for chief executives; even if "the buck stops here," the bulk of the work in preparing "the buck" and its possible alternatives will have been done by others.

Basic Honesty and Conformity to Law

The public servant is morally bound, just as are other persons, to tell the truth, to keep promises, to respect the person and property of others, and to abide by the requirements of the law. The law—the codes enacted or enforced by the legitimate organs of the state— usually embodies these basic obligations and provides sanctions for violating them. The law also includes many other requirements, and there is a moral obligation to conform, with arguable exceptions only in the most extreme circumstances. An orderly society cannot exist if individuals can choose to follow only those laws with which they agree; civil disobedience is an acceptable moral tactic only in very extreme situations. Conformity to law is especially necessary for public officials and employees.

These behaviors are basic requisites of an orderly society, which exists only if people can be reasonably secure in their persons and belongings, can normally rely on the statements and commitments of others, and can expect others to conform to the established norms of conduct. There are, of course, liars and thieves and law breakers in the public sector, as there are in the private, and it is transgressors of this kind which are particularly noticeable and which bring public condemnation.

In general, the difficulties in interpreting and following these broad mandates are about the same for public servants as for others. Definitional problems (What is true, and should all the truth be told? What property belongs to whom? What is the law?) which produce moral dilemmas are probably no more difficult in the public sector than in the private.

Even at this level of basic honesty and conformity, however, there may be some ways in which the public service differs. To some observers, the fact that public officials are vested with the power of the state produces more danger and more opportunity for transgression of the basic moral code. Power and access to public goods may provide more temptation; the necessity to communicate and deal regularly with the public may lead to more occasion to prevaricate or to ignore promises and commitments.

This may be more true in other societies than in America. In this country, two factors seem to limit, in some degree, the relative danger of official misbehavior at this basic level. One is the obvious fact that power is less concentrated in the government. The power that is lodged in private economic institutions and aggregations, especially apparent in the United States, may be no less corrupting than the power of public officials. Ours is a rather pluralistic society, with countervailing powers scattered quite widely. A second and related factor is the existence of stronger and more independent control mechanisms in our system, especially stemming from the judiciary and the press. In the United States, every public official (even the president, as the last decade demonstrated) lives in the shadow of the courthouse. And in America, unlike many other systems, it is the same courts which exercise control over public officials and private citizens. Moreover, the American press is particularly vigorous in trying to exercise surveillance over public officials and employees.

The visibility of public life may well make dishonest behavior less frequent there than in private life. Public officials are expected not only to be honest, but to appear to be honest. Both Caesar and Caesar's wife are to be above reproach, and departures from that norm are noticed. There is almost certainly less tolerance of public employees than of private employees who deviate from accepted standards in such personal matters as marital behavior, sexual [preference,] and use of alcohol and drugs, as well as in basic honesty.

Some would argue that the nature of political discourse in a democratic system greatly increases the likelihood of falsehoods and insincere and broken promises. There is no doubt that political campaigns and even the prospect of such campaigns produce statements that are at least selective in their veracity, and that neither the makers nor the recipients of campaign promises seem too surprised if the promised performances do not fully materialize. Without questioning the immorality of campaign falsehoods and forgotten promises, the fact that knowledgeable listeners to puffery of this sort do not really take it too seriously nor depend upon it for their own decisions may mitigate somewhat the seriousness of the sin. The same sort of salt is applied to advertising and salesmanship assertions in the private marketplace; regulations such as those of the Federal Trade Commission and the Securities and Exchange Commission attempt, without complete success, to limit inaccurate allegations and promises. Campaign puffery aside, however, the importance of veracity and adherence to direct interpersonal political commitments is widely recognized.

There are, on the other hand, areas of public official behavior where deceit and lying are not only condoned but approved. The police use undercover agents and "sting" operations to try to catch criminals. Official falsehoods and deception certainly accompany foreign intelligence and some national defense activities. In covert operations in particular countries, public employees may go well beyond falsehood in contravening the normal moral code. These are very difficult ethical areas, where the end is presumed to justify the means, and *raison d'etat* provides shaky moral justification.

Conflict of Interest

The ethical problems associated with conflicts of interest in the public service are even more complex and difficult. The general presumption is that the moral duty of an official or employee of a unit of government is to pursue the "public interest"—i.e., the needs and welfare of the general body of citizens of the unit. His own interests, and the interests of partial publics of which he may be a member, are to be subordinated if they differ from the broader, more general, public interest—as they almost inevitably will, from time to time.

In their cruder manifestations, conflict of interest transgressions fit clearly into the category of ethical problems discussed above, the obligation to respect the property of others and to conform to the law. Embezzlement of public funds, bribery, and contract kickbacks are all actions in which the offenders have pursued their personal interests in obvious contravention of the public interest, as well as in violation of the law. Expense account padding is a more common illustration of both conflict of interest and of theft. There are generally laws against the more obvious conflicts of interest.

But there are also very difficult and subtle conflicts of interest that are not so clearly theft, and which may not involve law violation. At a simple level, to take a trip at public expense which may not have been necessary but which enabled the official to visit family or friends, or just to have a semi-vacation without using leave time, presents a problem of conflict of interest ethics, as does nepotism, or the appointment or advancement of a relative or friend as is the case involving political party patronage systems.

Every public employee belongs to other groups than his governmental agency—church, professional, local community, racial, or ethnic group. Most officials and employees, especially those in public service for only limited periods, worked for and with someone else before the government service. It is very easy to presume that the interests of those groups coincide with the interests of the entire public, but others may disagree. Moreover, officials have private financial investments which can be affected by public policy decisions.

Among the most difficult and subtle of the public service conflict of interest problems are those relating to the obvious and inevitable interest of a person or group or party to win elections. This is related to the ethic of democratic responsibility, to be discussed later. It is sufficient to note at this point that to award contracts or jobs or shape public policies in such a way as to reward people or groups for their past or expected campaign support rather than on the merits of the public purpose involved raises tough ethical questions.[4] Campaign financing, for example, may be one of the most vulnerable spots in a democratic system. Those who contribute to a successful campaign usually expect, and get, a degree of access to decision makers that raises questions about the even-handedness of decisions and actions.

There are conflicts of interest in private life, as well. In the private as in the public sector, people are imbedded in collective entities—corporations, firms, associations—and they are confronted with many occasions in which their personal or small group interests may conflict with those of the larger entity. The potential, and frequently real, conflicts of interest between the management and the stockholders of a corporation, or between management and rank-and-file employees, are obvious illustrations.

There is a very important difference between conflicts of interest in public life and in private life. A basic cornerstone of the Western economic system is that the vigorous pursuit of self-interest by each participant is the most effective way to secure the general interest—the "invisible hand" of the market will transform selfish pursuits into the general welfare. The presumption that what is good for General Motors is good for the country, and vice versa, seems a truism in the capitalist economic system. In spite of the variety of exceptions that modern economists will make, the power and pervasiveness of this idea make conflicts of interest in the private sector considerably different than in the public.

This vigorous pursuit of self-interest by private economic entities, in their relationships to government, is one of the great causes of moral problems in the public sphere. Modern government has great impact on economic activity, and every economic entity needs and wants to influence public policy in its own interest. Government can give and withhold privileges of great value. The tendency to presume that what is good for the particular group or social segment of which a person is a member is also good for others is almost irresistible.

Lincoln Steffens, in attempting to explain the cause of corruption in government, compared the situation to that in the Garden of Eden. The trouble, he said, was not the serpent—that was his nature. Nor was it the weakness of Eve, nor of Adam—they also did what was natural. The fault, he said, was in the apple.

It can be argued that the political world is also a marketplace, where the pursuit of self-interest by all the varying segments will produce a harmonious general welfare. This was essentially the argument of Madison in Federalist Paper number 10, suggesting that the enlargement of the polity would increase the number of interests participating in the political market, and thus make more likely the achievement of a general rather than a particular interest.

But governmental decision making is far more institutionalized than the decision making of a free economic market, and the public official is always torn between acting as proponent of an interest (his own, or that of his department or agency, or of his profession, or of his faction or party) and acting as an arbiter among competing interests. Should U.S. Senator Richard Lugar support the interests of Indiana coal miners in dealing with acid rain? Can a real estate developer serve impartially on a zoning board? A longtime public servant coined an aphorism that is now widely known in the public administration community as Miles' Law: "Where a man stands depends on where he sits."

Since both potential and real conflicts of interest are so pervasive, major efforts are made to provide a degree of protection by procedural safeguards (a subject to be considered further in connection with the next level of public morality). Public acknowledgement of outside interests is required, and arrangements are made for officials to refrain from participation in matters where their interests may conflict. But these safeguards are far from sufficient to remove the moral responsibility of the individual employee or officer.

Here, as in so many other moral matters, degree may become crucial. Some conflict of interest is inevitable; the question becomes, how much conflict of interest taints a decision or action so much as to make it unethical?

Service Orientation and Procedural Fairness

We now enter areas where the overlap with private morality, while still present, is less noticeable, and where the problems are more peculiarly those of public officials and employees.

The purpose of any governmental activity or program is to provide service to a clientele, a public. This is true even if the activity is in some degree authoritarian, a regulation or control. It is sometimes easy for this moral imperative to be obscured by the fact that government and government officials, have and exercise power. The auditor at the Internal Revenue Service, the policeman on the beat, the teacher in the classroom, the personnel officer in an agency, are there to serve their clients, but they also exercise authority, and there is real moral danger in the possibility that the authority comes to overshadow the service.

Attitudes and the tone and flavor of official behavior are morally significant. Where power is being exercised, arrogance can easily replace humility, and the convenience of the official becomes more important than the convenience of the client. Delay and secrecy can become the norm. Procedures are designed for official purposes, not those of the public. This may well be the kind of corruption Lord Acton had in mind, rather than thievery or bribery, when he said that "power corrupts."

The effort to provide some degree of protection for the clients against the potential arrogance of officials is one of the reasons why procedural fairness is one of the central components of public morality. The concept that a person threatened with the power of the state has firm procedural rights is a very ancient one. The right to a trial, a public trial, or a hearing, with proper notice of what is alleged or intended, the right to counsel, the right of appeal, are built into administrative as well as judicial procedures. "Procedural due process" is a cornerstone of public morality.

In addition to the need to protect citizens against the corrupting power of the state officialdom, there is another important ground for a procedural component of public morality. As will be noticed further in connection with the sixth (highest?) level of morality to be discussed here, it is inevitable that in a complex society interests will be opposed to each other, that not all can be satisfied, and that often not even a Pareto optimality can be achieved. The public decision-making process may not be a zero-sum game, with inevitable losses accompanying all gains, but the difficulty of satisfying all makes it necessary that the *process* of arriving at decisions about action, or policy, be a fair one.

The Ethic of Democratic Responsibility

In our consideration of various types and levels of public morality, another transition occurs. The first three levels deal with the conduct of public officials as they go about their business, the last three with the content of what they do. It is upon these first three levels (particularly the first two) that attention is usually focused in discussions of official ethics. They are important, and difficult, but they may not be as central as the considerations affecting the moral choices involved in deciding what to do, in pursuing the purposes of the state and the society. The first set might be said to deal with collateral morality, the latter set with intrinsic morality.

The dogma of the morality of popular sovereignty is now general throughout the world, and nowhere is it more strongly entrenched than in the United States. To be democratic is good, undemocratic bad. Observers acknowledge the existence of elites, even power elites, but the suggestion is that their presence is unfortunate and, by implication at least, immoral. Citizen participation is encouraged, even required, in governmental programs. Hierarchy is suspect, participatory management the goal, even within an agency or institution.

In spite of almost universal adherence to the dogma, there are some reservations about its practice. Government by referendum does not always arouse enthusiasm. "Maximum feasible participation"—the legal requirement for several national programs—was certainly not an unqualified success; U.S. Senator Patrick Moynihan called it "maximum feasible misunderstanding." Many have reservations about open records, open meetings, public negotiations—all justified on the grounds of the public's "right to know." But these reservations are still only "reservations"; the basic notion of popular sovereignty is seldom challenged.

In its simplest logic, the legitimacy of popular control is transmitted to operating public servants through a chain of delegation. The legislature is supposed to do what the people want, while the public executive and administrator are to conform to legislative intent. The politically chosen official, either elected or appointed by someone who was elected, has the mandate of the people. The civil servant is ethically bound to carry out the instructions of these politicians, who derive their legitimacy from the people. The military is subordinate to politically chosen civilians. The "career" officials or employees are supposed to carry out Republican policies during a Republican administration, Democratic policies during a Democratic administration, because that is what the people want. For public employees to substitute their own judgments as to what the people want for the judgment of those who have the electoral or political mandate is unethical, according to this logic. They may advise to the contrary, but they are to carry out the instructions of their political superiors to the best of their ability. If they cannot conscientiously do so, their only ethical choice is to resign their posts.

In reality, of course, the situation is never as clear as the simple logic suggests. In this country (more, probably, than in other practicing democracies), there are usually multiple rather than single channels for the expression of the public will. Both the legislature and the executive (sometimes many executives), and sometimes the judges, are elected by the people, and they may emit different signals from their popular mandates. A statute, or a constitutional provision, may be considered to embody the popular mandate better than instructions from a political superior. Arrangements for direct citizen participation through hearings or advisory commissions may complicate the process further. A public employee may have considerable range of choice in choosing which popular mandate to respond to. But, in principle, strong support would be given to the concept that a public official in a democracy has a moral responsibility to follow the will of the people in his or her actions.

This ethic of democratic responsibility, the logic of which is quite powerful, produces difficulty for public employees. The employees may have goals and values which differ from those transmitted through the political channels. Or, the political superiors, even the people themselves, may not be fully informed. There may be no better illustration than the resistance which professors in a public university might make to instruction from a board of trustees or a legislature as to what to teach or who should teach it.

The conflict is particularly severe when the logic of democracy conflicts with the logic of science or of professional expertise. What is the ethical position of the public employee when the people, directly or through their properly elected representatives, insist on the teaching of "creationism" which is repugnant to the scientist? Or when the certified experts say that fluoride in the water is good for people but the people say no? To put it in the simplest terms, should the public official give the people what they want or what he thinks they ought to want?

To some, the voice of the people in a democracy may be equivalent to the voice of God. But, for most public officials in most circumstances, that axiom will not provide answers that allow them to escape their personal responsibility to make moral choices based on their own values. The voice of the people will not be clear, it will not be based on full knowledge, it will conflict in small or large degree with other persuasive and powerful normative considerations.

Here, as in other situations, the most popular (perhaps the best) answer may be a relativistic one. The official may be responsive to democratic control and his political superiors, but not too much. Democracy may be interpreted not as government by the people, but as government with the consent of the people, with professionals (either in a functional field or as practicing political leaders) making most decisions on the basis of standards and values derived from sources other than a Gallup poll, and submitting to only an infrequent exercise of electoral judgment as to the general direction of policy. Civil servants carry out the instructions of their political superiors with vigor and alacrity if they agree with them, and with some foot dragging and modification if they disagree. They try to give the people, and the politicial officials representing them, what they would want if they had full information, meaning of course, the information available to the particular person conducting the activity.

The Ethic of Public Policy Determination

Perhaps the most complex and difficult of all the moral levels is that involved in determining public policy, in making actual decisions about what to do. The problems of honesty and conformity to law, difficult as they sometimes are, are simple compared to those in decisions about public policy. And these are inescapably *moral* judgments; some policies, some actions, are good, some bad. Determinations about the nature of the social security program and how it is to be paid for, for example, turn only in minor degree on technical information; they depend chiefly on basic considerations about human values.

There can be no doubt that normative determinations are made at all levels of public service. They are not made just by legislatures, or by city councils, or school boards. They are also made by street level bureaucrats—the policeman on the beat, the intake interviewer at the welfare office, the teacher in the classroom. These individuals make decisions in their official capacity that involve equity and justice and order and compassion. Rules, regulations, and supervision of others in a chain of command stretching back, in theory, to the soverign people may provide a framework for decisions, but not a very tight framework. Every teacher knows that the department chairman and superintendent really have very little to do with how the job is actually done, and every cop knows the same thing about the police chief.

Though right and wrong certainly exist in public policy, they are frequently—usually—difficult to discern with confidence. There are degrees and levels of right and wrong. A medieval English verse about the enclosure of the village commons by the nobles goes:

> The law locks up both man and woman
> Who steals the goose from off the common,
> But lets the greater felon loose
> Who steals the common from the goose.[1]

The perpetrator of a regressive tax, or of a regulation which permits water or air used by thousands to be polluted, or of a foreign policy position which produces or enlarges armed conflict, may do far more harm to far more people than hundreds of common burglars. But is he or she a greater criminal? Or more immoral? The regressive tax may be better than leaving an essential public service unfunded; the pollution may result from a highway that allows people to go about their affairs expeditiously, or a power plant that permits them to air condition their homes; many reasonable people believe that the most effective way to deter armed conflict is to threaten, with serious intent, to initiate or escalate armed conflict.

The judgments that must be made about the rightness or wrongness of a public policy or decision involve at least two types of considerations—one, the benefit-cost calculation, and the other, the distributional problem (who gains and who loses).

Benefit-cost calculations are generally more difficult in the public arena than in the private, for two reasons. One is the matter of measurement—the currency in which the calculations are made. There are, of course, nonmonetary considerations in many private decisions, but they are more pervasive in public determinations. The other reason is the greater necessity for concern about externalities, for spillover costs and benefits. This is an important but usually secondary consideration in a private decision, but often central in a public one. The making of good benefit-cost calculations may be more a matter of wisdom (either analytical or intuitive) than of morality, but normative consideration in choosing factors to consider, and assigning weights to them, are inescapable.

Ethical considerations are particularly salient in determinations about *distribution* of benefits and burdens in a public activity or decision. Here we confront squarely the problems of equity and justice and fairness. Any attempt to define these concepts (which have occupied philosophers a long time) is obviously beyond the scope of this short paper. But a few reflections may be offered.

Policies which do not provide equal treatment for all, or which enhance rather than diminish inequality, are usually condemned as unjust and unfair. But equity is not synonymous with equality. Equity may require similar treatment for those who are similarly situated—but not all are similarly situated. Many public policies do, and should, discriminate. The critical question is whether the basis for differentiation and the kind of differentiation are appropriate. It is appropriate and desirable for public policy to reward desirable behavior and punish undesirable behavior. To give an A to one student and an F to another, and to admit one to graduate school and turn the other down, certainly discriminates, but it may be just and fair and equitable. A central purpose of social policy is to elicit desirable behavior, and discourage undesirable.

It is also usually considered appropriate to differentiate on the basis of need—to provide things for the widow and the orphan and the physically and mentally handicapped that are not provided for others. Compassion would certainly seem a morally defensible ground for such discrimination. And there are many other grounds for such discrimination, many of them morally debatable. For example, is it appropriate (ethical) to provide free education for a citizen but not for an alien? Is it ethical to deport a refugee from economic privation but not a refugee from political oppression? Is it inequitable, unjust, and unconstitutional to deviate from one man–one vote apportionment in electing members of state legislatures and city councils, but not in electing U.S. senators?

The Ethic of Compromise and Social Integration

To some, morality means uncompromising adherence to principle. To compromise with evil, or with injustice, is immoral. But "principle" and "evil" and "injustice" are not always certain, especially in complex social situations. One man's social justice may be another man's social injustice. Lincoln's classic formulation "with firmness in the right as God gives us to see the right," seems to carry with it the implication that God may give someone else to see the right differently, and that he also may be firm.

We must live with each other, adjust to each other, and hence make compromises with each other. This is a central feature of politics and of a bureaucratic world that is also political. We are all involved in politics—the only place without politics was Robinson Crusoe's island before Friday came. As every successful politician knows, it is necessary upon occasion to rise above principle and make a deal. Thus, compromise can be viewed as a highly moral act—without concessions to those who disagree, disagreement becomes stalemate and then conflict.

If sincere people hold to differing values, there must be institutional arrangements which legitimize courses of action which certainly can not satisfy all and may not fully satisfy any, and there is a moral obligation for both citizens and officials, but particularly officials, to participate and support such arrangements. These institutional arrangements are, in large degree, procedural, permitting and encouraging public policy discourse and mutual persuasion and, finally, resolution of differences. The needed public policy discourse is more than the discourse of a marketplace which involves bargaining between and among economic self-interests. It is somewhat different than the discourse in the "republic of science," in which evidence and proof (or at least disproof) can be marshalled. It may never attain the level of discourse which Habermas calls an "ideal speech situation." But it must be social discourse with a strongly moral component.

Since complete *substantive* due process, measured by the standards of a particular participant in the political process, can rarely be achieved, a large measure of *procedural* due process is a moral necessity, not only to protect an individual against the power of the state, but to make legitimate the process of public decision making. Complete reconciliation, or social integration, will always be elusive, but social cohesion, loyalty to and participation in a group, and in larger communities, is a moral goal of the highest order.

T. V. Smith put it this way: The world is full of saints, each of whom knows the way to salvation, and the role of the politician is that of the sinner who stands at the crossroad to keep saint from cutting the throat of saint. This may possibly be the highest ethical level of the public servant.

Notes

1. To the cynic, of course, the phrase Public Morality is an oxymoron—like Holy War, or United Nations, or Political Science.

2. This distinction has been pointed out by several writers—particularly by Wayne A. R. Leys in "Ethics and Administrative Discretion," *Public Administration Review* 3 (Winter 1943), and in *Ethics for Policy Decisions* (New York: Prentice-Hall, 1952).

3. This outline map differs from others—and they differ from each other. Several which have been particularly suggestive, even though they differ, are the items by Leys cited in note 2, above; a particularly rich two-volume collection of papers edited by Harlan Cleveland and Harold Lasswell, *Ethics and Bigness* and *The Ethic of Power* (New York: Harper, 1962); two substantial essays by Edmond Cahn, *The Sense of Injustice* (Bloomington: Indiana University Press, 1964) and *The Moral Decision*

(Bloomington: Indiana University Press, 1966); a volume by George A. Graham, *Morality in American Politics* (New York: Random House, 1952); and a lecture (unfortunately unpublished) given by Dwight Waldo at Indiana University in Bloomington in 1977.

4. A very sensitive description of such problems, based largely on his own experiences as mayor of Middletown, may be found in Stephen K. Bailey, "The Ethical Problems of an Elected Political Executive," in Cleveland and Lasswell (eds.), *Ethics and Bigness, op. cit.*, pp. 24–27.

5. I borrowed this from George Graham, *op. cit.*, p. 33. I'm sure he borrowed it from someone else.

Ethical laxity in government did not end with Watergate. Revelations surrounding the Iran-Contra scandal and the ensuing trials of top national security officials and a former general tainted the later years of the Reagan Administration. To address this issue, President George Bush, in his first Executive Order, created the President's Commission on Federal Ethics and Law Reform. This commission reviewed a number of issues that had been highlighted in the 1980s and suggested improvements to existing federal rules and procedures. The Commission's report, the introduction to which is included here, emphasizes that laws and rules are not adequate to ensure ethical behavior; the will of individuals and the example of leaders are also necessary. As you read this piece, compare its analysis with Kenneth Andrews' suggestions for the business setting and with David Flanagan's article on legal considerations. Then consider the implications of applying these ideas to the Land Ethic in a large organization.

CHAPTER THIRTY-FIVE

To Serve with Honor

PRESIDENT'S COMMISSION ON FEDERAL ETHICS LAW REFORM

We have approached the President's request to evaluate existing ethics rules with twin objectives: to obtain the best public servants, and to obtain the best from our public servants.

Ethical government means much more than laws. It is a spirit, an imbued code of conduct, an ethos. It is a climate in which, from the highest to the lowest ranks of policy and decision-making officials, some conduct is instinctively sensed as correct and other conduct as being beyond acceptance.

Laws and rules can never be fully descriptive of what an ethical person should do. They can simply establish minimal standards of conduct. Possible variations in conduct are infinite, virtually impossible to describe and proscribe by statute. Compulsion by law is the most expensive way to make people behave.

The futility of relying solely or principally on compulsion to produce virtue becomes even more apparent when one considers that there is an obligation in a public official to be sure his[1] actions appear ethical as well as be ethical. The duty is to conduct one's office not only with honor but with perceived honor.

We must start with a *will* at the top to set, follow, and enforce ethical standards. President Bush has given this Commission his first Executive Order, by which he has set initially the ethical tone he expects to pervade his Administration.

That order and this report are but a beginning. Each cabinet officer, head of an agency, subordinate official with supervisory authority—each must lead by example, by training and educating coworkers, by fair, just and persistent enforcement of the laws. This necessity of leadership applies with equal force to the legislature.

This article reprinted from *To Serve with Honor*, President's Commission on Federal Ethics Law Reform, March 1989, pp. 1–7.

What the Framers rightly considered the most powerful branch bids fair to be the least accountable branch. That is a dangerous combination, recognized by thoughtful Members themselves. We do not exclude the judiciary, although more rigorous and easily understood standards have obviated many problems there.

We believe that public officials want to follow ethical rules, and that they will do so if the laws are clearly delineated, equitable, uniform across the board, and justly administered. As Napoleon said: "There is no such thing as a bad soldier; there are only poor officers."

Ethical rules and statutes rest on moral standards. They are supposed to carry a certain moral authority, as are most laws. When the lawmakers prescribe laws for others but not for themselves, in the eyes of the public the essential moral authority is diminished. This is why it is essential to create ethical rules for the legislative branch as closely similar to those of the judiciary and the executive as is possible, given their differing functions. Instead of statutes applying to only one branch or two, standards should be applicable to all. No part of the Federal Government should be satisfied with a standard of less than absolute honesty in the conduct of public officials.

While our analysis is based on certain fundamental functions which conflict of interest restrictions are intended to serve, our analysis also incorporates the four key principles noted by the President when he signed Executive Order 12668 creating this Commission. One, ethical standards for public servants must be exacting enough to ensure that the officials act with the utmost integrity and live up to the public's confidence in them. Two, standards must be fair, they must be objective and consistent with common sense. Three, the standards must be equitable all across the three branches of the Federal Government. Finally, we cannot afford to have unreasonably restrictive requirements that discourage able citizens from entering public service. . . .

We begin by recommending that the conflicts of interest statute forbidding decision by an executive branch official on a "particular matter" in which he has a financial interest be extended to non-Member officers and employees of Congress and the judiciary, thus striving to achieve (at least in part) the level playing field desired in the ethics laws. Judges are already covered by very strict statutory standards, but there are difficult problems in applying a statutory standard to Members of Congress.

We favor centralizing the issuance of interpretive regulation for the executive branch in the Office of Government Ethics, and we suggest the creation of a similar centralized ethics authority within the legislative branch.

As the newcomer prepares to enter Government service, he or she must fill out various forms disclosing assets and income. This frequently leads to the realization that the prospective official has assets and sources of income which, if retained, would create a recurring conflict of interest with governmental duties. This conflict can be accommodated by recusal from decision-making or a waiver where the interest appears so small as to have no influence on the official's conduct, yet our strong recommendation is that the prospective official be encouraged to divest these troublesome assets at the very outset. If the official could do that by postponing the tax liabilities by a rollover of the troublesome assets into neutral holdings such as Treasury bills, municipal bonds, or bank certificates of deposit, then many more officials would do so. A divestiture of troublesome assets and reinvestment in neutral holdings is the single most important device we have encountered to eliminate completely or at least to mitigate greatly subsequent conflicts of interest. Many of the problems we discuss would never be problems at all, if such a change of holdings had occurred at the outset of the official's public service.

Since not all conflict problems can be erased by divestiture, we recommend that the Office of Government Ethics exercise a rule-making authority to deal with *de minimis* issues, pension plans, mutual funds, the investments of charitable organizations, and the industry-wide effects of some rulings on individual companies. With executive branch-wide standardized positions promulgated by the Office of Government Ethics, the compliance of individual public servants with the rules will be much simpler and easier, and the general public as a whole will have a vastly better understanding of exactly what holdings are permissible, and the nature of those retained by public officials filing annual disclosure reports.

This whole process should encourage seeking advice. One principle which helps to avoid a conflict of interest, and even an appearance of such conflict, is the age-old principle which should permeate the whole governmental ethical compliance system, *i.e.*: "No one shall be a judge in his own case." The possibility of a conflict of interest can be put to rest by submission to an impartial ethics authority, and following the advice received.

We have made specific and uniform recommendations in regard to the thorny problems of augmentation of Government income by private sources, a cap (with the exact percentage yet to be determined) on outside earned income of senior officials of all three branches, a ban on honoraria, and on outside boards and directorships. We believe there must be some cap on most types of outside income earned by the public servant, otherwise many public servants would slowly edge into private activity as a disproportionate source of income, to the detriment of their expected public service. This danger is enhanced every year that governmental salaries lag further and further behind those which can be obtained by the same individuals in private enterprise. We would propose, however, that the President be authorized to exempt from the cap any category of income he determines to be generated by a type of activity which did not pose ethical issues or detract from full performance of official duties.

The disgracefully low compensation for public service affects also the increasingly notorious problem of honoraria. Executive branch rules now prohibit executive branch officials from receiving honoraria for any speeches, writings, or other actions undertaken in their official capacity. In stark contrast, in the legislative branch honoraria for speeches, writings, public appearances at industry meetings, or even at breakfast with a small group, have become a staple and relied-upon source of income up to, in some cases, the limit of thirty or forty percent of the Member's salary. Some Members earn several times their governmental salary by honoraria, and give the excess to charities of their choice. This practice, which some Members of the legislative branch defend as both inevitable and proper in an era of recognized meager and unfair compensation, obviously produces several evils: first, the Member is diverted with increasing frequency from the performance of official duties; and second, the honoraria all too frequently come from those private interests desirous of obtaining some special influence with the Member. While we believe that no position of a Member of the legislative branch could be changed by a $2,000 honorarium, the honorarium is often perceived to guarantee or imply special access to the particular Member by the granting organization.

In the judicial branch, the apparent evil is simply the diversion of time from judicial duties. The groups before whom judges appear are usually professional groups who are genuinely interested in the judge's views on legal problems, and whose members have no expectation whatsoever of any special influence with a court. Individual lawyers or members of the public do not feel free to pick up the phone and talk to them; contact between bench and bar is in the decorum of a public courtroom.

Outside boards and directorships represent another diversion of a public official's time and energy. Sometimes they produce actual conflicts of interest, many times they create an appearance of a conflict of interest. Our recommendation is that senior public officials not serve on the boards of commercial enterprises, and that participation by such individuals in the management responsiblities of charitable organizations be carefully guarded, particularly if there is a danger of abuse of the public official's name for fundraising purposes.

Gifts, travel, entertainment, and simple meals have caused ethical problems for years. We encountered the most amazing diversity of interpretation within the executive branch and a stark disparity between the three branches. Our recommendation is a uniform and, we hope, reasonable policy in all branches as to what may be accepted without creating an appearance of impropriety.

Negotiation for future employment while the public official is still serving has always created a disruptive effect on the work of the agency involved, and frequently real problems of a conflict of interest. Our recommendation is designed to make clearer and brighter the lines of what may be done and not done, and to minimize any disruptive effect.

Post-employment restrictions center on the question of the period of time during which former employees should be barred from contacting certain parts of the Federal Government with which they were previously associated. At the present time there is no statutory bar in the legislative and judicial branches, but there are four separate bars in the executive branch. Our recommendations attempt to equalize and simplify these post-employment restrictions. We recommend the extension of the one-year cooling-off period now applicable to senior employees of the executive branch to comparable positions in the legislative and judicial branches. Separate inconsistent and duplicative post-employment restrictions should be considered for repeal. The existing lifetime bar in the executive branch regarding representation on particular matters handled personally and substantially by the former employee should be extended to the judiciary. To thwart the former employee who wants to "switch sides," there should be a two-year ban on using or transmitting certain types of carefully defined non-public information.

We think compartmentalization in each branch in applying these restrictions has its place, but it should not be abused. We recognize compartmentalization in limiting the ban for the legislative branch to one House only, and for the judicial to the specific court with which the former employee was affiliated. In the executive, we make no recommendation changing the responsibility of the Office of Government Ethics to prescribe compartmentalization in large departments, where a person working in one of numerous large sub-agencies would have little contact or influence with other agencies, but we do recommend that compartmentalization within the Executive Office of the President be abolished, as not consonant with the standards and objectives of this device.

We have carefully studied the question of whether the validity of the actions of the former employee, in making representations on behalf of others to the part of the branch with which he was previously affiliated, should depend on whether the employee receives compensation for such representational activity. We find that the injection of the element of compensation would vitiate the worthwhile objectives of the prohibition itself, and that there is no logical or constitutional justification for it. As a practical matter, requiring the element of compensation to make the representation improper would create a large loophole, which would inevitably be exploited to the Government's detriment. If representation of another person is improper, it is improper whether compensation is involved or not. We believe, however, that there should be no restriction whatsoever on a former employee's right to present his views on policy issues in any form—testimony, press articles, speeches, interviews—and indeed personally to represent his own interests anywhere.

Financial disclosure has been variously described as the linchpin of the ethical enforcement system, as the disinfectant sunlight which makes possible the cleaning up of abusive practices. We have made three basic recommendations on financial disclosure, two to strengthen it, the other to simplify it. To strengthen financial disclosure, to make it more meaningful, we would raise the highest limit describing assets to "1,000,000 and over," and the highest limit describing income to "250,000 or over." We believe the present intermediate bands of assets and income are so close together as to be somewhat meaningless, and would leave to the Office of Government Ethics the responsibility of recommending the number and size of the categories between the $1,000 and $1,000,000 or $1,000 and $250,000 marks. To strengthen disclosure for political appointees we would eliminate the home mortgage and family debt exceptions, because in our view the existence of heavy obligations, no matter what the laudable purpose of assuming those obligations may be, is something about which the public should be informed. We are also recommending a simplification and increased uniformity to the extent possible of the disclosure forms required by the Senate committees, the White House, and the FBI, and propose that a coordinating committee with representatives from all three branches be convened to address this task.

We now turn to the structure of federal ethics regulation. In the executive branch we recommend the strengthening of the Office of Government Ethics for its advice, consultation, and rulemaking function. We would continue the principal investigative, enforcement and compliance responsibility in the individual departments and agencies for several reasons. We think there should be one overall set of government regulations interpreting the ethics statutes, promulgated by the Office of Government Ethics. Variations from these uniform regulations should be permitted only on a showing by the individual department or agency that such is necessary for its particular mission, and on the approval of the Office of Government

Ethics. Likewise, the final authoritative interpretation of the regulations should rest with that office, to prevent the divergent standards and interpretations now current. The Office of Government Ethics should promote uniformity by an enlarged and strengthened training mission for ethics officers in the individual agencies, and for an annual review of their plans and programs.

In contrast to the rulemaking centered in the Office of Government Ethics, we recommend that the investigative enforcement and compliance function remain principally the responsibility of the individual departments and agencies. The Cabinet secretary or the head of an independent agency is the person whom the President should hold responsible for ethical standards in his agency. He must therefore have the responsibility for investigation, enforcement, and producing compliance with the overall executive branch standards within his department. Furthermore, the agency inspectors general and ethics officers are closer to the facts of any violation, and would probably do a better job of investigation. Likewise, training and education within the department or agency is the responsibility of that agency, although assisted, reviewed, and checked by the Office of Government Ethics.

In the legislative branch we recommend the establishment of a joint ethics officer with an adequate investigative staff to investigate alleged offenses and bring the results of the investigation and recommendation for enforcement to the appropriate Congressional ethics committee.

The Commission recommends additional enforcement mechanisms. In this area we are especially indebted to the work already done by the Congress and embodied in the legislation introduced in the last Congress and in current proposed legislation. These new enforcement measures were singled out for praise by President Reagan in his message vetoing the legislation for other reasons. We believe that in addition to continuing the felony level penalties, there should be civil penalties, which can be in some instances more persuasive and better tailored to the offense than criminal penalties. The Attorney General should have authority to seek injunctions to restrain conduct violative of ethical standards. We pass no judgment on the wisdom of the independent counsel device for investigating and prosecuting ethical violations, but we do urge that if the independent counsel device is retained, it apply to both the executive and legislative branches. If the device is retained, we also urge that it be strictly limited to the very highest officials in each branch, as this is the only area in which there may be justification for it.

In closing, we would emphasize that in addition to serving specific functions, our recommendations are also offered to stimulate the continuing development of the ethics system within the government. We believe that the implementation of these recommendations will serve both the public interest in protecting the integrity of the government, and the federal employee's interest in preserving an individual sense of pride and honor in serving the public good.

Notes

1. Masculine or feminine pronouns appearing in this report refer to both genders unless the context indicates another use.

Resource management professionals are often asked to serve on local planning boards and commissions because of their technical knowledge. In planning work, decisions often have major impacts on the lives of individuals. Cumulatively, these decisions can affect the quality of life of an entire community. The responsibilities are serious and the pressures can be intense. These cases, presented by Carol Barrett, show how general ethical principles come into play in actual situations. You might find it useful to review and discuss your reaction to these cases with a colleague or mentor. When this article was written, Ms. Barrett was Manager of the Community Development Bureau at the Greater Washington Board of Trade in Washington, DC.

.

CHAPTER THIRTY-SIX

Ethics in Planning: You Be the Judge

CAROL D. BARRETT

The specific ethical concepts that underlie the Code of Professional Conduct adopted by the American Institute of Certified Planners in 1981—public interest, conscientiousness, pursuit of excellence, fair treatment of colleagues, consideration, equity, integrity—are the stuff of apple pie and the flag. While their statement in the AICP code evokes little controversy, their day-to-day application may be quite problematical.

How one makes certain ethical decisions is usually more dependent on family and religious training than on the view of professional colleagues. Planning schools devote little time to the subject, and the sessions on ethics at recent national planning conferences have not drawn nearly as well as those featuring computers. The upshot is that planners often feel they have no guide to evaluating their own behavior.

Most planners would agree, of course, that no one should accept bribes left in shoeboxes. However, the subtleties that characterize life with public and private decision makers make it difficult to achieve such clear-cut agreement as to whether a particular course of action is ethical.

Published research by University of Wisconsin-Madison planning professors Elizabeth Howe and Jerome Kaufman on the subject of ethics finds agreement among planners that certain kinds of behavior *are* acceptable.

- Dramatizing a problem or issue to overcome apathy.

- Making tradeoffs in negotiating situations.

This article reprinted by permission from *Planning*, November 1984, pp. 22–25. Copyright © 1984 by the American Planning Association, 1313 E. 60th St., Chicago, IL, 60637.

- Assisting citizen groups, on a planner's own time, to prepare a counterproposal to an official agency decision.

Likewise, there is consensus that the following tactics are *not* acceptable:

- Making threats.
- Distorting information.
- "Leaking" information.

[In 1983], the AICP Ethics Committee published *Ethical Awareness in Planning*, a report designed to encourage more systematic attention to planning ethics. The 66-page report was distributed to planning schools and APA chapters for use in their classes, newsletters, meetings, and AICP exam preparation sessions.

Seventeen "scenarios," all developed by planners, are included in the report. Six of those scenarios are reprinted here in slightly edited form. Readers are invited to consider each and then to offer their own "answers. . . ."

1. Potential Conflict of Interest

The county council is considering an ordinance that would drastically increase the water and sewage fees for rental units. The county's housing planner has analyzed the proposal and feels that the proposed fees are excessive because the amount of water consumed by apartment units is far less than that of single-family houses. The planner also feels the rate hikes will exacerbate the county's existing rental housing shortage by encouraging the conversion of rental units to condominiums.

The planner prepares a staff report that recommends that the revised fee structure not be approved. However, the planner does not declare a potential conflict of interest, even though her husband owns a small rental property.

The behavior of the planner who prepared the staff report was: ethical, probably ethical, probably unethical, unethical? Not sure?

2. Release of Development Information

The staff of a state planning agency is reviewing a development proposal. Most of the data it has assembled show the project in an unfavorable light. The state's policy is that all working files should be open to the public, but the staff planners are concerned about releasing information in a piecemeal fashion because it could be misconstrued.

The president of a citizens group opposed to the project has requested an appointment to see the file. The president has also stated her intention to seek the state's help in organizing opposition to the project. The state's director of planning decides to remove the single most critical document and keep it in his desk for "further study" during the time when the leader of the citizens group is reviewing the file.

The behavior of the planner who edited the file was: ethical, probably ethical, probably unethical, unethical? Not sure?

3. Letter to the Editor

A city planner writes a letter to the editor of a local newspaper. The letter compliments the county's planning commission on its refusal to approve a rezoning request that would have allowed further industrial development. The planner signs the letter with his name and home address only. The city's planning director agrees with the planner's conclusions and even notes that the comments expressed are of a professional, not a political nature.

The letter to the editor provokes behind-the-scenes activity in which pressure is put on the planning director to fire the planner. The director refuses. Instead, he inserts a memo in the office file listing several "legitimate vehicles"—going to meetings and giving speeches—through which staff planners can express themselves publicly. The planner also is told to use more discretion in the future and never to sign his own name to such a letter. (This scenario is excerpted from a 1971 Planning Advisory Service Report, No. 269, *Dissent and Independent Initiative in Planning Offices*, by Earl Finkler.)

The behavior of the planner who wrote the letter was: ethical, probably ethical, probably unethical, unethical? Not sure?

4. Gag Order

Several city planners oppose a freeway system plan that was adopted by a regional planning agency. They contend that the original staff plan has been emasculated and that the final product discredits the profession.

The city's planning director, who supports the freeway system plan, refuses to allow her staff to express public opposition to the plan, either as professionals or as citizens. She threatens to fire any planners who disobey her orders in this matter.

The planners draft a statement for presentation at the local APA chapter meeting, but then receive word from a reliable source that pressure will be put on the planning director to fire them if such a statement is presented. Fearing for their jobs, the planners do not make any statements in opposition to the freeway system. But they do tell the local APA chapter, at a meeting attended by the director, that they have been forbidden from taking a public position on the freeway system plan. (This scenario is also excerpted from Planning Advisory Service Report No. 269).

The behavior of the planning director in threatening to fire her employees was: ethical, probably ethical, probably unethical, unethical? Not sure?

5. Employment Opportunity

A small city of 25,000 on a lovely lake is being wooed by several hotel entrepreneurs. In evaluating the various proposals, the city's planning staff has asked for information about the number and types of jobs to be made available and, also, how many of these jobs would be targeted to city residents.

In reviewing the data submitted, the staff notices that the jobs are segregated by sex. For example, women are to be employed in the coffee shop as waitresses and men are to work in the main restaurant as waiters.

A member of the planning staff meets with a planning commissioner to discuss this matter, and the commissioner volunteers to contact the developer and challenge the hotel's policies. A debate develops among the planners, with some arguing that the management of the hotel is outside the purview of their responsibilities.

How do you regard the behavior of the planner who contacted the commissioner: ethical, probably ethical, probably unethical, unethical? Not sure?

6. Save the Wetlands

A regional planner who worked on a wetlands preservation study gives certain findings to an environmental group, without receiving authorization from the director of the agency. The planner took this action because he felt the director had purposely left out of the study report those findings that did not support the agency's official policies. The findings that were deleted had been well documented. (This scenario is used with permission of Jerome Kaufman.)

The behavior of the planner who released the information was: ethical, probably ethical, probably unethical, unethical? Not sure?

Winston Churchill observed that "our system is the worst in the entire world—except for all the others." This article from *The Economist,* a British newsweekly, offers a brief survey of the problems that arise when companies from wealthy Western nations bribe officials from developing countries to obtain contracts, exemptions, or privileges. It concludes that bribery reflects on the ethics of the officials and the people offering the bribes. In one of the more vivid scenes in the film, "The Untouchables," Elliot Ness (Kevin Costner) flings an envelope full of money back into the face of a corrupt alderman, fiercely lecturing him on what the Romans did to those seeking to bribe public officials. It was this rectitude that earned the Untouchables their nickname—they were not for sale like most of the rest of the Chicago law enforcement machinery. Most cases of bribery are not so clumsy or dramatic. You will probably never be offered a bribe. You may be offered a gift, an expensive meal, or another gratuity that would be either contrary to rules or imprudent to accept. This article may help sharpen your perceptions of such situations.

CHAPTER THIRTY-SEVEN

On the Take

There are parallels between bribery and nuclear weapons. A bribe can win a contract, just as a nuclear bomb can win a war. But to offer bribes and to make nuclear weapons invites rivals to do the same. When all companies bribe, none is sure of winning the contract, but each must pay so as not to be outdone. As bribers bid against each other, the cost rises, bribery's effectiveness does not. All companies—and the countries whose officials are corrupted—would gain from an agreement to scrap bribes.

In 1975 the United Nations began work on an international ban on bribery. Progress is even slower than on arms control. Frustration has bred an urge for unilateralism; but here the nuclear comparison stops. Unilateral nuclear disarmament would hardly serve the interests of a country like the United States. But America is bribery's unilateralist, and its experience indicates that renouncing bribery need not damage the fortunes of a country's businessmen.

In 1977 America passed the Foreign Corrupt Practices Act, which forbids American companies from making payments to foreign officials. Companies are liable to a fine of $1 million for each violation; individuals to a fine of $10,000 and five years in jail. Prison terms are the more powerful half of the deterrent, since the potential revenues from some bribes make a $1 million fine look like loose change. The PEMEX scandal in the 1970s, in which Mexico's national oil company received bribes from a Texan businessman, involved contracts worth $293 million.

After the anti-bribery legislation was passed, American businessmen complained that they were losing orders to Japanese and European competitors for whom bribery was sometimes not merely legal but tax free, since it could be counted as a business expense.

Business lobbyists have repeatedly demanded that the act be repealed or diluted, citing the country's $150 billion trade deficit as a reason for urgent action.

America's law suffers from being vague. It does not, for instance, forbid "facilitating payments" to government employees "whose duties are essentially ministerial or clerical." Only a handful of companies have been prosecuted under the Foreign Corrupt Practices Act. But though the act has its faults, damage to American exports is apparently not one of them.

Studies by Mr. John Graham of the University of Southern California and Mr. Mark McKean of the University of California at Irvine suggest that the businessmen's cries of pain are exaggerated. Using information from American embassies in 51 countries that together account for four-fifths of America's exports, Mr. Graham divides the countries into two groups; one where bribery is endemic, the other where it is not. He then checks the embassies' impressions against American press reports of bribery, which broadly confirm the corrupt/non-corrupt classification. He has found that in the eight years after the Foreign Corrupt Practices Act was passed, America's share of the imports of corrupt countries actually grew as fast as its share of the imports of non-corrupt ones (see table). His findings are convincing even though, over the period studied, a few of the baddies may have become goodies (and vice versa).

Virtue rewarded

	US share of imports of corrupt countries, %	US share of imports of uncorrupt countries, %
1977	17.5	13.8
1978	17.8	13.6
1979	18.6	12.9
1980	18.6	13.7
1981	19.1	14.8
1982	18.9	14.8
1983	18.7	14.6
1984	18.2	14.4

Source: John Graham and Mark McKean.

Shady Folk in Sunny Places

The need to pay bribes to win business is, it seems, overestimated. Bribes are awkward to distribute; it is not always clear in a foreign country whom should be bribed, or with how much. "Commissions" are sometimes not passed on. Sometimes they are, but the enriched official then awards the contract on the basis of merit. Costs are incurred and risks are run for uncertain benefits. As well as being expensive, bribes can be embarrassing if exposed. Many man-hours are therefore spent fudging accounts and keeping things quiet. Low prices and high quality are often an easier way to win contracts.

The success of the Foreign Corrupt Practices Act ought to have encouraged other governments to copy America's virtuous example. None has: nearly all countries have laws against the bribing of their own officials, but only America forbids the bribing of other people's. Despite the evidence from America, bribery is still thought of as a necessary part of doing business in the third world. Anthropologists' studies of gift-giving are wheeled in to show that bribery is part of the culture of many poor countries; non-bribers are presented as cultural imperialists as well as naive businessmen. The way share-ownership is becoming more international is cited as another reason for business managers to bribe freely; whatever their personal moral scruples, they should not impose them on shareholders to whom such morals might be alien.

Such attitudes once prevailed in America, too. Lockheed, an American aircraft manufacturer, admitted in 1975 that it had paid out $22 million in bribes since 1970; but it protested that: "Such payments . . . are in keeping with business practices in many foreign countries." Yet the Lockheed scandal—along with the humiliating revelations of corrupt political practices that came with the Chilean-ITT and Watergate hearings—helped to bring about a change of mood among America's politicians, even if not among its businessmen.

In 1977 the Senate was told that the Securities and Exchange Commission had discovered that more than 300 American companies had paid bribes abroad. The image both of American government and of American business was suffering, and so were America's relations with friendly foreign governments. Lockheed's bribes to Mr. Kakuei Tanaka when he was Japan's prime minister in 1972–74 led to his arrest in 1976 and a protracted trial. . . . The Senate report also made a point that has grown with the fashion for *laisser-faire* economics. A free-market economy is based on competition—which corruption subverts.

The report's result was the Foreign Corrupt Practices Act. Far from being patronising, the act's proponents argued, it enforced American compliance with other countries' anti-corruption laws. Even Saudi Arabia, renowned for the lush bribing that goes on there, has anti-corruption legislation on its books. Indeed, it may often be developing countries' standards that are brought down by multi-national firms, rather than the other way round.

Innocents Abroad

Two researchers from the University of Western Ontario, Mr. Henry Lane and Mr. Donald Simpson, argue that foreign business-men on brief visits to Africa presume corruption too easily, and so make it worse. If they fail to win a contract, they prefer to believe that the rivals won with larger bribes than that their own products were not up to scratch. Once sown, rumours of corruption spread quickly among the expatriates of an African capital. This leaves westerners with the impression that they have little choice but to bribe: the rumours are self-fulfilling.

The style of western business also encourages bribery. Execu-tives from head office spend fleeting days in a poor country's capital. Few know their way about, or understand the workings of the cumbersome local bureaucracy. The foreignness of foreign cities makes it hard to resist the speakers of excellent English who hang around the hotel bars: a westerner gets conned, and quickly spreads the news that the city is corrupt. Alternatively, his lack of time makes him impatient with local bureaucratic rules. The simplest solution, so it seems, is to cut through the rules with bribes.

Mr. Lane and Mr. Simpson base their views on private talks with officials and businessmen. None, for obvious reasons, wants to be named, so the theories cannot be checked. But they fit with Mr. Graham's conclusions. First, the moral justification for bribery abroad—that it is part of local custom—is sometimes spurious. Second, the business justification does not stand up either: since bribery is not expected of foreign firms, contracts can be won without it. Yet European governments show no signs of heeding such research and legislating against bribery abroad.

Their mistakes need not be repeated by European companies, which could also learn from their counterparts in the United States. More and more American companies are telling their employees to act ethically as well as profitably. Managers have three standard weapons in their armory. Company codes of practice lay down general ethical guidelines. These are fleshed out with training courses, based mainly on case studies. Then there are ways of catching offenders by encouraging colleagues to report them. One is to create an ethics ombudsman to whom employees may report anonymously. Another is IBM's "skip level" management reporting, whereby everybody spends periods working directly for his boss's boss, and so has a choice of two familiar superiors to report to.

According to the Ethics Resource Centre, a Washington-based research group, 73% of America's largest 500 companies had codes of ethics in 1979; by 1988 the figure had risen to 85% of the 2,000 biggest. In 1980 only 3% of the companies surveyed had ethics training for their managers; now 35% do. In 1985 the centre knew of no company that had an ethics ombudsman; by 1987 more than one in ten had created the post.

Most American business courses now include a training course in ethics. At Harvard nearly a quarter of the business-school students opt for the ethics course. More European business schools are also starting to teach business ethics. Last year an umbrella body, the European Business Ethics Network, was set up in Brussels.

Down to Self-defense

Yet the fight against corruption remains a peculiarly American concern. The Europeans and Japanese (whose own country is pretty corrupt) hurt themselves by their complacency, but they hurt developing nations more. In the end it is up to poor countries to defend themselves from foreigners' corruption—as well as from their own.

Sheer poverty makes this hard to do. By 1900 Britain had beaten the worst of its corruption. But in 1900 the average Briton had a yearly income (GDP per person in today's prices) of $4,000—more than ten times that of the average person in the developing world today. Britain had acquired a middle class, whose belief in reward for hard work was the antithesis of corruption. Few of today's poor countries have a sizable middle class; the rest are sat upon by elites accustomed to acquiring money through inheritance and other gifts.

Poverty goes with a weak state, which makes corruption worse. If the state cannot enforce laws, nobody will respect it. Disrespect quickly breeds disloyalty among civil servants: corruption seems eminently sensible, since it involves robbing from the state in order to give to relatives and friends who provide the security that the state is too feeble to deliver. Thus impoverished, the state's strength diminishes further; the rival authority of the clan is consolidated.

Though hampered by their poverty, developing countries can dent the worst of their corruption. The first step is to admit corruption exists. It hides behind respectable masks. Mexican policemen ask for "tips." Middlemen in business deals demand "consultancy fees" and "commissions." A favourite trick in Pakistan is for the post-office teller to be out of stamps. Terribly sorry, but there happens to be a street vendor just outside the post office who sells them—at a premium. Not everybody guesses that half the premium goes to the teller.

It is also necessary to admit it is damaging. The Mexican policeman gets the national minimum wage (a bit over $3 a day), so it may seem natural that he should supplement his pay. The bribes accepted by an official before he awards a government contract do not necessarily distort competition among rival tenderers: sometimes, all are accompanied by a similar bribe, which serves as an entry fee. Equally, a judge may offer the plaintiff with justice on his side the first chance to make a "contribution." For businessmen, a modest bribe may seem an efficient way to secure a license quickly.

Even these apparently mild examples of corruption are harmful. Mexican policemen refuse to investigate crimes reported by those who cannot afford to pay the tip: access to public services, which should be equal, is thus restricted to the better off. While refusing to investigate crimes that do not pay, the Mexican police assiduously tackle non-crimes that do: innocent motorists are stopped to extract a bribe.

The poor and innocent suffer, but there is wider damage too. The tip for a quickly issued license encourages officials to invent new licenses. The tangle of lucrative red-tape strangles would-be entrepreneurs—and the economy suffers. The state's venality diminishes its standing in the eyes of its citizens. No sane Mexican respects the police. South Africa's supposedly independent home-lands are made all the more despicable because their rulers are thieves.

By weakening the state, corruption can even prompt—or at least provide the excuse for—political violence, as when Nigeria's President Shehu Shagari was deposed in 1984. Honest regimes, by contrast, are generally strong enough to get even their unpopular policies accepted. Ghana's Flight-Lieutenant Jerry Rawlings, who overthrew his civilian predecessors because of their corruption, has imposed an awesome dose of economic austerity on his people, but still survives in power.

Once corruption's harmfulness is acknowledged, train civil servants to spot and stop it. The polite silence that surrounds corruption often blocks the passing on of useful tips on how to tackle it. The story is told of an engineer responsible for an irrigation system in India. The rich farmers in the area bribed a local politician, who in turn ordered the irrigation engineer to divert water from poor farms to rich ones. The engineer agreed to do as he was told, so long as the politician would speak his order into the engineer's tape recorder: whereupon the politician backed down. If this was made a case study for trainee water engineers, India's water might be better managed.

As well as instructing the virtuous on how to beat corruption, training should explain to the not-so-virtuous why corruption is so damaging. It may not be a bad idea to explain the benefits to a country of an honest civil service, much as student lawyers learn some jurisprudence. The same goes for businessmen. Some Latin countries—Mexico, Chile—are making ethics training part of their business-school curriculums, which should make businessmen aware of the harm that corruption does to the economy, and to the standards of their firms.

Next, let journalists and other snoops help in exposing corruption. It is not enough for governments to break their silence on the subject; general openness is essential for having corruption discussed. In the Soviet Union, parts of which are pretty poor, Mr. Mikhail Gorbachev is allowing more press freedom than before partly in order to expose the corruption that festered under the secretive rule of Leonid Brezhnev.

Greater openness is the first step towards increased accountability. Mr. Gorbachev also wants some party officials to be exposed to elections, so that they can be judged on the records that *glasnost* has made known. Elections are one good way of holding people to account. Another is the separation of powers. Independent executives, judiciaries and legislatures can keep tabs on each other.

Even strong and open states have difficulty retaining civil servants' loyalty, so the wise ones reduce bureaucrats' discretion: fewer licenses will mean fewer bribes. In famously corrupt Indonesia, the government's economic-reform program includes the burning of red tape. To build a hotel only one license is now required; once, an entrepreneur needed 33.

In particular, do away with economic controls that create black markets. If the state fixes the exchange rate artificially high, foreign currency will be scarce, and distributing it will be the task of bureaucrats. Bribes will flow, because businessmen who need to import spare parts will pay generously for dollars or import quotas. The same happens when state food-marketing boards force farmers to sell their crops at artificially low prices: farmers are encouraged to bribe the board's officials to overlook their grain, and then to bribe customs officials to allow it across the border into a country where it will fetch more. [In 1983] Ugandan coffee could be sold in Kenya for ten times its domestic price.

Slimming down the state will make possible the next corruption-beating move that is sometimes needed: pay public employees more, so that they no longer depend on "tips." The Indonesian government is likely to find its anti-corruption policies damaged by the freeze it has put on civil servants' pay. It may do wonders for Indonesia's budget, but it will probably encourage civil servants to find pay of their own. Along with better training, better pay will improve morale. The more pride that officials take in working for their governments, the less likely they are to subvert them by accepting bribes.

Another way to raise the professional morale of bureaucrats is to make the civil service meritocratic. Competitive entry examinations and promotion on merit helped diminish corruption in nineteenth-century England. In Mexico today, the relatively high professional standards of the Finance Ministry, Bank of Mexico and Foreign Ministry go with their relatively clean reputations. The Indian civil service has competitive examinations but, in some states at least, civil-service jobs are known by the size of bribe needed to obtain them. So long as that persists, those who do the jobs will see them as an instrument of plunder, not as a chance to serve the state.

Tolerated, corruption spreads easily. The civil servant who buys his job will reimburse himself corruptly. In the Philippines corruption has even infected the body investigating corruption under the country's deposed ruler, Mr. Ferdinand Marcos. Because it is so hard to beat, and because all societies and institutions develop taboos against snitching on colleagues, corruption is too often met with defeatism or indulgence. That is an unkindness to bureaucracies and businessmen, whether poor or rich.

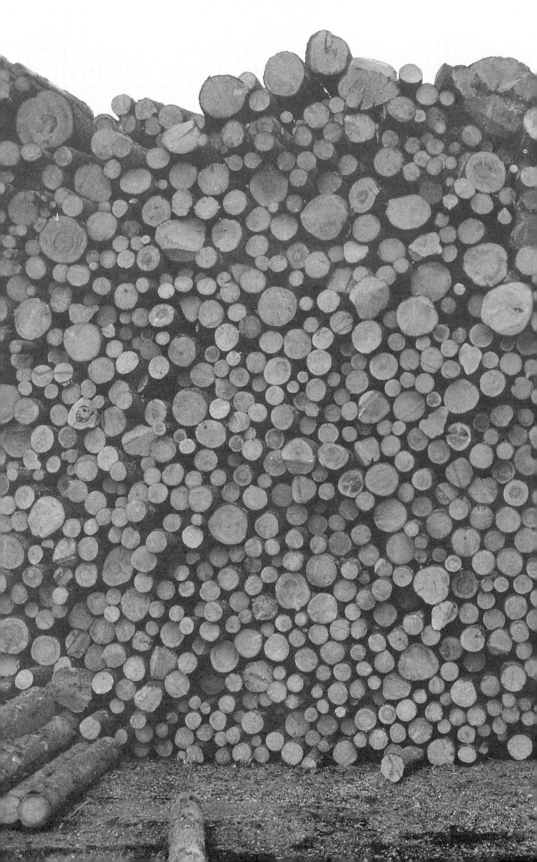

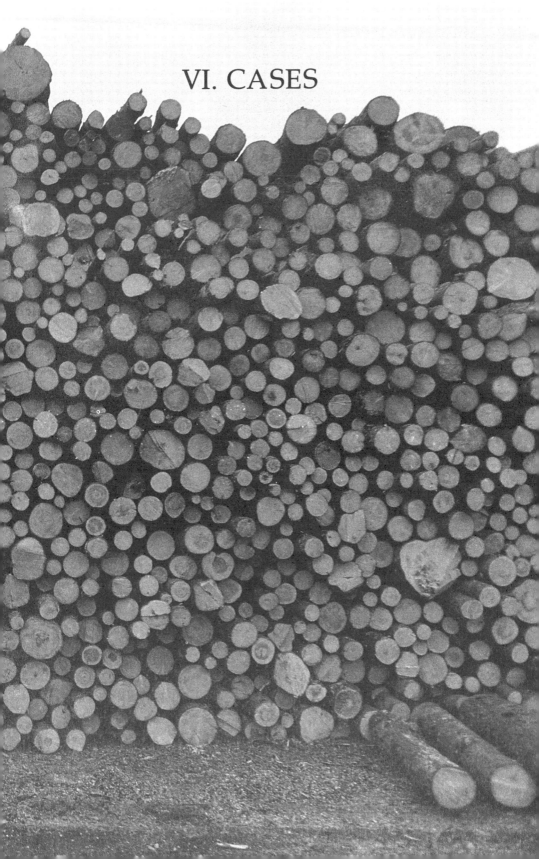

VI. CASES

The cases offered in this section will help you to visualize the kinds of situations that raise ethical questions and to polish your ethical thinking skills. They are adapted from cases contributed for this purpose by Ted Stuart, Keville Larson, and Greeley McGowin.

In particular, the cases and accompanying questions are intended to help you focus on the following key ethical skills:

- detecting emerging problem situations before they lead to an actual lapse in behavior;

- understanding the many ambiguities and conflicts of loyalties that a career in resource management brings;

- knowing when to consult a friend, mentor, or colleague for advice;

- developing skills in constructing alternatives that will successfully resolve a situation;

- applying general ethical concepts to particular situations to help you make decisions you'll feel good about explaining, even to your nine-year-old son.

In addition, a valuable article about the Society of American Foresters Canons is included, the authors of which served on the SAF Ethics Committee. It offers both explanation and questions for reflection concerning several important canons of the ethics code. Banzhaf was president for many years of a leading consulting firm, and now serves as executive director of the SAF. Burns is an attorney active in SAF activities, and Vance is a longtime Forest Service official. These diverse backgrounds help enrich the article.

CHAPTER THIRTY-EIGHT

Unethical Professional Behavior?

EDWARD STUART, JR.

Let us examine some situations that foresters can expect to encounter regarding *conflict of interest,* a term that covers many situations and activities. Foresters must know what constitutes a conflict of interest—or even the *appearance* of a conflict of interest—so that they can maintain their professional standards. Some of the following situations have occurred in the past; others could occur at any time. Relevant sections of the Society of American Foresters and Association of Consulting Foresters Codes of Ethics (as of Summer 1993) are noted in each heading.

Conflict of Interest in Buying Timber (SAF-9, ACF-13)

A management forester, employed by a lumber company to assist private landowners, marks a block of timber for harvesting, estimates the volume, determines the market price, and submits this to the landowner. The employer then purchases the timber based on the forester's appraisal, though the timber is never exposed to the open market. There is no question that this type of practice is a conflict of interest or, at the least, an appearance of conflict, regardless of how unbiased the forester is in determining the timber value.

The decision of the North Carolina Supreme Court, *Roberson v. Williams* (No. 98, October 13, 1954), should be a warning to those timber buyers who induce sellers to use the buyer's estimate of value as a basis for the purchase price, especially when the seller is an uninformed party. It seems that Mr. Williams, a sawmill operator, approached Mrs. Roberson, a widow, offering to buy her timber. She replied that she had no idea of its worth. Williams offered to have his

---

Mr. Stuart is a consulting forester in Virginia. This article reprinted by permission.

cruiser go over the timber and appraise it. He found 250,000 board feet and said that he would pay her $10,000 for the timber. The timber cruiser assured Mrs. Roberson that the cruise was correct, and the sale was closed at $10,000.

A short time later Williams resold the timber for $19,000. Subsequent timber cruisers employed by Mrs. Roberson showed there were over 700,000 board feet of timber. Mrs. Roberson sued Williams, eventually reaching the North Carolina Supreme Court, which decided in her favor. The Court "held that the evidence showed that statements made by defendant-buyer and his companion as to amount and value of timber were false, that they were made under circumstances which would permit inference that purpose was to deceive, and that plaintiff was thereby induced to part with her property for less than half of its value." A lower court, after another trial, awarded monetary damages to Mrs. Roberson.

Many large wood-using firms in different parts of the country have developed landowner assistance programs. Most believe that because of their financial assistance to the landowner over the years, they should have the right to purchase the wood when it matures, either by an option or a right of first refusal and then based on their own appraisal. However, if they would allow the timber to be exposed to the open market or if they had the timber appraised by an unbiased third party, the appearance of a conflict of interest could be avoided.

Some company foresters dispute the contention that these programs create an ethical conflict and a danger of conflict of interest. How would you argue this side of the question? What do you conclude?

Many professional foresters are employed in wood procurement work for sawmills, paper mills, and other wood-using firms. Think of some of the other ethical conflicts they may face on a day-to-day basis.

Conflict of Interest: Contingent Fees (SAF-9, ACF-18)

A landowner is having some property taken through condemnation and asks his attorney to retain appraisers. Because the property had considerable timber on it, the attorney hires a forester. The attorney indicates that his own fee is contingent on 50% of any money obtained over the condemning agency's offer. In view of the fact that he is receiving his fee on a contingency basis, he suggests that the forester's fee should also be on a contingency basis. The forester agrees and then makes an excellent appraisal followed by a court presentation. All seems to go well until, under cross examination, the forester is asked how he is being paid. The forester explains that his fee is contingent upon the amount of the award. On the motion of the condemning agency's attorney, the judge immediately disqualifies the forester.

Why is this contingent fee situation a conflict of interest? If offered such an engagement, what should you do? If shown an appraisal prepared on such a fee arrangement, would you rely on it?

A timber owner wishes to retain a consulting forester to prepare a management plan. The consultant accepts the assignment with the understanding that his fee will be paid from the first timber-cut.

Is there a conflict of interest or the appearance of a conflict? The consultant could easily be accused of setting the year of the first harvest too early so that his fee would be paid at an early date.

A forester is asked to appraise a tract of timber. The landowner asks what his fee will be. The forester states that if he makes a 5% cruise, the fee will be 5% of the appraised value. If he makes a 20% cruise, the fee will be 10% of the value. In short, the fee will be contingent upon the value placed on the timber.

There is no question that there is a conflict of interest in this situation: in formal appraisal reports, appraisers are required to state that their compensation is not contingent on the value reported.

Misrepresentation: Duty to Report Unethical Conduct (SAF-5, SAF-16, ACF-3, ACF-24)

An 80-year-old widow, living in New York City, owns 600 acres of timber in a state some distance away. She receives a letter from a firm of foresters in a far-off state stating that her timber is badly infested by the Southern Pine Bark Beetle and that the only remedy is to sell the remaining timber immediately. The firm would be glad to act as her agent. She is understandably upset and is ready to allow the firm to sell her timber. However, she has the presence of mind to get a second opinion from a consulting forester in the area of her timber. He reports that the timber is in excellent growing condition and that there is no insect infestation. This finding is confirmed by a letter from the local state forester. The first firm is guilty of misrepresentation and also certainly of a conflict of interest. The widow does not want to get involved in filing charges and, as a result, no ethical charges are brought against the forestry firm.

The foresters in the first firm and the consultant who gave the second opinion are members of the same professional organization. Is the consultant guilty of unethical behavior for not reporting the situation to his professional organization?

A forester is retained by a prospective timberland buyer to examine and appraise a tract of timberland. The forester reports that the asking price is too high and recommends that the client not purchase the property. With this information, the client loses all interest in the property. Later on, the client has occasion to look at some records at the county court house and discovers that the property was sold at the original asking price and that the deed was recorded less than 30 days after he received the report from his forester. He notices that his forester is one of the new owners.

This situation is not only unethical, but also violates criminal law.

Loyalty to Employer (SAF-5, ACF-10)

A federal forester is responsible for awarding forestry service contracts. He receives an invitation from a local contractor to attend a professional football game at a city some distance away. The contractor will provide the tickets, transportation, lodging, and meal costs.

Should the forester accept?

Forester B is employed as a service forester by the state. Her responsibilities include assisting landowners in getting government subsidies to help finance their reforestation projects and engaging site preparation contractors for the landowners. Her brother-in-law is in the business of site preparation with heavy equipment.

Is it unethical for the forester to engage her brother-in-law to do site preparation on private lands?

The same forester owns land of her own and applies for government funds to reforest it at a time when funds are limited. She feels, as a landowner, entitled to a share of government subsidies. She seeks your advice.

Is there a conflict of interest between Forester B's personal goals and professional duties? If another landowner is denied assistance because the funds have run out, has Forester B violated her trust?

A forester who resides in a county that has large county-owned timber acreage has, for a number of years, unsuccessfully attempted to get the county to retain him to manage the land and to act as agent for any timber sales. He learns that the county is planning to sell the timber at $200 per acre. However, he knows from casual observation that the timber is worth at least $1,000 per acre. At a public hearing he is undecided whether or not he should speak against the sale because of the appearance of a conflict of interest.

Should the forester formally withdraw his offer of services to the county before speaking at the hearing?

Plagiarism (SAF-13, ACF-22)

Professor X requests and receives research assistance from a graduate student. The student performs all of the research and writes the paper for publication, while Professor X edits the paper and submits it to the printer. After publication, however, the only author's name that appears on the paper is that of Professor X. The entire student body and faculty of the school of forestry knows about this lack of credit, but no charges are brought, even though many of the faculty are members of their professional organization.

Does this situation illustrate a violation of any other canon?

Confidentiality (SAF-8, ACF-12)

A lumber company is retained by a consulting forester to cruise a tract of timber being offered for sale and pays him for his cruise. No sale is made, but in six months another lumber company offers to pay the forester for the results of the original cruise. Because the forester's relationship with the first lumber company had concluded some six months earlier, he feels he has the right to sell the cruise to the second lumber company.

If he sells his earlier cruise, does he violate the confidentiality of his first client? What should he do?

A graduate forestry student is doing research on the relationship between timber appraisals and eventual sales prices. He contacts numerous consulting foresters for information, but to assist in assembling the data, the student suggests that he personally go through the foresters' files.

Is the suggestion unethical? If a consultant allows the student access to such files, could this be seen as unethical behavior?

A forester has been employed by Company X for a number of years doing research on fertilizing costs. She is hired by Company Y at a higher salary and upon reporting for work, is asked to perform the same cost analysis research she had been doing, using the knowledge obtained at her previous job.

Can she proceed, using the information developed while employed by Company X? Explain the reasons for your answer.

A forester appraises some timbered property for a landowner who subsequently leaves the country. The forester receives a phone call from a caller who identifies himself as an agent of the I.R.S. and requests a copy of the cruise. The forester cannot locate the landowner, and, realizing the information is several years old, provides the I.R.S. with the data requested.

Has the forester acted unethically? Should he have refused the I.R.S. the information?

After 30 years of private practice, a forester decides to sell his business. His one stock in trade, besides some miscellaneous equipment, is his clients and his files. He considers these as assets when determining an asking price. He is fortunate in selling his business, lock, stock, and barrel, including his files, to another forester.

Does he violate the canon on confidentiality when making the sale? How can he avoid the charge of unethical conduct if this is true?

Advertising (SAF-3, ACF-3)

Professional codes now allow advertising in a dignified and truthful manner. The problem arises, what is a dignified manner? There is no question of what the word "truthful" means. The following situation occurred in a local newspaper:

A forester employed by a lumber company advertises publicly that landowners will receive more money for their timber if they sell directly to the company, based on the company's scale, rather than having a consulting forester cruise and handle the sale.

Is this ad dignified and truthful? This proposition raises issues regarding at least two other canons. Which canons do you think apply?

Deviations from Professional Standards (SAF-7, ACF-11)

A consulting forester is retained by a landowner to block out a section of timber for a clearcut. Upon examination of the area to be harvested, he notices that the timber is on a steep slope with fragile soil; any clearcutting will cause considerable erosion to the property and stream sedimentation on other landowners' property. The consultant informs the landowner of these effects, but the landowner insists that that is the way he wants his timber harvested. He instructs the forester to proceed with the cruise of the timber and the sale.

Will the forester be committing any breach of professional ethics by continuing with the sale? Should he walk away?

Using the same tract, assume the landowner is a lumber company and the forester is the company's employee. The orders are to mark the area and put in crews to cut the timber. The forester informs his boss of the consequences but is still ordered to proceed. To keep his job the forester follows the orders.

Are the forester's actions correct? Has he abided by the SAF Code?

In the first case, the consultant can walk away and should do so.

The second case places a burden on the forester—his job is at stake. *Can he hide under the umbrella of his supervisor's responsibility?*

Reputation of Another Forester (SAF-12, SAF-15, ACF-23)

Forester B has qualified as an expert witness in a court case. He gives the results of his timber cruise, which is pertinent to the case, under direct examination. Prior to his testimony, the plaintiff placed Forester A on the stand. A's cruise was triple B's, and B is asked to explain the difference in the volume estimate, as well as what he knows about Forester A's competence. B has heard from a number of sources that Forester A is incompetent. He is under oath and has been required by the judge to give his opinion of Forester A.

B's knowledge of Forester A is based solely on hearsay, but he feels A's cruise of the tract is completely inaccurate and good evidence of incompetence. What should he answer?

Loyalty to Employer (SAF-5, ACF-10)

The defendants in a partition suit retained Forester E to appraise a tract of timber. He estimates a value of $150,000. When the suit is settled less than six months later, the plaintiff retains ownership and immediately places the timber on the market. Forester E is then retained by a prospective buyer to make a cruise and appraisal. The owners are extremely disappointed in the buyer's bid and ask if they can compare Forester E's two cruises. The difference between them is almost $100,000 worth of timber.

Is Forester E merely incompetent in his appraisals? If he made his original appraisal according to the wishes of his client, has he breached any SAF codes?

A company-employed forester has been assigned to cruise a property some 100 miles from his home. At four o'clock he finds that he needs less than an hour to complete the job, but he and his wife are giving a dinner party for some friends that evening. If he finishes the cruise that day, he will be late for the party. If he leaves immediately, he will have to spend the best part of the next day travelling to and from the tract, including another hour's work. He quits and makes the party on time.

Has he demonstrated loyalty to his employer?

A publicly employed forester is asked to give a talk to a local civic organization. During the talk, the forester criticizes some of the current forest policies, thinking of the remarks as constructive criticism. He did not clear the remarks with superiors.

Is this unethical? Does it violate any SAF Canons?

Conclusion

The SAF's Code of Ethics is not worth the paper it is printed on unless it is strongly enforced. Both the organization's members and their governing bodies must be alert to violations and have the fortitude to file charges when a violation occurs. Investigating committees have the responsibility of determining the facts of cases in a fair and impartial manner, and the council that passes final judgment must have the will and the ability to enforce its decisions. All SAF members must have a clear understanding of ethics to ensure that the integrity of the profession is maintained; this is clearly stated in SAF Canon 15.

All too often charges have been dropped in spite of overwhelming evidence, either because the violation seemed to be minor or because the governing body did not have a complete understanding of its own code. Canon 15 places on members' shoulders the task of presenting evidence of a violation, but it does not specifically spell out the responsibility of the council. Here is the weak link in the enforcement of the code.

The importance of foresters' understanding the code and conducting themselves accordingly cannot be overstressed. The entire forestry profession is affected by the unethical behavior of even a few of its members. Foresters will be suspect unless they can demonstrate professional competence combined with personal integrity.

CHAPTER THIRTY-NINE

Company Assistance Programs

J. GREELEY McGOWIN, II

These cases raise questions of ethics, and also questions of business policy and prudence. Notice that it is not always easy to tell which are which.

Stumpage Pricing Policy (SAF-5, ACF-1, ACF-10, ACF-13)

Charley Stewart is a prominent local businessman who lives in a small town a few miles from an older sawmill that is owned by the region's largest lumber company. Charley owns about 2,000 acres of well-stocked timberland, most of which has been in the Stewart family for over 50 years. Whereas the average logging distance to the mill is 25 miles, Charley's land is located 15 miles from the company's sawmill, has a pretty good road system, and is a little easier to log than most outside tracts. In addition, the average size and quality of his timber compares favorably with most of the short-term leases the company has been buying. Charley's father had been a great "friend of the company" and had always traded with the company whenever he sold his timber. Charley inherited the land almost 20 years ago and has made three timber sales to the company—in each case he simply negotiated a price and did not invite competitive bids.

Bill Jones is the company's procurement manager and has been personal friends with Charley for many years. Bill had always been glad to mark Charley's timber, do a little road work, and give him forestry advice. Four years ago, Bill made Charley a formal participant in the company's assistance program and with Charley's help, Bill has signed up several smaller landowners for the program. Charley is now in a bit of a financial bind and wants to sell about $100,000 worth of timber. The timber market in the area has become much more competitive, and prices have escalated. Although he

---

Mr. McGowin is Vice President of the Union Camp Corporation in Savannah, GA. This article used by permission.

doesn't have access to hard inside information and is not an expert on timber values, Charley does try to keep up with current market prices. He feels strongly that he wants to get all he can for his timber, but he values his relationship with Bill and the company, and he really prefers not to go out for competitive bids. He has always trusted Bill and he believes he has been treated fairly in past timber sales. However, he is the type of person who would probably never forgive if he felt someone had really taken advantage of him, and he would be inclined to let everyone know his feelings.

Charley phones Bill, and asks him to come by to discuss a timber sale. He lets Bill know that he is well aware of the fact that stumpage prices are way up from what they were last time he sold timber.

Bill's short-term lease inventory is low, and he needs to buy more wood at reasonable rates. Last week he bid top stumpage prices on 600 acres about 30 miles from the mill because he was bidding on both land and timber.

Bill decides he had better discuss this one with his boss before he goes to see Charley. Bill is enthusiastic about the company program, has done a good job of promoting it in his area, and believes it can play a more important role in the future in his difficult wood procurement job. He is beginning to think that he needs some clear policy guidelines on how to handle this kind of situation. He knows he must shoot straight with Charley, but he also knows that he is expected to get his wood cost down.

What should Bill's boss tell him to do in his negotiations with Charley?

Owner vs. Employer Needs (SAF-9, ACF-1, ACF-10, ACF-13)

Charley's land has a pretty good road system on it, except that the 300 acres of timber that he just sold to Bill are cut off by a small creek. The old logging bridge over the creek has fallen in, and there are no passable roads on the 300 acres. Bill can buy a right-of-way from an adjoining landowner so he doesn't have to cross the creek to get the logs out, but the haul distance will be five miles further.

Charley wants a permanent creek crossing (either bridge or culvert) and would like to have about a mile and a half of usable roads on the 300 acres. He also needs a bit of maintenance work done on his existing road system. Charley goes to see Bill and asks what he can do to help.

Bill figures it would cost about $10,000 to do what Charley wants, but thinks it would only cost about half this much to get the logs out via the right-of-way.

What should Bill tell Charley?

Landowner Assistance: Costs vs. Regeneration
(SAF-5, SAF-9, ACF-1, ACF-10, ACF-11, ACF-13)

Sam Wood sold his 100 acres of timber by sealed bid. The company was the high bidder at $36,000, but left too much money "on the table," bidding $5,000 higher than the next bidder. Bill is unhappy about it, figuring his cruise must have been high. Bill has orders to take the company's assistance program more seriously, and is expected to always try to get landowners to replant when he buys their timber. Bill is having a tough time getting all of the company's own land regenerated this year (he had to carry over 2,000 acres last year), so he really doesn't have his heart in trying to sell Sam on letting the company help him replant his land. So Bill goes to his boss, and asks how hard he really should try (if at all) on a deal like this.

What does the boss tell Bill? If Bill has already paid too much for the wood, can they afford planting costs? What is the tradeoff between completing replanting on company lands versus private cooperators in the assistance program? How much of this question is ethics, and how much is business policy?

Objective Advice; Reputation of Others
(SAF-4, SAF-12, ACF-23)

Charley wants to sell the timber on 200 acres within the next year or so. He intends to cut it selectively this time, and may or may not clearcut the 200 acres next time. He goes to see Bill to discuss it.

Charley points out that there could be a "conflict of interest" involved in selective cutting. It is to the company's advantage to take a heavier cut per acre and to concentrate on the bigger and better trees, while Charley is more interested in the spacing of the remaining trees, the future growth he will get on the stand after cutting, and the removal of the lower quality trees regardless of size. Charley doesn't seem concerned, perhaps because he trusts Bill; he asks Bill if this apparent "conflict of interest" creates any particular problem for him.

In the course of the conversation, the subject of consultants comes up. Charley observes that the company seems to be offering "free" consulting services in the assistance program. Charley likes the idea of getting "something for nothing," but he wants to know what Bill, who has a high regard for several consultants in the area, thinks. It's not at all clear to Charley whether consultants and assistance programs are competitors or not, so he asks Bill to explain in some detail when a private landowner like himself ought to use a consultant and when he should go the assistance program route.

How should Bill respond to Charley?

Competing for a Client
(SAF-4, SAF-9, SAF-14, ACF-10, ACF-25)

Charley has a sudden heart attack and dies, leaving his widow, Marie, and three kids, two in college, one in high school. His estate consists of 2,000 acres of timberland and $500,000 in other assets. Marie will get an income of about $12,000 a year from stock dividends and figures she can maintain her standard of living and get the kids through school if she can get $20,000-$25,000 a year income from timber sales.

Charley stated in his will that Lawyer Jackson would tend to his estate and help Marie handle her finances, but Jackson does not know the timber business. The estate tax is going to be about $350,000, so Jackson suggests to Marie that they hire John Smith, a consultant in the area who has a good reputation, to handle a $350,000 timber sale and to continue representing Marie as she makes annual timber sales. Marie agrees.

John comes to see Marie to discuss the situation. Marie explains that Charley had always negotiated his timber sales with Bill, and she knows that Charley would have wanted her to continue the practice. She is comfortable with the idea, as she is aware that Bill did a lot of favors for Charley and helped manage the land. She knows Bill would like to keep the land in the company program. However, she makes it clear to John that she is going to need all the dollars she can possibly get from her timber sales. John calls on Bill, and explains the situation. John is perfectly willing to negotiate with Bill for the sale of Marie's timber, but he and Lawyer Jackson must protect Marie's interest, and they have to be satisfied that the "price is right." He points out that Marie's timber is somewhat better in size and quality than what Bill is buying on average, that it is relatively close to the mill, and that logging conditions are somewhat better than normal. Therefore, he thinks a fair price would be "top dollar" less the out-of-pocket cost of special services rendered as part of the company assistance program. He points out that while this is no bargain, it is still a good deal for Bill as he will be paying no more than top dollar and will be getting a continuing flow of timber, year after year. He can also have a lot of flexibility about when he cuts it. The big problem, John tells Bill, is that Bill must convince him that the price Bill is offering is a fair one.

What might Bill say to convince John? What would be some of the ethical and unethical arguments he might use?

Clearcut vs. Partial Cutting (SAF-4, SAF-7, ACF-11)

Jack Johnson is a bachelor in his early fifties who owns the best liquor store in town, does a lot of hunting and fishing, and is satisfied with his lot in life. He and Bill are hunting buddies of long standing.

Jack owns 300 acres of timberland that are in the company assistance program, but he is just not interested in spending money to regenerate his land—he'd rather just sell timber from time to time and enjoy spending the proceeds.

Jack tells Bill that he could use about $50,000 and wants to sell some timber. He and Bill have no trouble agreeing on unit prices, and he asks Bill's advice on how and where to cut. Bill would prefer to clearcut a little over 100 acres, but he knows there's no chance of talking Jack into spending $15,000 to regenerate. Bill could selectively cut the whole 300 acres, generate the $50,000 stumpage, and leave the whole tract in reasonable growing condition, but it costs money to mark the timber, and he would have to supervise the logger. Bill wants to recommend the clearcut, and he feels sure that Jack would go along as long as Jack is not required to regenerate the land. But Bill is well aware that a major objective of the company program is to improve the productivity of non-industrial private timberland—and there is no doubt in his mind that, under the circumstances, the 300 acres will remain more productive if he selectively cuts the timber.

What should Bill tell Jack to do?

Competing Loyalties (SAF-5, ACF-10)

The Southern Pine Beetle epidemic is bad this year. Bill is having a hard time getting all the beetle-damaged wood harvested from company land. John has just checked Marie's 2,000 acres, and found half a dozen beetle outbreaks. He is very much concerned, and immediately goes to see Bill. John fully understands Bill's problems, but he knows that Bill has certain obligations to Marie as part of the company assistance agreement. John estimates that, if they could get started right away, about 250 cords (including the green buffers) need to come out. He asks Bill what he will do, when he will do it, and what stumpage price would be fair.

What can Bill do to remain loyal to all parties?

John also wants to understand the company's policy for suppression of wild fires on timberland owned by others, and in particular for participants in the assistance program. He asks Bill what the company would do if he found out about a wild fire on Marie's land.

Would he respond the same way as if the fire were on company land? *Having recently reread SAF Canon 5, what would Bill answer?*

Competing Loyalties (SAF-5, ACF-10)

Doctor Randle has 500 acres of timberland in the assistance program. He has just purchased an adjoining 50 acres of open land, and wants to get a pine plantation started on it as soon as possible. He has read about the company's "supertrees," and wants to be sure that he gets the same "supertree" stock that the company plants on its own land.

Randle goes to see Bill in early January. Randle expects to pay for the cost of establishing the new plantation, but he has no tree planting contractor contacts and doesn't have the time or inclination to get involved in the details. He wants Bill to get the trees in the ground this planting season and to give him the best deal available. As usual, Bill is a little behind with his planting responsibilities for the company and may not have enough genetically improved seedlings to go around.

What should Bill tell Doctor Randle?

CHAPTER FORTY

Ethics and Forestry

WILLIAM H. BANZHAF, ANN FOREST BURNS,
and JOHN VANCE

Hypothetical Case 1

SAF member Sam Deer is a graduate forester employed by a large forest landowning corporation. He is the sole supporter of his mother and four children, aged 14 to 20. He has been employed by the same company since graduating from college 23 years ago. Approximately 20 percent of the foresters in Sam's area are unemployed.

Because of a recent economic downturn, the executives of Sam's corporation—most of whom are trained in business rather than resource management—have decided to cut the company's timber as rapidly as possible. This will require scrapping the company's 10-year-old sustained-yield projections. In some situations, the company will be in technical, but not actual, violation of state-legislated regeneration requirements. Because of the way the law is written, the state will not be able to take enforcement actions for at least five years.

The executives hope that the new rapid cutting schedule will improve the company's financial position enough to attract new investment capital. The new money would go toward a planned program of forestland stocking improvement, forestry research, and commercial development of some of the company's suburban holdings. If this plan does not succeed, the executives will try to sell the company's forestry operations.

Sam is not in an enviable position. He must make a number of ethical considerations while trying to avoid jeopardizing his family. *If*

Mr. Banzhaf is currently Executive Director of the Society of American Foresters. Ms. Burns is an attorney. Mr. Vance is with the Forest Service. All have served on the Society's Ethics Committee. This article reprinted from *Journal of Forestry*, April 1985, pp. 219–223, by permission of the Society of American Foresters. It was published before the 1992 revision of the SAF Canons, which added the Land Ethic as Canon 1.

Sam considers what his company proposes to be harmful to the forest resource, what are his options?

Unethical Practices

In the February 1968 *Journal of Forestry*, J. O. Lammi defined unethical practices as "deviations from progress toward professional goals. They arise from intentional misconduct, from lack of knowledge, or from lack of resources" [see "Professional Ethics in Forestry" in this volume].

As the case above shows, a forester such as Sam who is not backed by established professional standards—or who is lacking in knowledge of them—could be cast rudderless in a moral sea. In a more general sense, ethics are the foundation of an organized society. Without ethical standards—whether they be religious in origin or based on professional need—society as we know it would not progress, if, indeed, it were able to survive at all.

Hypothetical Case 2

Sylvia Swan is a graduate forester licensed by her state. She is one of 20 employees of a consulting forestry firm that locates forest properties for prospective buyers and manages other forestland for landowners, charging a percentage of the proceeds from management.

Sylvia's grandmother owns a sizable tract of wooded property, one of a number of similar tracts owned by various parties that might be of interest to a real estate developer. Sylvia is strongly opposed to any consideration of developing her grandmother's property. The grandmother has suggested that Sylvia manage the land and keep any proceeds.

Is Sylvia obliged to disclose to her firm her grandmother's offer of outside employment? If Sylvia is asked by her firm to evaluate the property for a prospective developer, how should she respond? If she knows that a developer is interested in an adjoining piece of property which is for sale and which her grandmother could purchase first, what should she do?

Suppose Sylvia were to falsify cruise data from the tract in an effort to keep it out of the developer's hands, and you, a colleague at another firm, found out. *What would you do?* Suppose the employer also found out and fired Sylvia. *What would you do if she then asked you for a letter of reference to another firm?*

Code of Ethics

As early as 1914, the Society of American Foresters was interested in maintaining the high ethical standards of its members and providing guidance in cases such as Sylvia's. It wasn't until 1948, however, that the SAF Code of Ethics was enacted. A standing Committee on Ethics was subsequently formed; the Code was revised most recently in 1976 [the 1992 revision is reprinted in this book].

The 15 [now 16] canons that comprise the Code, though reasonably straightforward, are occasionally misunderstood, especially in ambiguous cases such as the two described above. In order to alleviate any misunderstanding, late last year [1984] the Committee on Ethics completed work on an ethics handbook, which will soon be on sale to members through the national office. The Society's goal is to use the handbook to increase the ethical knowledge of its members. To this end, the following material, excerpted from the handbook, helps interpret the code. When reading the comments, keep in mind the hypothetical cases.

Canon 1

A Member's knowledge and skills will be utilized for the benefit of society. A Member will strive for accurate, current, and increasing knowledge of forestry, will communicate such knowledge when not confidential, and will challenge and correct untrue statements about forestry.

This underscores members' ultimate responsibility to serve the long-term interest of society as a whole. The canon states the duty of the profession to "apply our specialized knowledge toward making the country's forests yield their fullest contribution to the economic and social welfare of the nation," as the *Journal*'s editorial in March 1937 put it.

Further, this canon commits members to continuing education, and to cooperate with others in sharing information and knowledge (except where there is need to respect confidentiality).

Canon 2

A Member will advertise only in a dignified and truthful manner, stating the services the Member is qualified and prepared to perform. Such advertisements may include references to fees charged.

The canon recognizes that the manner in which individual members present and conduct themselves reflects on other members and the profession as a whole.

Canon 3

A Member will base public comment on forestry matters on accurate knowledge and will not distort or withhold pertinent information to substantiate a point of view. Prior to making public statements on forest policies and practices, a Member will indicate on whose behalf the statements are made.

Members are encouraged to make public comment on forestry-related subjects, and in fact are expected to take initiative in correcting inaccurate and misleading statements (see Canon 1). However, it is imperative that members themselves do not engage in the same type of subjectivity or inaccuracy that they intend to correct. Care must be exercised to maintain objectivity, and provide an accurate and fair portrayal of the facts. Conjectures or opinions should be identified as such.

Canon 3 also obliges members to make clear for whom they are speaking. They must guard against the appearance, intended or unintended, of representing the view of the Society, or other members, when not authorized to do so. The probable effect of one's statements, and the way in which they will be perceived by the public, must be carefully evaluated beforehand.

Canon 4

A Member will perform services consistent with the highest standards of quality and with unqualified loyalty to the employer.

Loyalty to the employer refers to the responsibility of the member to render high-quality advice and service, as free as possible from personal bias. The term "unqualified" does not mean a blind loyalty that is oblivious to the larger interests of society and to other canons of the Code of Ethics. It does mean that members will faithfully perform professional services to the full extent of their ability. Where a member is unable to reconcile loyalties to self, employer, profession, and society, it may be wise to disengage from service.

Canon 5

A Member will perform only those services for which the Member is qualified by education or experience.

See Canon 6 for comment.

Canon 6

A Member who is asked to participate in forestry operations which deviate from accepted professional standards must advise the employer in advance of the consequence of such deviation.

SAF members should know the accepted professional standards or methods applicable to whatever task or operation is being contemplated. The employer or client may not have equal expertise. It therefore is imperative that the member explain what could occur to the detriment of the employer, or society, if accepted methods or practices are not implemented.

Canon 7

A Member will not voluntarily disclose information concerning the affairs of the Member's employer without the employer's express permission.

This canon deals with the concept of confidentiality and ownership of information. Information acquired during the performance of work may, if disclosed, put the employer at a competitive disadvantage. Depending on the circumstances in which the information was developed, it may be the property of the employer. Before disclosing information the forester should consider whether it is public or private information.

This canon also implies that in some situations, such as in a court of law, there may be no other choice but to disclose confidential information.

Canon 8

A Member must avoid conflicts of interest or even the appearance of such conflicts. If, despite such precaution, a conflict of interest is discovered, it must be promptly and fully disclosed to the Member's employer and the Member must be prepared to act immediately to resolve the conflict.

A conflict of interest implies a relationship or situation where one's judgment may be influenced by potential personal gain. It may also imply serving two masters where the gain of one means another's loss.

When a conflict is not real, but nevertheless perceived, it will still harm the credibility of the profession and of the individual involved.

Canon 9

A Member will not accept compensation or expenses from more than one employer for the same service, unless the parties involved are informed and consent.

This canon describes a particular example of conflict of interest, in which an individual could potentially produce information that might give one party an advantage over the other. Full disclosure to both parties is the first and most critical step in removing a real or seeming conflict of interest.

Canon 9 once again raises the issue of information ownership. Having been paid by one client to produce information, if you resell it to another client you may be receiving compensation for something you do not own.

Canon 10

A Member will engage, or advise the Member's employer to engage, other experts and specialists in forestry or related fields whenever the employer's interest would be best served by such action, and Members will work cooperatively with other professionals.

It is sometimes difficult to admit to an employer that you are not fully qualified to perform a particular task. Nevertheless, the forestry profession is based on a wide range of disciplines, some of which require a great deal of specialized study. That being the case, it is important that we learn to work cooperatively within the interdisciplinary teams sometimes required to accomplish our professional objectives.

Canon 11

A Member will not by false statement or dishonest action injure the reputation or professional associations of another Member.

This canon clearly prohibits any member from lying or making deliberately false or misleading statements about the professional expertise, ability, or reputation of another member. Certainly, everyday ethics prohibits such conduct against *any* person.

The highest standards of human conduct also require us to guard against the temptation to engage in gossip, or to pass along stories and anecdotes that might hold another up to ridicule or lower that person in the eyes of others. Questioning the accuracy or truth of a statement does not excuse the perpetuation of gossip.

Honest, direct criticism and airing of differences between professionals often serves to advance the profession and safeguard the public interest, and is in no way a violation of the canon.

Canon 12

A Member will give credit for the methods, ideas, or assistance obtained from others.

This canon recognizes the basic property value of ideas, of intangible concepts. In the final analysis, taking or using another's ideas or methods without compensation is theft. Like plagiarism, such conduct is unethical.

Canon 12's importance is most obvious in forestry research. A forester who uses another's research results should obtain permission of the original investigator and give due credit. This is particularly vital where the original investigator's findings have not been established as their own by publication.

Canon 13

A Member in competition for supplying forestry services will encourage the prospective employer to base selection on comparision of qualifications and negotiation of fee or salary.

Canon 13 was the most significant modification to the Code when it was revised in 1976. The former Code, in common with that of many other professions, prohibited competitive bidding for professional services. That restriction was recognized as being contrary to the legitimate public interest in open and free competition in the marketplace and was discarded.

The new canon recognizes that decisions about forestry services should not be made on the basis of price alone. Potential clients should be informed of the qualifications of those seeking to supply services such as education, equipment, experience, and access to specialists. Fees or salaries can reasonably be negotiated once these factors are known and recognized.

Canon 14

Information submitted about a candidate for a prospective position, award, or elected office will be accurate, factual, and objective.

Canon 14 forbids making false or misleading statements about one's own or another's qualifications. Circumstances in which misstatements might occur are in an application for employment, scholarship, fellowship, or grant; in presenting qualifications of one's firm for a contract, license, or certificate; in seeking an award or recognition for oneself or another (within and outside the Society of American Foresters); and in running for public office or such position in a club or organization (including, of course, SAF).

Canon 15

A Member having evidence of violation of these canons by another Member will present the information and charges to the Council in accordance with the Bylaws.

Article VIII of the Society's Bylaws governs the procedures for dealing with violations of the Code of Ethics. Charges against a member must be made in writing. The charges must cite the specific canon or canons believed to have been violated, along with the evidence known to those bringing the charges.

As a safeguard against unwarranted accusation, charges must be signed by five or more voting members of the Society. It is preferred, but not required, that members signing the charges have firsthand knowledge of the alleged unethical conduct.

The discipline of those engaged in unethical conduct protects the public and the profession from unscrupulous and harmful activities. The bringing of charges against a member is a grave responsibility to be undertaken only after thorough and thoughtful consideration.

Can the Code Work?

In the matter of ethical conduct, the act of questioning is almost as important as the answers themselves. If we as members continually question the ethical nature of our professional conduct, the chances of violating the Code of Ethics will be greatly reduced.

Hypothetical Case 3

Chris Byrd is an SAF member who is not a graduate forester. He is employed as a soils specialist by a government agency that manages public forestlands.

Chris's employer has issued a finding that no significant environmental impact will result from a road maintenance program proposed for one of the areas being managed. Chris strongly disagrees with this finding, believing that seeding along the roadbanks will result in a significant increase in the deer population, leading to an influx of hunters during the hunting season and probable damage to forest regeneration during the winter as the animals try to find alternative forage. Chris has aired his views at all levels within his agency, but no one has seemed to listen or to present data to refute his ideas.

Chris is a member of the Humane Wildlife Management Society (HWMS). If this society became aware of and agreed with Chris's analysis, it would probably bring suit against the government agency, challenging its findings and seeking to have the road maintenance program stopped. Stoppage, however, might result in streambank erosion and degradation of water quality.

Should Chris share his knowledge with HWMS? If he is asked to testify by HWMS at a legislative hearing concerning agency wildlife practices, can he list his membership in the SAF as a qualifying credential? His employment by the agency? Could Chris ethically testify at all in such a hearing?

VII: APPENDIXES

Appendix 1: Exercises in Case Study

Reading articles and studying specific cases are valuable ways to enhance your ethical awareness and your ability to exercise ethical judgment, but they are not enough. The following exercises can be used as class assignments, discussion topics, or even test questions. They are designed to help you actively employ the tools of ethical thinking.

Exercise: Common Themes from the Cases

Several common themes are illustrated by these cases. Review the cases and identify the ones that illustrate each of these themes.

1. In many instances, people will disagree over what is ethical.

2. Many situations involve more than one of the canons.

3. By their very nature, ethical canons cannot include all the detail or be specific enough to provide guidance in every situation.

4. Some canons *mandate* specific actions, while others *prohibit* specified actions.

5. In the examples of conflict of interest, the forester involved usually failed to take one obvious step that would have satisfactorily resolved the problem in many instances. What was this step?

6. Many of these situations could have been avoided, and the problems thus easily prevented, by a higher level of awareness and information.

7. Ethical decisions can have high stakes.

Exercise: Cases Not Covered

Think of a number of cases involving specific kinds of behavior that are not covered by any SAF Canon. How would you resolve these situations?

Exercise: Tough Case

Invent a realistic example of the toughest ethical bind you can imagine getting yourself into. Consult three friends for their advice. Do they all agree? What would your solution be and why?

Exercise: Real Life Cases

Ask three foresters or resource managers to identify the toughest ethical choice they ever had to make. Get them to be very specific. Ask them what they decided to do about it and why. Would they act differently now? Why?

Exercise: Introspection

1. Think of a situation in which you think you did something unethical or at least dubious. Analyze it fully, along the lines of the approach to reflective thinking described in the introduction to this book. What should you have done?
2. Think of an ethical conflict that you feel you handled well, or a problem you avoided by sound ethical thinking. Analyze the situation. What did you learn?
3. Think of the sources of your own personal views on professional ethics. To what extent are they based on (*a*) religious upbringing or beliefs; (*b*) parental example; (*c*) guidance of a strong teacher or respected "mentor;" (*d*) organizational rules; (*e*) example of colleagues; (*f*) professional society ethical codes; (*g*) introspection or instinctive feel for what is right; (*h*) focused study of the subject of ethics; (*i*) study of men and women who have made great accomplishments.

Is there any reason to prefer any one of these over the others? Do some of them have elements of risk?

Exercise: Train Yourself to Seek Advice

Think of a situation you are now facing that will require you to make a decision that takes ethics into consideration. Seek the advice of a trusted associate and discuss the matter thoroughly. Analyze what you learn from this.

Appendix 2: Questions for Reflection

Questions of Ethics

1. Think of an event in your own experience or that you have observed that corresponds to each of the canons of the three codes of ethics reprinted at the beginning of this book. How might that situation have been handled differently?

2. Do these canons as written provide all the guidance needed for individuals to make sound decisions? If not, why not? What else is needed?

3. Why do you suppose that the three codes of ethics are so different?

4. Are public officials held to a different standard of ethical conduct than private citizens? How? Why?

5. Do any provisions of these codes disturb you in any way? Why? How would you modify the code to address your concern(s)?

6. Many large corporations have adopted ethical codes to guide their policies and their employees. Select a firm with which you are familiar, obtain its code by writing the corporate secretary, and review it carefully. Does the code provide enough clear guidance that employees will know what is expected of them? Does the code balance the claims of the company's owners and those of society responsibly? How does the code differ from the professional codes of ethics reprinted above? Read the firm's annual reports and follow press coverage on its operations. Does the company seem to be following its own code of ethics? Is publishing a code enough?

Professional Ethics Questions

Questions on Patterson (Chapter 4)

1. How does Patterson view the role of ethical codes in fostering ethical behavior?

2. What are Patterson's four guides to behavior? Are they enough? Can you think of any others?

Questions on Lammi (Chapter 5)

1. Lammi defines sources of unethical practice as intentional misconduct, lack of knowledge, or lack of resources. Think of examples of each from your own experience. Are there other sources of misconduct beyond these?

2. Lammi connects professional ethics to philosophical tenets put forth by Dag Hammarskjold and Franklin D. Roosevelt, among others. Review his discussion of these points.

Questions on Merchant (Chapter 6)

1. What does Merchant recommend as a way of upgrading one's ethical awareness? Do you think it would work?

2. Merchant is an extension agent in Maine. Imagine some of the ethical problems that extension agents might face in their daily work.

Questions on Flanagan (Chapter 7)

1. Define and discuss Flanagan's distinction between acts that are bad in and of themselves and those that are bad because they are prohibited. What does this distinction imply for professional conduct?

2. Flanagan states that a code of conduct brings with it legal obligations for professionals. What are the three specific legal liabilities that he discusses?

3. Flanagan comments on the role of foresters in compliance with land use laws. Review his observations and comment.

4. Summarize his discussion of the possibility of malpractice actions being brought against foresters.

Questions on Irland (Chapter 8)

1. What are the obstacles to making sound decisions about using pesticides in forestry? Have any important considerations been omitted?

2. What is an ultrahazardous activity? What are its implications for legal liability and professional practice?

3. In your understanding of the Land Ethic, where would pesticide use fit in forest management, recalling that there are a range of uses besides silvicultural ones. Make the fullest list you can and comment on how their use might relate to your vision of a land ethic.

4. Citizens affected by pesticide use have two specific rights according to this article. What are they? Do you agree with them? Can you think of any others?

5. Review the list of recommendations at the end of the article. Is it complete? Are there any points you would dispute? Why?

Questions on Johnson (Chapter 9)

1. Johnson quotes Liddell Hart on how eagerness for promotion reduces senior officers' willingness to challenge their superiors. Do you see parallels in forestry? What canons of the SAF Code of Ethics address such behavior?

2. What are the four pressing ethical issues suggested by Johnson? Describe their relevance to forestry.

3. If you have ever served in the Armed Forces or any large organization, make a list of situations in which you felt that your personal ethical standards were compromised by organizational pressures. What did you do about it? Make a list of situations in which you felt that professed policies were being violated routinely. What effect did this have on you?

4. What does Johnson urge organizations to do to foster a higher level of ethical behavior?

Questions on Klemperer (Chapter 10)

1. Define the concept of sustained yield.

2. What is Klemperer's argument about the ethical importance of the sustained yield concept? Do you agree with his view? How does this view accord with your sense of a land ethic?

Questions on Irland (Chapter 11)

1. According to the article, why do organizations want to keep secrets? Are these reasons legitimate?

2. Possessing files full of clients' confidential information creates a serious ethical responsibility for foresters. How should foresters prepare themselves for this eventuality?

3. Read, consider, and discuss with a mentor or friend the four questions posed toward the end of the piece.

Questions on Ladd (Chapter 12)

1. Summarize Ladd's definition of ethics.

2. Summarize his view of the objectives of professional ethics codes.

3. Compare Ladd's view of the objectives of ethics codes with those of previous authors.

4. What are Ladd's objections to professional ethics codes? Do you agree with him?

Questions on Marshall (Chapter 13)

1. Outline Howard Wilshire's case against the USGS. Comment on its relevance to professional forestry.

2. Interagency conflict over management decisions is a daily occurrence in forestry. Discuss how the Wilshire case illustrates many of the ethical dilemmas that arise. From an ethical perspective, should any of the participants in the Wilshire case have acted differently? How?

3. How do the demands of organizations and the standards of individuals conflict in these cases?

Questions on the Chapter by the Committee on the Conduct of Science, National Academy of Sciences (Chapter 14)

1. Identify three situations in which practicing foresters engage in scientific or technical work similar to the work outlined in this report.

2. Recall the principal areas in this selection in which ethical concerns arise, and think of examples from your own experience.

Questions on Koshland (Chapter 15)

1. Restate the two principles with which Koshland begins the editorial, and describe the balancing act between these two principles.

2. Can you think of examples of this balancing act in a forestry context?

Questions on Marshall (Chapter 16)

1. What conflicts can arise from the extensive business relationships in academia depicted in this article? What do you think about these issues?

2. How do you think academic scientists in forestry and related fields should handle situations like this?

Business Ethics Questions

Questions on Marks (Chapter 17)

1. Summarize the differing views of the business school students quoted in this article concerning their interest in and need for training in ethics. Which view do you share?

2. Think of two situations in public or private forestry that would give rise to situations similar to examples given in the article.

3. In the article's closing section, Marks notes that instructors have a variety of views on the value of teaching ethics. Which makes the most sense to you?

Questions on Boulding (Chapter 18)

1. List and describe Boulding's four ethical principles.

2. What does Boulding mean by a "value system"?

3. What does he say about ethical relativism?

4. What are the three sets of problems Boulding defines toward the end of his essay?

5. How does he characterize the relationship between the market system and the wider society?

Questions on Barry (Chapter 19)

1. Compare Barry's view of the place of business in the social system with Boulding's.

2. Note that Barry defines conflict of interest in the context of an employee in an organization. Would it make a difference if an independent professional were engaged for a specific task by a client?

3. What issues does Barry identify as "use of official position"? Can you think of any other issues that could be included in this category? Link each of these points to a canon of the SAF or ACF code.

4. Think about cases in which you might be faced with ethical choices regarding gifts and entertainment. Which of the SAF and ACF canons provide guidance for such situations?

5. Review how Archie Patterson's four questions could help resolve the types of ethical conflicts described in this article.

Questions on Kokus (Chapter 20)

1. Think of five situations in which foresters become involved in the types of appraisal activities reviewed by Kokus.

2. Carefully read the positive and admonitory canons. Explain in your own words what each one means, and connect each one to a similar one in the SAF or ACF codes.

3. Review Kokus' discussion of qualifications and limits of experience. Think of situations in which a forester would face these issues. Which SAF or ACF canons apply?

4. Summarize the relationship between appraisal practice and the various regulations affecting the real estate industry.

5. In the section "The Client," the author reviews an appraiser's duty to the client and the potential for conflict between professional standards and client loyalty. Summarize the points made and apply them to situations in forestry.

Questions on Dorsey (Chapter 21)

1. Consulting foresters often appraise forest properties or standing timber for owners or potential buyers. Have a chat with some acquaintances in this line of work and ask them to review with you the most difficult ethical question they ever faced doing appraisals and how they dealt with it.

2. Toward the end of the article, Dorsey notes, "The appraiser cannot act contrary to the interests or requirements of the client. At the same time, the appraiser cannot perform in a manner which conflicts with the dictates of conscience, with ethics, and as evidence of both, with professional standards." Does Dorsey leave an unresolved conflict here? Specifically, what should be done if the client's "interests" plainly seek a high value, as, for example in a condemnation case? Could it be argued that an objective value placed on the property conflicts with the client's interest in obtaining a high one? Reflect further on the issue embodied in Dorsey's sentences quoted above.

3. Make a list of practical situations in which a client may have an interest in obtaining a particular outcome in an appraisal.

4. In any given area, appraisers are known by reputation as being "conservative," "middle of the road," or "liberal" in appraisal values they reach. Depending on their needs, clients may select one whose predilections fit their needs of the moment. Comment on the ethical questions involved for both client and appraiser.

Questions on Harrison (Chapter 22)

1. Describe the ethical problems in the paper industry identified by Harrison. How would these concerns arise in forestry?

2. Are the situations Harrison describes only a matter between the individual participants or are there higher social stakes in these situations and their outcomes?

Questions on Stuart (Chapter 23)

1. Define conflict of interest, and review the applicable SAF and ACF canons.

2. Recall a conflict of interest from your own experience. How did it arise? What did it say about the ability of others to rely on your judgment? What should have been done differently? What lessons did you learn?

3. Review the advertisement in the article. What do you think about it?

Questions on Andrews (Chapter 24)

1. How does Andrews describe the responsibility of organizations toward the moral development of employees? What specifically does he call upon senior managers to do?

2. What role does Andrews see for existing codes of professional ethics?

Questions on Case (Chapter 25)

1. Review these four stories, the summaries of how others responded, and what actually happened as described in Case's article.

2. Offer your own comment on what you would have done in the same situation and why.

3. Identify the canons of the SAF and ACF codes most directly relevant to these situations.

Environmental Ethics Questions

Questions on Leopold (Chapter 26)

1. What experiences do you think led Leopold to perceive and articulate the Land Ethic in the way he did?

2. How does Leopold view the Community Concept in relation to his Land Ethic?

3. How does Leopold define conservation? Do you agree with this definition? How would his definition apply to forestry today?

4. How does Leopold characterize the role of economic motives in contrast to ethical motives in conservation?

5. What does he say about the public programs of his day and their effectiveness in light of the Land Ethic?

Questions on Flader (Chapter 27)

1. How does Flader describe the evolution of Leopold's thought on wildlife ecology?

2. How did Leopold's scientific thinking influence his ethical thinking over the years?

3. According to Flader, what was Leopold's influence on the development of the Land Ethic?

Questions on Worrell (Chapter 28)

1. This piece originally appeared in a volume concerned with forest soils. How does Worrell describe the significance of soils in forestry?

2. How does Worrell summarize conservation problems in relation to the market system?

3. Review and comment on Worrell's six-fold summary of the land ethic. Where does he agree with and where does he differ from the way Leopold would state the matter?

4. What does Worrell see coming to the fore in forest policy as he concludes the essay? Define and discuss this concept.

Questions on Fritsch (Chapter 29)

1. Fritsch notes that many writers hold Judeo-Christian religious precepts responsible for the misuse of land in western societies. He argues, however, that these traditions also contain mandates to care for nature. He notes three in particular. What are they?

2. How does Fritsch's concept of prophetic witness apply to forestry? Can you think of a mentor or forestry leader who has exemplified the ideal of prophetic witness?

3. How does the concept of exemplary community apply to forestry?

4. What are the elements of Fritsch's concept of stewardship? How has this concept developed over history?

5. Are there common points between Fritsch's concept of stewardship and Leopold's community concept?

Questions on Selle (Chapter 30)

1. Summarize Selle's argument.

2. How does Selle's approach compare to that of Leopold and Fritsch?

3. How do you connect your religious beliefs with your professional and environmental ethics in your life?

Questions on Heilbroner (Chapter 31)

1. How does Heilbroner answer the question posed in his title?

2. What alternatives to that answer does he suggest?

3. How does this apply to forestry?

4. How would Klemperer respond to Heilbroner's argument?

Questions on Sagoff (Chapter 32)

1. Describe Sagoff's description of the economic model for environmental policy.

2. How does Sagoff view the emphasis of economics on consumer welfare?

3. What is the Lockean view of land use as summarized by Sagoff?

4. Summarize Sagoff's view of the concept of "internalizing" externalities.

5. Compare Sagoff's discussion of rational self-interest to Leopold's.

6. Summarize Sagoff's closing argument. Do you agree with it? How might this view apply to forestry?

Questions on Coufal (Chapter 33)

1. Comment on Aldo Leopold's influence on professional foresters as shown in Coufal's essay.

2. Coufal discusses Milbrath's New Social Paradigm. Do you think most foresters share this view or not?

3. Comment on Coufal's discussion of objectivity and values in relation to ethics.

4. Summarize Coufal's case for including a Land Ethic Canon in the ACF's Code of Ethics. Do you agree that such a canon should have been included in the Code?

Government Service and Public Policy Questions

Questions on Willbern (Chapter 34)

1. Review Willbern's six types and levels of public morality. Describe how they would apply in forestry.

2. What does Willbern mean by his "ethic of democratic responsibility?" Think of an example in forestry.

3. What are the implications of the "ethic of compromise and social integration" for forest management on public lands? On private lands?

Questions on the Chapter by the President's Commission on Federal Ethics Law Reform (Chapter 35)

1. What are the four key principles offered by the Commission?

2. Summarize the Commission's points about conflict of interest.

3. The Commission proposes extending certain ethics rules to cover the Congress. What are the pros and cons of this proposal?

4. The Commission refers to the question of honoraria for members of Congress and the judiciary. Why should such officials be able to accept honoraria while executive branch employees may not?

5. Can you think of instances in public forestry or resource management that involved the issues discussed in this report?

Questions on Barrett (Chapter 36)

1. In each of these six cases, do you see a course of action that the individual could have taken that would have eliminated the ethical problem?

2. What are the legitimate ethical claims on both sides of each case?

3. Apply Archie Patterson's four questions to the individual in each situation.

4. Invent a forestry example as an analogue to each of these cases. Discuss how it would best be handled.

Questions on "On the Take" (Chapter 37)

1. What are the three steps the article suggests to deal with public corruption?

2. Are the solutions to corruption provided by the Foreign Corrupt Practices Act likely to work in your opinion?

3. Have you ever encountered bribery or other misbehavior of the sort described in this article? If you ever were in a situation in which someone attempted to bribe you, what would you do?

Index

This index locates important concepts and terms. The references cited in the articles and the questions for reflection are not indexed.

455